Solid State Physics

The Manchester Physics Series

General Editors

F. MANDL : R. J. ELLISON : D. J. SANDIFORD

*Physics Department, Faculty of Science,
University of Manchester*

Published

Properties of Matter: B. H. Flowers and E. Mendoza

Optics: F. G. Smith and J. H. Thomson

Statistical Physics: F. Mandl

Solid State Physics: H. E. Hall

In preparation

Electromagnetism: I. S. Grant and W. R. Phillips

Atomic Physics: J. C. Willmott

Electronics: J. M. Calvert and M. A. H. McCausland

SOLID STATE PHYSICS

H. E. Hall

Department of Physics,
University of Manchester

John Wiley & Sons Ltd.

LONDON   NEW YORK   SYDNEY   TORONTO

Library of Congress Cataloging in Publication Data:

Hall, Henry Edgar, 1928–
Solid state physics.

(The Manchester physics series)
Bibliography: p.
1. Solids. I. Title.

QC176.H24   530.4'1   73–10743
ISBN 0 471 34280 7 Cloth bound
ISBN 0 471 34281 5 Paper bound

Set on Monophoto Filmsetter and printed by
J. W. Arrowsmith Ltd., Bristol, England

To FRANZ,
who talked me into it,
and PAT,
who made me finish it.

Editors' Preface to the Manchester Physics Series

In devising physics syllabuses for undergraduate courses, the staff of Manchester University Physics Department have experienced great difficulty in finding suitable textbooks to recommend to students; many teachers at other universities apparently share this experience. Most books contain much more material than a student has time to assimilate and are so arranged that it is only rarely possible to select sections or chapters to define a self-contained, balanced syllabus. From this situation grew the idea of the Manchester Physics Series.

The books of the Manchester Physics Series correspond to our lecture courses with about fifty per cent additional material. To achieve this we have been very selective in the choice of topics to be included. The emphasis is on the basic physics together with some instructive, stimulating and useful applications. Since the treatment of particular topics varies greatly between different universities, we have tried to organize the material so that it is possible to select courses of different length and difficulty and to emphasize different applications. For this purpose we have encouraged authors to use flow diagrams showing the logical connection of different chapters and to put some topics into starred sections or subsections. These cover more advanced and alternative material, and are not required for the understanding of later parts of each volume.

Since the books of the Manchester Physics Series were planned as an integrated course, the series gives a balanced account of those parts of physics which it treats. The level of sophistication varies: '*Properties of Matter*' is for the first year, '*Solid State Physics*' for the third. The other volumes are intermediate, allowing considerable flexibility in use. '*Electromagnetism*', '*Optics*', '*Electronics*' and '*Atomic Physics*' start from first year level and progress to material suitable for second or even third year courses. '*Statistical Physics*' is suitable for second or third year. The books have been written in such a way that each volume is self-contained and can be used independently of the others.

Although the series has been written for undergraduates at an English university, it is equally suitable for American university courses beyond the Freshman year. Each author's preface gives detailed information about the prerequisite material for his volume.

In producing a series such as this, a policy decision must be made about units. After the widest possible consultations we decided, jointly with the authors and the publishers, to adopt SI units interpreted liberally, largely following the recommendations of the International Union of Pure and Applied Physics. We did not outlaw physical units such as the electron-volt. Nor were we pedantic about factors of 10 (is 0.012 kg preferable to 12 g?), about abbreviations (while s or sec may not be equally acceptable to a computer, they should be to a scientist), and about similarly trivial matters.

Preliminary editions of these books have been tried out at Manchester University and circulated widely to teachers at other universities, so that much feedback has been provided. We are extremely grateful to the many students and collegues, at Manchester and elsewhere, who through criticisms, suggestions and stimulating discussions helped to improve the presentation and approach of the final version of these books. Our particular thanks go to the authors, for all the work they have done, for the many new ideas they have contributed, and for discussing patiently, and frequently accepting, our many suggestions and requests. We would also like to thank the publishers, John Wiley and Sons, who have been most helpful in every way, including the financing of the preliminary editions.

Physics Department   F. MANDL
Faculty of Science   R. J. ELLISON
Manchester University   D. J. SANDIFORD

Author's Preface

In keeping with the general aims of the Manchester Physics Series I have tried to write a short book containing all the solid state physics that an honours student, with some special interest in the subject, should learn— *but no more*. This book is therefore very far from comprehensive in its coverage; the amount of factual information presented has been pruned with extreme severity. The facts presented have been selected mainly on the criterion of usefulness in explaining basic principles, and I have tried to make these explanations clear and complete. I have simplified the arguments as far as possible, so that the essentials may be grasped; but, as befits the intellectual ability of honours students, I have not shirked the discussion of difficult ideas or dubious assumptions. I have particularly had in mind that students using this book will have to cope with physics thirty years hence, and have therefore largely avoided the description of currently fashionable calculation methods, but have emphasized basic assumptions. The central theme of this book is the large variety of *qualitatively different* ground states that an assembly of atoms can have; I hope I have said sufficient to warn the reader to watch out for surprises!

The general plan of the book is shown on the flow diagram inside the front cover. This plan has been determined largely by two factors:
(1) After several years teaching a course for honours physicists based firmly on reciprocal lattice and Brillouin zone theory, and an attempt to give a much simpler course to honours chemists, largely avoiding these topics, I have come to the conclusion that they are the 'Maxwell's equations' of solid

state physics. The reciprocal lattice and Brillouin zones are an elegant formal structure from which much of solid state theory can be deduced; but they are much better appreciated as a unifying principle after some simpler things have been done without their aid. For this reason the general discussion of waves in periodic structures in Chapter 6 is preceded by one-dimensional discussions of phonons in Chapter 2, electrons in Chapter 3, and magnons in Chapter 5.

(2) In the new course structure at Manchester the second half of the solid state course is optional; I therefore wanted the first half of the book to constitute a suitable introductory course, containing some important applications. It is for this reason that the whole treatment of semiconductors is in Chapter 3. It seemed logically preferable to reverse the usual order and treat semiconductors before metals, since the former are conceptually easier because the electron gas is dilute; also, the discussion of divalent metals requires the extension of energy bands to two dimensions, and should therefore come later.

Another feature worthy of comment is that I have put more chemistry and less crystallography than is usual in Chapter 1. The crystal structures one actually needs to know are few; the hydrogen molecule is not only fundamental to atomic binding, but is also an easy example with which to illustrate both the simplicity and the doubtful validity of the independent particle approximation—a point I wanted to make right from the start. The structure of Chapter 1 has been rather carefully organized so that, by the judicious use of optional starred sections and appendices, this key introductory chapter may be read at a wide range of levels, according to the tastes and abilities of the student.

I hope that these considerations have given the book a structure that will make it suitable for courses of honours degree standard that vary considerably in length, breadth, and depth. The chapters are of rather unequal length, but as a rough general guide I would recommend four or five 50-minute lectures per chapter as an adequate allowance; the whole book thus contains material for over fifty lectures, more than would usually be given in any undergraduate course. The flow diagram inside the front cover will enable a suitable selection to be made. Thus, a short introductory course at full breadth can be based on Chapters 1–5; but a narrower course of similar length directed specifically at the band structure of metals could be based on Chapters 1, 3, 4, 6, 9 and 10. As a further aid to selection, sections and subsections that may be omitted without loss of continuity are indicated by a star ★ at the left of the heading, and paragraphs in this category are printed on a grey background; these optional sections are often rather harder. Also, the appendices may be omitted if the student is prepared to take results on trust. It is thus by the omission or inclusion of starred section and appendices that the level at which this book is read may be most conveniently adjusted.

As used at Manchester this book presupposes courses based on Willmott's '*Atomic Physics*' and Mandl's '*Statistical Physics*',[2] but one could if necessary get away with considerably less preparation: for quantum mechanics, Heitler's '*Wave Mechanics*';[9] and for statistical mechanics, familiarity with the Boltzmann factor and acceptance of the Bose and Fermi distribution functions. In electromagnetism, Maxwell's equations are assumed known, and some idea of magnetic fields in matter on the lines of Vol. 2 of the Berkeley Physics Course.[4]

In accordance with editorial policy and current educational practice I have used SI units in this book as fully as seems reasonable. An important exception to standard SI practice occurs in electromagnetism where, to avoid confusion in discussing fields in matter, I have defined **H** so that it is measured, like **B**, in tesla, with other consequential changes. A full description of units in electromagnetism is given in Appendix E.

On the few occasions where formulae from atomic physics occur I have written them also in terms of the fine structure constant $\alpha = (e^2/4\pi\varepsilon_0 hc)$; this gives formulae that are not only more concise, but are also independent of the system of units. I have also departed from strict SI in not eschewing such useful units as the Angstrom, the electron volt, the Rydberg, and the Bohr radius.

I should perhaps conclude with an apology for writing this book at all. I look like a solid state physicist only when viewed from a coordinate system centred on the Physics Department at Manchester, which is dominated by nuclear physics and radioastronomy; because I am not a professional I have merely treated the topics that interest me in a way that I can understand. I have therefore probably used methods that cannot be developed into a more proper treatment, and I hope the professionals will forgive me; I also hope they will tell me, as kindly as possible, of the errors they find.

I would finally like to thank the many people who have helped me with comments on the preliminary edition. I am especially grateful to Drs. F. Mandl, P. G. J. Lucas, and D. J. Sandiford for their detailed and constructive criticism of successive versions of this book. I should also like to thank Dr. Lucas for allowing me to use some of his problems, and him and Drs. J. R. Hook and I. S. Mackenzie for checking the solutions. I am grateful to the copyright holders for permission to use many published figures, as specifically acknowledged in the text, and especially to the authors for providing original photographs for Figs. 1.28, 1.29, 10.12 and 12.5.

April 1973   H. E. HALL

---

[2],[4],[9] : References by number are to the bibliography.

Contents

★ Starred sections may be omitted as they are not required later in the book.

2   LATTICE VIBRATIONS

3   MOBILE ELECTRONS

4   METALS

8 THERMAL CONDUCTIVITY OF INSULATORS

9 REAL METALS

CHAPTER

Crystal structure

1.1 INTRODUCTION

Solid state physics is concerned with attempting to account for the properties of ordinary matter as found on Earth. For almost all purposes the properties of such matter are expected to follow from Schrödinger's equation for a collection of permanent atomic nuclei and electrons interacting with electrostatic forces. In this sense, solid state physics is no longer fundamental; nowadays it is only in cosmology, astrophysics and high energy physics that we believe the fundamental laws are still unknown.

It is commonly said that most solids are crystalline; in the sense of the traditional solid–liquid–gas classification of matter this is true. Solids that have a sharp first order transition* to a liquid or vapour phase *are* crystalline in the sense that their atomic structure is based on a regularly repeated pattern, a sort of three-dimensional wallpaper, even if external crystalline form is not manifest. But many of the commonest solids—glass, plastics, wood, bone—are not so highly ordered and are not crystalline.

In this book we shall be concerned almost entirely with crystalline solids composed of one or a few types of atom, simply because this is the easiest problem, and the one with which most progress has been made. More complicated materials, particularly those of biological origin, we have

---

* i.e. one with a latent heat (see Mandl,[2] p. 233).

hardly begun to understand at the level of seeing how they can arise as solutions of Schrödinger's equation.

Even in the very restricted field of crystalline solids the most remarkable thing is the great variety of *qualitatively different* solutions to Schrödinger's equation that can arise. We have insulators, semiconductors, metals, super-conductors—all obeying different macroscopic laws: an electric field causes an electric dipole moment in an insulator, a steady current in a metal or semiconductor and a steadily accelerated current in a superconductor. Solds may be transparent or opaque, hard or soft, brittle or ductile, magnetic or non-magnetic.

In discussing various types of solid we shall have occasion to speak of many different types of force or binding: ionic, metallic, covalant, van der Waals, exchange. But we must always remember that these terms are inventions of the human imagination, introduced as an aid to thought. *They are all consequences of the electrostatic interaction* between nuclei and electrons obeying Schrödinger's equation.* That these simple principles can lead to such a variety of consequences for the very simple systems we shall study may perhaps make it less incredible that the complex everyday world can also be a consequence of these same basic principles.

Our purpose in this chapter is to show how quantum mechanics leads to various types of chemical binding and a consequent variety of types of crystal structure. In section 1.2 we introduce a form of quantum mechanics that will be generally useful in this book and apply it to the simplest problem of chemical binding, H_2^+. The results are applied to a more qualitative discussion of the various types of binding in section 1.3, while the fundamen-tals are taken to greater depth in the optional section 1.4. Section 1.5 intro-duces the basic ideas of crystal geometry; these are not essential until Chapter 6, but some acquaintance with them is helpful in understanding the description in section 1.6 of various crystal structures we shall refer to later. We conclude this introductory chapter with a brief outline of x-ray crystallography in section 1.7, sufficient to suggest the experimental basis of the structures we have described.

1.2   INTERATOMIC FORCES

We know the stationary state solutions of the Schrödinger equation for an isolated atom (see, for example, Heitler[9]). What happens when two atoms approach each other closely? What do we mean by an interatomic force? As with any force, we infer its existence from the motion of the particles it acts on; since most of the mass of an atom is in the nucleus, it is essentially the motion of the atomic nuclei that concerns us.

---

* Also, of course, an invention of human imagination?

At first sight this is a very difficult problem, requiring the solution of the many particle Schrödinger equation for both atomic nuclei and all the electrons. Fortunately, the fact that nuclei are much more massive than electrons enables us to make a great simplification, both conceptually and mathematically. Because of the large mass ratio the nuclear motion is very much slower than that of the electrons, so that the electrons at any instant behave almost as if the nuclei were stationary in their instantaneous positions. In other words, we can to a very good approximation think of an electronic wavefunction which is an eigenstate for nuclei fixed in their instantaneous positions; as the nuclei move, this wavefunction smoothly adjusts itself to the changing boundary conditions, but remains an eigenstate. Such a slow perturbation of the boundary conditions (ideally, infinitely slow) is called an **adiabatic perturbation**, and it is a principle of quantum mechanics that such a perturbation does not cause transitions between quantum states (compare Mandl,[2] p. 86, Born, *Atomic Physics*, Blackie, London, 4th edn., 1946, p. 109). The wavefunction and energy alter during an adiabatic perturbation, but the quantum state does not.

This enables us to split our calculation into two stages. First we calculate the electronic energy $E(R)$ for two *fixed* nuclei separated by a distance R, for various values of R. We then make the adiabatic approximation described above and assume that $E(R)$ so calculated is the electronic contribution to the total energy of the system when we allow the nuclei to move. The total energy of the two atoms is now given by

$$E_{\text{tot}} = \frac{p_1^2}{2M_1} + \frac{p_2^2}{2M_2} + \frac{q_1 q_2}{4\pi\varepsilon_0 R} + E(R), \qquad (1.1)$$

where q_1 and q_2, M_1 and M_2, p_1 and p_2 are the nuclear charges, masses and momenta, respectively. The first two terms in Eq. (1.1) are nuclear kinetic energy, the third term is the nuclear electrostatic potential energy, and the last term is the electronic energy. We see that $E(R)$ appears like an extra potential energy of interaction between the two nuclei; it is as if there were a force $\partial E(R)/\partial R$ between them in addition to their mutual Coulomb repulsion.

For many purposes the nuclear motion can be treated classically, but it is almost equally easy to replace the momenta in Eq. (1.1) by operators $-i\hbar\nabla$ to obtain a two particle Schrödinger equation; this is easily reduced to a single particle equation by separation of the centre-of-mass motion.

Our use of interatomic forces in this book will be confined to the case where the effective potential is the electronic ground state energy. But it is worth noting here that nothing in our arguments requires $E(R)$ to be the ground state; we have assumed only that it is an eigenstate. The idea of interatomic forces and potentials is therefore equally applicable to excited states, but of course both the magnitude and dependence on R will be different from that for the ground state.

1.2.1 Coupled quantum states

We are used to seeing quantum mechanics expressed in the form of the Schrödinger equation

$$i\hbar\frac{\partial\Psi(\mathbf{r}, t)}{\partial t} = H\Psi(\mathbf{r}, t), \tag{1.2}$$

where H is the hamiltonian (kinetic plus potential energy) operator. The stationary states are the eigenfunctions $\psi_n(\mathbf{r})$ of this equation such that

$$H\psi_n(\mathbf{r}) = E_n\psi_n(\mathbf{r}), \tag{1.3}$$

where the eigenvalues E_n are the energy levels of the particle, usually an electron. For an electron in a stationary state we can write

$$\Psi_n(\mathbf{r}, t) = c_n(t)\psi_n(\mathbf{r}). \tag{1.4}$$

Substitution of Eq. (1.4) in Eqs. (1.2) and (1.3) shows that c_n obeys the equation

$$i\hbar\frac{dc_n}{dt} = E_n c_n, \tag{1.5}$$

which has the solution

$$c_n \propto \exp\left(-iE_n t/\hbar\right). \tag{1.6}$$

Note that Eq. (1.6) is just a time-dependent phase factor; $|c_n|$ is independent of time. $c_n(t)$ is called the **probability amplitude** for the state ψ_n, because, if ψ_n is normalized, $|c_n(t)|^2$ is the probability of finding the electron in the state ψ_n at time t; in the present case of an electron known to be in the state ψ_n, this is unity, independent of time.

When several atoms are brought together to form a molecule or solid the atomic wavefunctions with which we are familiar are no longer stationary states, and Eqs. (1.3)–(1.6), describing the behaviour of stationary states, are no longer adequate. The atomic quantum states are now coupled together, and an electron which is in a particular state at $t = 0$ may be found in another state at a later time. To describe this situation of coupled probability amplitudes we need a set of coupled equations in place of the independent Eqs. (1.5). The obvious generalization of Eqs. (1.5) is

$$i\hbar\frac{dc_n}{dt} = \sum_m E_{nm}c_m, \tag{1.7}$$

in which the rate of change of each probability amplitude is coupled to other probability amplitudes by the coefficients E_{nm} for $n \neq m$. We show in

Appendix A that a result of the form of Eq. (1.7) does indeed follow from the Schrödinger equation, Eq. (1.2), by substituting a wavefunction of the form

$$\Psi(\mathbf{r}, t) = \sum_l a_l(t)\psi_l(\mathbf{r}). \tag{1.8}$$

The coupled probability amplitude equations, Eqs. (1.7), are the form of quantum mechanics which we shall use to treat electrons in molecules and solids. This formulation of quantum mechanics, in terms of probability amplitudes, is taken as basic by Feynman,[1] who uses it to make the Schrödinger equation plausible (Chapter 16), the converse of our proof in Appendix A.

1.2.2 The H_2^+ ion

We now apply Eqs. (1.7) to the simplest possible problem of atomic binding, the binding of two protons by a single electron. In accordance with the adiabatic approximation we shall proceed by considering two protons a distance R apart and calculating the electronic energy $E(R)$.

We simplify the problem by considering only two probability amplitudes: c_a, the amplitude for the electron to be in ψ_a, the atomic ground state function for proton A at position R_a; and c_b, the amplitude for the electron to be in ψ_b, the ground state for proton B at R_b. These ground state functions are sketched in Fig. 1.1(a). We neglect the amplitudes for the electron to be found in atomic excited states, since to include them would complicate the algebra without adding anything to our physical understanding of the problem.

With this simplification Eqs. (1.7) become

$$i\hbar\frac{dc_a}{dt} = Ec_a - Ac_b,$$

$$i\hbar\frac{dc_b}{dt} = Ec_b - Ac_a; \tag{1.9}$$

these two equations differ only in the interchange of the suffixes a and b, because the two protons are identical. The energies E and A are of course functions of the separation R between the two protons. To calculate these energies would require more detailed quantum mechanics (see Appendix A), but this does not concern us here. The minus sign in front of A is arbitrary; we shall see later that this choice makes A positive.

We notice that Eqs. (1.9) look rather like the coupled oscillator equations that one gets, for example, for two identical pendulums coupled together; the main difference is that the differential operator is id/dt instead of d^2/dt^2. Because of this similarity the same technique can be used to solve them.

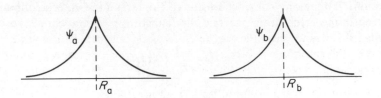

(a) The wavefunctions of two isolated atoms.

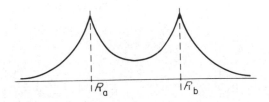

(b) The bonding wavefunction $(\psi_a + \psi_b)$.

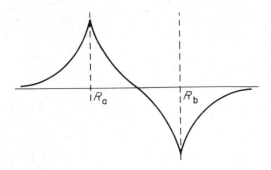

(c) The antibonding wavefunction $(\psi_a - \psi_b)$.

Fig. 1.1.

Taking the sum and difference of Eqs. (1.9) we obtain *uncoupled* equations for the new variables $c_+ = c_a + c_b$ and $c_- = c_a - c_b$:

$$i\hbar\frac{dc_+}{dt} = (E - A)c_+,$$

$$i\hbar\frac{dc_-}{dt} = (E + A)c_-.$$

(1.10)

These uncoupled equations are of the form Eq. (1.5), and the amplitudes

c_+ and c_- therefore refer to stationary states of the new problem—an electron in the field of two protons. To find the wavefunctions of these new stationary states we note that if the electron is definitely in the state represented by the amplitude c_+, then $|c_-| = 0$ and $c_a = c_b$. Because ψ_a and ψ_b are the same function centred on different points, Eq. (A.2) of Appendix A shows that $c_a = c_b$ if $a_a = a_b$ in Eq. (1.8). Eqs. (1.8) and (1.10) therefore imply that

$$\Psi_+(\mathbf{r}, t) \propto [\psi_a(\mathbf{r}) + \psi_b(\mathbf{r})] \exp{[-\mathrm{i}(E - A)t/\hbar]}. \tag{1.11}$$

Similarly, the amplitude c_- refers to a state with the wavefunction

$$\Psi_-(\mathbf{r}, t) \propto [\psi_a(\mathbf{r}) - \psi_b(\mathbf{r})] \exp{[-\mathrm{i}(E + A)t/\hbar]}. \tag{1.12}$$

The spatial variation of these wavefunctions along the internuclear line is sketched in Figs. 1.1(b) and (c) respectively.

Knowing these wavefunctions, we can say something about the energies of the states, without calculating E and A. The electron energy consists of two contributions: electrostatic potential energy in the field of the two protons, and kinetic energy proportional to $|\nabla\psi|^2$. Comparison of Figs. 1.1(b) and (c) shows that both these contributions are lower for the symmetric state $(\psi_a + \psi_b)$ in Fig. 1.1(b); the potential energy is low because $|\psi|^2$ is large in the region of low potential between the two nuclei, and the kinetic energy is low because $|\nabla\psi|^2$ is lower between the nuclei and much the same elsewhere. The energy A is therefore positive.

Our wavefunctions are not exact because of our neglect of atomic excited states in Eqs. (1.10), but in spite of this we can obtain some exact energies by noting the symmetries of our solutions. The symmetric state in Fig. 1.1(b) is nodeless, like an atomic 1s state. As the separation R between the two protons decreases to zero we therefore expect that this state will become the 1s state of the ion He$^+$ (nuclear charge $2e$), which has an electronic energy of -4 Rydberg (1 Rydberg is the ground state binding energy of hydrogen, 13.6 eV). On the other hand the antisymmetric state $(\psi_a - \psi_b)$ shown in Fig. 1.1(c) has a single nodal plane perpendicular to the internuclear line. This is the symmetry of an atomic 2p state, and we therefore expect that in the limit $R \to 0$ it will become the 2p state of He$^+$, with an electronic energy of -1 Ry. In the opposite limit $R \to \infty$ both states tend to the ground state energy of atomic hydrogen, -1 Ry.

These limiting energies are seen to be correct in Fig. 1.2(a), which shows the results of exact calculations of the symmetric and antisymmetric states of H$_2^+$. In other words, the curves in Fig. 1.2(a) show $E(R)$ in Eq. (1.1) for the ground state and first excited state of H$_2^+$. If we add to this the third term in Eq. (1.1), the internuclear Coulomb repulsion, we obtain the total effective internuclear potential energy curves shown in Fig. 1.2(b). We see that only the lower curve shows a potential minimum, giving the possibility of a stable nuclear separation. For this reason the symmetric wavefunction is

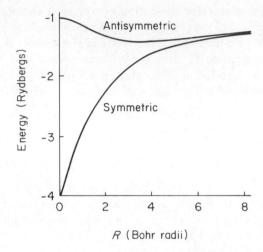

(*a*) Electron energy as a function of nuclear separation.

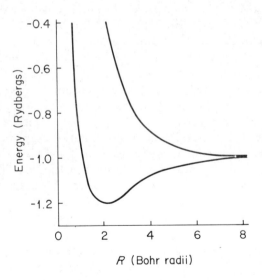

(*b*) Total energy including nuclear repulsion.

Fig. 1.2. {after *Quantum Theory of Molecules and Solids*, Vol. 1, by John C. Slater. © 63 McGraw-Hill Book Company Inc. Used with permission}.

called a **bonding orbital** and the antisymmetric wavefunction an **antibonding orbital**.

1.3   TYPES OF BINDING

1.3.1   Covalent binding

The binding together of two atoms by shared electrons is called a covalent or homopolar bond. The binding of two protons by a single electron, which we have just considered, serves to illustrate the basic effect—reduction of energy by concentration of electrons near the internuclear line—but it is not quite typical because the resultant molecular ion, H_2^+, is charged.

When neutral atoms are bound by equally shared electrons this necessarily involves taking one electron from each to form the bond. Consider the simplest example, the neutral hydrogen molecule H_2. If we neglect the mutual Coulomb repulsion of the two electrons they can be treated independently and our results for H_2^+ are immediately applicable. The lowest energy state is obtained by putting both electrons in the bonding orbital, and this is allowed by the exclusion principle provided that the two electrons have opposite spin. More precisely, both electrons in the bonding orbital $\psi_+(\mathbf{r})$ are represented by a simple product wavefunction

$$\psi(\mathbf{r}_1, \mathbf{r}_2) = \psi_+(\mathbf{r}_1)\psi_+(\mathbf{r}_2), \tag{1.13}$$

which is symmetric under interchange of the electron coordinates; to make the total wavefunction, including spin, antisymmetric, the spin wavefunction must be antisymmetric, implying antiparallel spins. (Heitler,[9] Chapter V).

This independent particle approximation, neglecting electron interaction, is one that is very widely used in solid state physics. This is not so much because it is a good approximation as because it is the only tractable approximation for problems involving many electrons. It is therefore instructive, in the simple case of the hydrogen molecule, to examine the sort of errors to which it gives rise, so as to obtain an idea of how far it is to be trusted.

To do this we examine the structure of the two electron bonding wavefunction, Eq. (1.13), using our approximate form for $\psi_+(\mathbf{r})$ from section 1.2.2,

$$\psi_+(\mathbf{r}) = \psi_a(\mathbf{r}) + \psi_b(\mathbf{r}).$$

Eq. (1.13) then becomes

$$\psi(\mathbf{r}_1, \mathbf{r}_2) = \psi_a(\mathbf{r}_1)\psi_a(\mathbf{r}_2) + \psi_b(\mathbf{r}_1)\psi_b(\mathbf{r}_2)$$
$$+ \psi_a(\mathbf{r}_1)\psi_b(\mathbf{r}_2) + \psi_b(\mathbf{r}_1)\psi_a(\mathbf{r}_2). \tag{1.14}$$

If we remember that ψ_a is large near nucleus A and ψ_b large near nucleus B we can see that the first two terms in this wavefunction have a rather different physical interpretation from the last two. Thus, the first term has a large amplitude when both electrons are near nucleus A, and the second term has a large amplitude when both electrons are near nucleus B; but the last two terms in Eq. (1.14) have a large amplitude when one electron is near each nucleus. Thus, for large nuclear separations the last two terms give the probability amplitude for finding two neutral hydrogen atoms, whereas the first two terms give the probability amplitude for finding a bare proton H^+ and a negative hydrogen ion H^-. Within the independent particle approximation these states have the same energy, and it is quite legitimate to mix them, because an electron has a binding energy of 13.6 eV to a proton, independently of whether another electron is already bound. In fact, however, an electron is bound to a neutral hydrogen atom to make H^- by only 0.7 eV, because of the Coulomb repulsion of the first electron. Therefore, at large separations, when a clear distinction can be made between the states $H + H$ and $H^+ + H^-$, the independent particle approximation is bad; the higher energy of the state $H^+ + H^-$ means that we are almost certain to find one electron near each proton. In contrast, the independent particle wavefunction, Eq. (1.14), gives equal amplitude to the states $H + H$ and $H^+ + H^-$.

At smaller internuclear separations when ψ_a and ψ_b overlap appreciably this clear distinction between the two types of state can no longer be made and it is no longer obvious that the independent particle approximation is bad; it does not, however, follow that it is good, and we should always treat it with caution. Nevertheless, even if Eq. (1.14) is an inadequate wavefunction, we expect the true wavefunction for a covalent bond to share with it the property of symmetry under exchange of the electron coordinates r_1, r_2, leading to antiparallel spins for the two electrons.

This has the important consequence that a *pair* of electrons are an essential feature of a covalent bond; one electron from each atom gives one covalent bond. Consequently hydrogen can form only one covalent bond, and in general an atom cannot form more covalent bonds than it has electrons outside closed shells. Because of this limited number of bonds covalent binding is said to be **saturable**.

Some new features arise in an atom with several electrons outside a closed shell. Let us consider the case of carbon (ground state configuration $(1s)^2(2s)^2(2p)^2$), which is of particular importance to us because the semiconductors silicon and germanium have the same outer electron configuration. When the atom has formed four covalent bonds with its outer electrons it is in an environment that is no longer spherically symmetric; we expect there to be *four* equivalent directions—the bond directions—but we do not expect *all* directions to look alike.

The s and p wavefunctions are appropriate to a spherically symmetric situation because an s state (Fig. 1.3 (a)) has that symmetry and the three independent p states $xf(r)$, $yf(r)$, $zf(r)$ can be taken in linear combinations

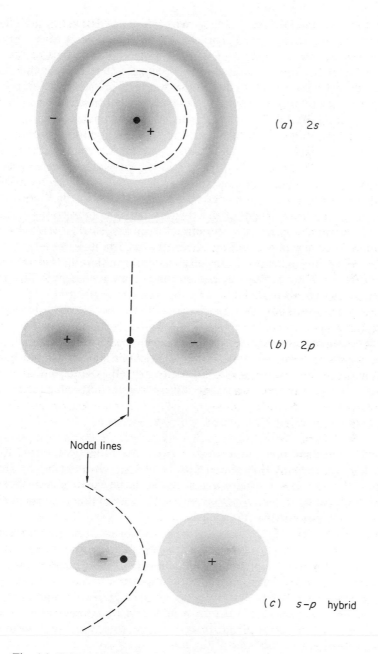

(a) 2s

(b) 2p

Nodal lines

(c) s-p hybrid

Fig. 1.3. Schematic electron wavefunctions in s and p states and an
s-p hybrid. Broken lines are nodal lines of ψ.

to give a p wavefunction (Fig. 1.3(b)) with its symmetry axis in any orientation whatever. For an environment with four equivalent special directions we need to take linear combinations of these four wavefunctions to produce four equivalent functions oriented in these special directions. Such a linear combination, called an s-p hybrid, is shown in Fig. 1.3(c); an important feature of such a hybrid is that there is a large concentration of wavefunction amplitude in a particular direction, which is favourable to overlap with a wavefunction from a neighbouring atom to give a bonding orbital in that direction. To obtain s-p hybrids oriented in four equivalent directions we vary the relative proportions of p_x, p_y, and p_z in the mixture to produce wavefunctions like Fig. 1.3(c) oriented in the four equivalent directions towards the corners of a regular tetrahedron, as shown in Fig. 1.4(a) (this is the only possible arrangement of exactly four equivalent directions in space). This tetrahedral arrangement of bonding directions is called sp^3 hybridization, and is the basis of the organic chemistry of carbon.

Although sp^3 hybridization is the most usual, carbon can form fewer than four bonds. Thus the s state can be combined with p_x and p_y to give three equivalent s-p hybrids in a plane, in the directions shown in Fig. 1.4(b). This lower symmetry is known as sp^2 hybridization, and leaves the p_z state unmodified.

We see from the foregoing that when an atom makes more than one covalent bond, these bonds occur at quite well-defined angles with respect to each other. Covalent bonds are thus **directed** as well as **saturable**. These two properties are crucial in determining the type of crystal structure to which they give rise.

1.3.2   Ionic and metallic binding

When a covalent bond is formed between two identical atoms it is a symmetry requirement that the electrons in the bond should be shared evenly, so that the resulting diatomic molecule has no electric dipole moment. But when the atoms are not identical, as for example in the hydrogen halides, this symmetry requirement no longer operates, because the two nuclei are now distinguishable. There may then be partial or complete transfer of charge from one atom to another. Thus, we can imagine hydrogen and fluorine forming a covalent bond from the hydrogen electron and the unpaired electron on the fluorine. Or in the opposite extreme we can imagine the hydrogen electron transferred to the fluorine giving the ion F^- with a rare gas electronic structure; the ions would then attract each other to form molecules, giving an ionic bond. In fact, an intermediate situation usually prevails. An objective measure of the ionicity of a bond is provided by the electric dipole moment of the resulting diatomic molecule, which would be the electronic charge multiplied by the internuclear separation for a fully ionic bond. These dipole moments can be deduced from measurements of

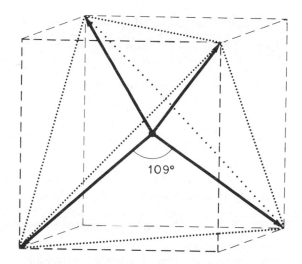

(a)   Tetrahedral directions of sp^3 hybrid, inscribed
      in a cube.

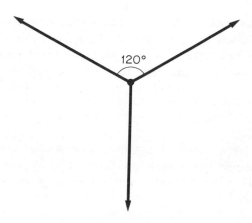

(b)   Triangular directions of sp^2 hybrid.

Fig. 1.4.

dielectric constant as a function of temperature (compare paramagnetism, section 5.2.1), and ratios of the measured value to the fully ionic value for the hydrogen halides are shown in Table 1.1.

We see that there is a continuous gradation from a bond that is about half ionic in HF to an almost purely covalent bond in HI. Hydrogen does not form fully ionic bonds because the bare proton has to penetrate the electron

Table 1.1. Ratios of measured electric dipole moments of the hydrogen halides to the theoretical value for a fully ionic bond.

HF	0.43
HCl	0.17
HBr	0.11
HI	0.05

cloud of the negative ion somewhat to attain electrostatic equilibrium. The most ionic bonds are formed between alkali metals and halogens; this is because each ion has a rare gas electron configuration so that the exclusion principle prevents much interpenetration—the ions behave almost like charged spheres of definite radius.

With such ions the forces involved in crystal formation are electrostatic attraction and hard core repulsion. The structure is therefore determined largely by the packing of spheres so as to minimize the electrostatic potential energy. In contrast to covalent structures, the bonds are neither saturable* nor directed.

The metallic bond can be thought of as a limiting case of the ionic bond in which the negative ions are just electrons. Thus, sodium chloride contains equal numbers of Na^+ and Cl^-, and metallic sodium contains equal numbers of Na^+ and e^-. The crucial difference is that the very small mass of an electron means that its zero-point motion is large, so that it is not localized on a lattice site. A metallic structure is therefore determined largely by the packing of the positive ions alone; the electron fluid is just a sort of negatively charged glue.

Alternatively, metals may be regarded as a special case of a covalent structure in which it is not possible to make enough bonds to produce a rigid structure. Thus, most metallic atoms have fewer than four electrons in the outermost shell, yet a minimum of four covalent bonds per atom is necessary to build a three-dimensional structure from a single type of atom. With fewer than four electrons per atom some of the bonds have to be left out, and there are many ways of doing this. We can imagine a ground state wavefunction for the crystal which is a linear combination of all possible ways of leaving out the prescribed fraction of bonds. Note that this picture also leads to the idea of electrons that are not localized, as is necessary for electrical conductivity.

---

* Do not confuse the saturation of covalent bonds with the saturation of nuclear forces. The latter concept means only that there is a finite limiting density of nuclear matter with a constant binding energy per particle—analogously to a solid or liquid. In this weaker sense *all* the forces in solid state physics are saturated.

1.3.3 Weak binding

The covalent, ionic, and metallic bonds are all quite strong, with binding energies of the order of the Coulomb energy of two electrons a few Angstroms apart, a few eV/atom. There are also some much weaker forces between atoms that are responsible for the liquefaction and crystallization of neutral molecules or rare gas atoms. The most important for our purposes is the van der Waals force, so called because it is responsible for the deviations from ideal gas behaviour studied by van der Waals. The force arises because even a spherically symmetrical atom or molecule has a *fluctuating* electric dipole moment due to electronic zero-point motion. This can *induce* a dipole moment in a neighbouring atom or molecule, and the two dipoles will attract each other. In other words, the fluctuating dipole moments in neighbouring molecules are correlated because they are coupled via the electromagnetic field, and this correlation leads to an average net attraction. Like the metallic bond, this force is undirected and not saturable. The rare gas solids are therefore, like metals, largely determined by the packing of spheres; and organic molecules crystallize in such a way as to pack together their shapes most compactly.

Another weak bond that is worth mentioning for completeness is the hydrogen bond. Table 1.1 illustrates that a hydrogen atom is usually a somewhat positively charged region of a molecule. This can, by electrostatic attraction, form a weak bond to a negatively charged region of another (or the same) molecule. An interesting example is that of water. In H_2O the outer electron configuration of the oxygen is four sp^3 hybrid orbitals as illustrated in Fig. 1.4(a); two of these are used for covalent bonds to the hydrogen atoms, while the other two are doubly occupied by oxygen electrons. Thus, of the six outer oxygen electrons, two are used in bonding and the other four are in 'spare' orbitals. As a result of this the water molecule is rather like a tetrahedron in shape with the two hydrogen occupied corners positively charged and the other two corners negatively charged. Hydrogen bonds are then formed between the positive corners of one molecule and the negative corners of another, leading to a rather open structure, because each molecule can form hydrogen bonds with only four neighbours.

Another interesting example, outside the scope of this book, is the spiral form of the DNA molecule, which is due to hydrogen bonding between different parts of the same long molecule.

1.4 QUANTUM MECHANICS OF THE COVALENT BOND

We now extend our discussion of the hydrogen molecule to include excited states. This enables us to relate the independent particle approximation to the alternative Heitler–London approximation (Heitler,[9] Chapter

IX), and also makes possible a fuller introduction to the idea of exchange energy than we give in section 5.4.

Consider all two particle wavefunctions that can be constructed as products of the bonding and antibonding wavefunctions of H_2^+, ψ_+ and ψ_- respectively. The wavefunction for the lowest such state was given in Eqs. (1.13) and (1.14). It is convenient to have a pictorial representation of such a function, but this is difficult because it is a function of six variables (the vectors $\mathbf{r}_1, \mathbf{r}_2$). We therefore restrict our attention to the variation of ψ along the internuclear line, which we take as the x coordinate, and plot the amplitude $\psi(x_1, x_2)$ as a contour diagram in the (x_1, x_2) plane. The various possible product wavefunctions derived from ψ_+ and ψ_- are plotted schematically in this way in Fig. 1.5; the solid circles are contours, the $+$ and $-$ signs indicate positive and negative amplitude peaks of ψ, and the broken lines are nodal lines of ψ.

The wavefunction for both electrons in a bonding orbital, given in Eq. (1.14) is depicted in Fig. 1.5(a); the peak labelled aa is the first term in Eq. (1.14), the peak bb is the second term, and the peaks ab and ba are the third and fourth terms respectively. Fig. 1.5(b) is the wavefunction for both electrons in an antibonding orbital and Figs. 1.5(c) and (d) are wavefunctions for one electron in the bonding orbital and one in the antibonding orbital.

The last two wavefunctions are not physically acceptable because they do not satisfy the condition

$$|\psi(x_1, x_2)|^2 = |\psi(x_2, x_1)|^2$$

or

$$\psi(x_1, x_2) = \pm\psi(x_2, x_1), \qquad (1.15)$$

which is required by the indistinguishability of electrons: the statement that electron 1 is a bonding orbital and electron 2 in an antibonding orbital is meaningless; we can only say that there is one electron in each orbital. Two wavefunctions such as those in Figs. 1.5(c) and (d) that differ only in the labelling of the electrons are said to be **exchange degenerate**. To obtain physically acceptable wavefunctions from them we have to construct linear combinations that are symmetric or antisymmetric under the exchange of coordinates (x_1, x_2), so as to satisfy Eq. (1.15). Such linear combinations are shown in Figs. 1.6(b) and (c); they are even or odd under reflection in the line $x_1 = x_2$. The independent particle approximation thus yields the four states shown in Fig. 1.6: (a) is two electrons in a bonding orbital with energy $2(E - A)$; (b) and (c) are degenerate states for one electron in a bonding orbital and one in an antibonding orbital, with total energy $2E$; and (d) is two electrons in an antibonding orbital, with energy $2(E + A)$. There is one additional requirement we have to satisfy: the total electron wavefunction must be antisymmetric under exchange of space and spin coordinates; a

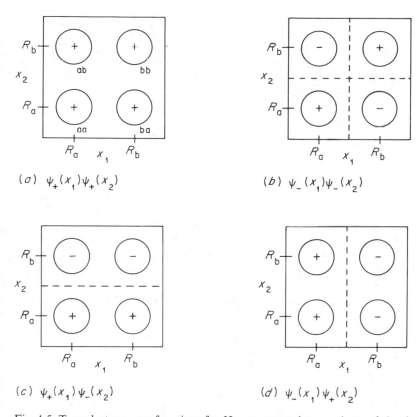

(a) $\psi_+(x_1)\psi_+(x_2)$

(b) $\psi_-(x_1)\psi_-(x_2)$

(c) $\psi_+(x_1)\psi_-(x_2)$

(d) $\psi_-(x_1)\psi_+(x_2)$

Fig. 1.5. Two electron wavefunctions for H_2 constructed as products of simple electron wavefunctions for H_2^+. The functions (c) and (d) are not physically acceptable because $|\psi(x_1,x_2)|^2 \neq |\psi(x_2,x_1)|^2$.

symmetric space wavefunction must be associated with an antisymmetric spin function and vice versa. Since symmetric and antisymmetric spin wavefunctions are associated with parallel spin (triplet) and antiparallel spin (singlet) states respectively, (see Heitler,[9] Chapter V), we thus obtain the spin states shown in Fig. 1.6.

We can obtain a qualitative idea of the effect of the Coulomb repulsion of the two electrons on the energies of these states by noting that the energy will be raised by the largest amount for those states which have a large amplitude for $x_1 \approx x_2$. Thus, states containing the peaks aa and bb of Fig. 1.5(a) will be strongly affected, and states containing the peaks ab and ba will not be much affected; this agrees with our conclusion in section 1.3.1 that ionic states with both electrons near the same proton are higher in energy. The state shown in Fig. 1.6(c) will therefore be raised most in energy and the state in Fig. 1.6(b) will be raised least, removing the exchange degeneracy

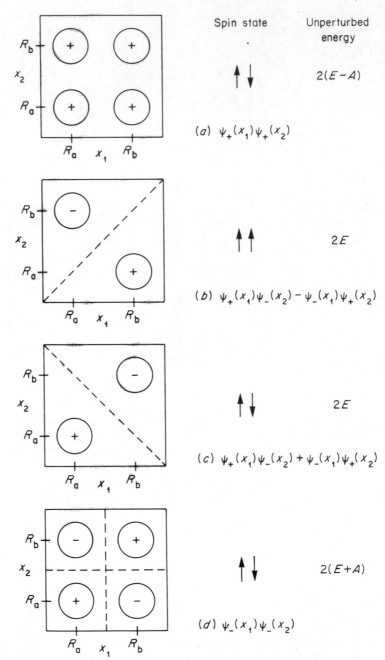

Fig. 1.6. Physically acceptable two electron wavefunctions for H_2, derived from the single electron wavefunctions of H_2^+ in the independent particle approximation.

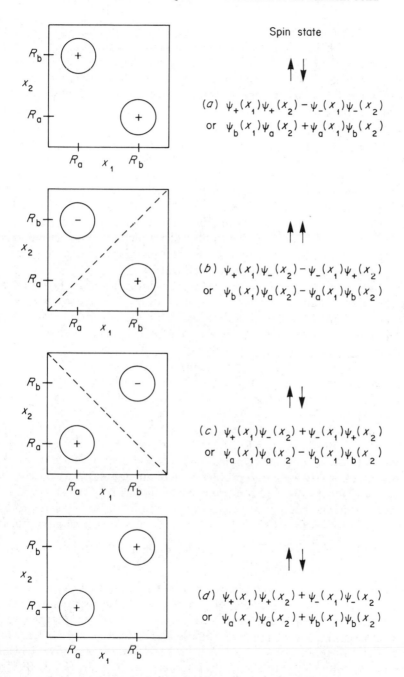

Spin state

(a) $\psi_+(x_1)\psi_+(x_2) - \psi_-(x_1)\psi_-(x_2)$
or $\psi_b(x_1)\psi_a(x_2) + \psi_a(x_1)\psi_b(x_2)$

(b) $\psi_+(x_1)\psi_-(x_2) - \psi_-(x_1)\psi_+(x_2)$
or $\psi_b(x_1)\psi_a(x_2) - \psi_a(x_1)\psi_b(x_2)$

(c) $\psi_+(x_1)\psi_-(x_2) + \psi_-(x_1)\psi_+(x_2)$
or $\psi_a(x_1)\psi_a(x_2) - \psi_b(x_1)\psi_b(x_2)$

(d) $\psi_+(x_1)\psi_+(x_2) + \psi_-(x_1)\psi_-(x_2)$
or $\psi_a(x_1)\psi_a(x_2) + \psi_b(x_1)\psi_b(x_2)$

Fig. 1.7. Physically acceptable two electron wavefunctions for H_2, after taking some account of the mutual repulsion of the electrons.

between these two states. The states in Figs. 1.6(a) and (d), which contain both types of peak, will be raised an intermediate amount. In fact, as we argued in section 1.3.1, these states are not correct at large nuclear separations, because they are linear combinations of states with different energy— $(H + H)$ and $(H^+ + H^-)$. Instead we should take linear combinations of the states in Figs. 1.6(a) and (d) so that they contain only the peaks ab and ba (corresponding to $H + H$) or only the peaks aa and bb (corresponding to $H^+ + H^-$). Such states are shown in Figs. 1.7(a) and (d). All four states that now result are shown in Fig. 1.7, in order of increasing energy, from (a) to (d). Our arguments suggest that the wavefunctions of Fig. 1.7 should be closer to the truth than the independent particle wavefunctions of Fig. 1.6.

It is interesting that if we set $\psi_+ = (\psi_a + \psi_b)$ and $\psi_- = (\psi_a - \psi_b)$, as in section 1.2.2, the wavefunctions in Fig. 1.7 can alternatively be expressed simply in terms of ψ_a and ψ_b, the single atom functions. You should check for yourself that the alternative wavefunctions given in the captions to Fig. 1.7 are equivalent, apart from normalization.

The wavefunctions in Figs. 1.7(a) and (b) are in fact the basis of the Heitler–London approximation for calculating the binding of the hydrogen molecule (Heitler,[9] Chapter IX). They are the acceptable linear combinations of the exchange degenerate states $\psi_a(x_1)\psi_b(x_2)$ and $\psi_b(x_1)\psi_a(x_2)$, and are quite a good approximation to the lowest two states. The wavefunctions in Figs. 1.7(c) and (d) are not such a good approximation because $|\psi|$ is reduced along the line $x_1 = x_2$ by the Coulomb repulsion between the electrons. Accurately calculated energies for states corresponding to those shown in Fig. 1.7 are shown in Fig. 1.8. These accurate calculations show that at the equilibrium internuclear separation the ground state wavefunction of H_2 is much closer to Fig. 1.7(a) than Fig. 1.6(a); the Heitler–London approximation is therefore to be preferred. Unfortunately the Heitler–London method is almost impossible to extend to problems involving more than two electrons, whereas the independent particle approximation is readily extended. It is because of this, and because of its conceptual simplicity, that the independent particle approximation is so widely used in solid state physics, despite the weaknesses discussed above.

1.4.1 Exchange interaction

The removal of degeneracy between states such as those in Figs. 1.6(b) and (c) is said to be due to **exchange interaction**. This is an essentially quantum mechanical effect resulting from the requirement of Eq. (1.15) that the wavefunction be symmetric or antisymmetric under exchange of particles. Thus, the product wavefunctions shown in Figs. 1.5(c) and (d) both give the same expectation value of the Coulomb interaction $e^2/4\pi\varepsilon_0|\mathbf{r}_1 - \mathbf{r}_2|$. But the correctly symmetrized linear combination in Figs. 1.6(b) and (c) give quite different expectation values; we have already remarked that the state in

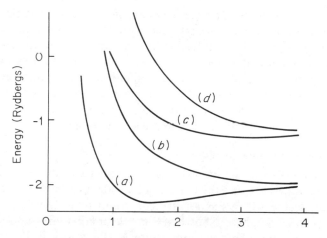

Fig. 1.8. Energies of H_2 for wavefunctions roughly like those of Fig. 1.7. The upper two curves tend to the energy of $H^+ + H^-$ at large R; the lower curves tend to the energy of two neutral H atoms. The letters on the curves indicate the corresponding wavefunctions in Fig. 1.7.
{after *Quantum Theory of Molecules and Solids*, Vol. 1, by John C. Slater. © 63 McGraw-Hill Book Company Inc. Used with permission}.

Fig. 1.6(c) has its energy raised the most because of the larger $|\psi|$ for $x_1 \approx x_2$.

Because of the relation between space and spin symmetry these two states therefore behave as if there were an energy difference depending on whether the spins are parallel or antiparallel. It is convenient to express this by a term $-2J\boldsymbol{\sigma}_1 \cdot \boldsymbol{\sigma}_2$ in the energy, where the quantity J is called the exchange energy; its form, which is considered in Appendix B, need not concern us here. This is of the same algebraic form as a magnetic dipole–dipole interaction energy, but it is *not* a magnetic effect. As we have seen, it is an electrostatic effect resulting from the symmetry of wavefunctions; it is *much stronger* than the magnetic effect, a typical energy being of order 1 eV. Because an antisymmetric space wavefunction has $\psi = 0$ for $x_1 = x_2$, this state will always feel Coulomb repulsion less. Consequently, Coulomb repulsion always gives a lower energy for parallel spins, as in Figs. 1.6(b) and (c); this effect is the basis of Hund's rule in atomic spectroscopy.

There may, however, be other contributions to an exchange interaction. Thus, we may also regard the states in Figs. 1.7(a) and (b) as split by an exchange interaction, for they are the acceptable linear combination of product states formed from ψ_a and ψ_b. In this case the state with antiparallel spin is lower. This can occur because the electron on each atom is perturbed not only by the repulsion of the other electron, but also by the attraction of the other proton; a result of either sign is possible for this type of exchange inter-action. It is this type of exchange interaction that is responsible for spin align-ment in ferromagnetic and antiferromagnetic substances (see section 5.4).

Finally, we remark that there is an element of arbitrariness in deciding which part of the Coulomb interaction is labelled exchange interaction. Thus in Fig. 1.7, if we start from ψ_+ and ψ_- it is natural to say that states (b) and (c) are separated by exchange interactions; but if we start from ψ_a and ψ_b it is natural to say that states (a) and (b) are so separated.

1.4.2   Transition to ionic binding

All the states in Fig. 1.7 are even or odd under exchange of the nuclear coordinates; this is because the two nuclei are identical. For a heteronuclear diatomic molecule AB this symmetry constraint is no longer present, and the states become roughly as shown in Fig. 1.9, for fairly large nuclear separa-tions. Fig. 1.9(c) represents the ionic state $A^- B^+$ and Fig. 1.9(d) the ionic state $A^+ B^-$; these now have different energies as the internuclear separation $R \to \infty$. The important point to note about Fig. 1.9 is that states (a), (c) and (d) are all singlets (antiparallel spin) with ψ positive everywhere. As R decreases so that ψ_a and ψ_b overlap, we expect the ground state to remain nodeless, but the other two states must acquire negative regions to remain orthogonal to it. There is, however, no symmetry constraint to say whether the ground state wavefunction at the equilibrium value of R will resemble Fig. 1.9(a), (c) or (d) more closely. A wavefunction can continuously evolve from one form to the other as R is decreased, because all three functions have the same symmetry. Thus, the final ground state may be ionic, covalent, or intermediate, for a heteronuclear molecule, and only detailed calculation can give the answer. On the other hand, a homonuclear diatomic molecule is always covalent by symmetry.

1.5   CRYSTAL GEOMETRY

We have already mentioned that a crystal is a regularly repeated structure on an atomic scale, a sort of three-dimensional wallpaper. In this section we explain some of the basic geometrical notions used to describe such struc-tures, and in section 1.6 we go on to describe some of the typical atomic arrangements that result from the various types of binding discussed in section 1.3. We conclude in section 1.7 with a brief introduction to x-ray crystallography, which provides the experimental evidence for the structures we shall describe.

1.5.1   The crystal lattice

The idea that a crystal structure could be built up from a fundamental unit originated before the direct evidence of x-ray diffraction; it arose from

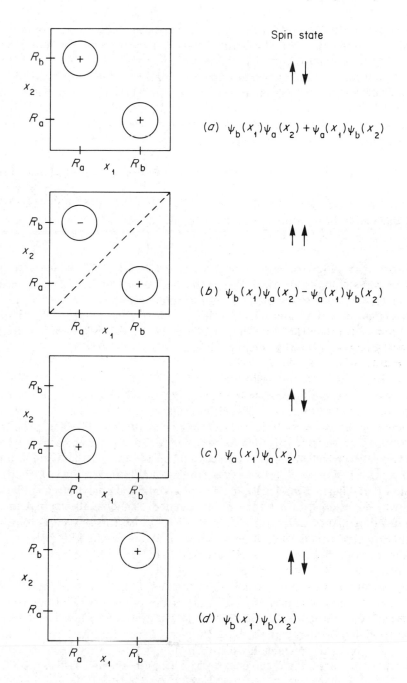

Fig. 1.9. Schematic wavefunctions for a diatomic molecule AB.

the observation of certain regularities in the angles between faces of natural crystals. Fig. 1.10 is an illustration from an early book on mineralogy showing how macroscopically plane faces in various orientations can be built up by stacking microscopic cubelets. In general, the unit is a parallelopiped defined by non-coplanar vectors **a**, **b**, **c**. This parallelopiped is the **unit cell** of the structure, and the complete structure is specified by giving both the unit cell and its contents, specified by **basis vectors** describing the atomic positions. If the basis vector from the origin of the unit cell to an atom is $\alpha\mathbf{a} + \beta\mathbf{b} + \gamma\mathbf{c}$, it is usual to shorten the notation and say that the atom is at (α, β, γ); the complete set of basis vectors is called the **basis** of the structure.

The repetitive nature of the atomic pattern is expressed by the fact that displacement through the crystal by a vector

$$u\mathbf{a} + v\mathbf{b} + w\mathbf{c}$$

where u, v and w are integers, brings us to an exactly equivalent point; an infinite crystal looks identical, and in the same orientation, from two points connected by such a vector. If the **translation vectors a, b, c** are chosen to make the unit cell as small as possible, both translation vectors and unit cell are said to be **primitive**. The set of points defined by all integral linear combinations of primitive translation vectors.

$$\mathbf{r} = u\mathbf{a} + v\mathbf{b} + w\mathbf{c}, \tag{1.16}$$

is called a **Bravais lattice**; it is the 'scaffolding' on which any crystal structure is built. A Bravais lattice in two dimensions is illustrated in Fig. 1.11(a), together with three possible choices of primitive unit cell. The primitive unit cell of a given lattice is therefore not unique; it is not even necessary that it should be a parallelopiped. A choice of primitive unit cell that is sometimes convenient is illustrated for our two dimensional lattice in Fig. 1.11(b). It is called the **Wigner–Seitz* cell,** and is the **coordination polyhedron** (polygon in two dimensions) of a lattice point, defined as the smallest polyhedron bounded by planes which are the perpendicular bisectors of vectors to other lattice points. From the definition, the interior of the coordination polyhedron is the locus of those points which are nearer to the given lattice point than to any other.

A Bravais lattice may have higher symmetry than just the translation symmetry implied by Eq. (1.16); we illustrate this point for a two-dimensional lattice. Such a lattice is specified by $|\mathbf{a}|$, $|\mathbf{b}|$, and the angle α between **a** and **b.** The special cases that can arise are shown in Fig. 1.12. First, we may have $\alpha = 90°$, giving the **rectangular lattice** in Fig. 1.12(a). Alternatively, with a general value of α we may have $|a| = |b|$, giving the **rhombic lattice** shown in

---

* After those who first used it for a quantum mechanical problem.

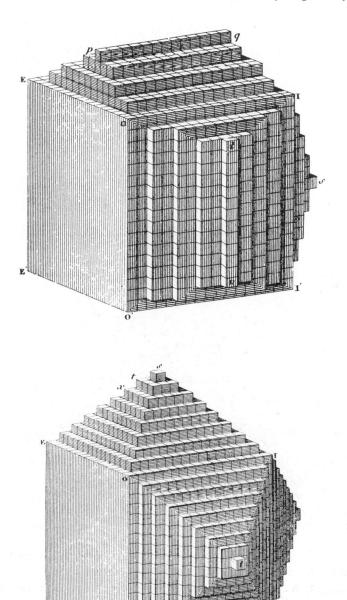

Fig. 1.10. The development of crystal faces from elementary
cubelets (Haüy, *Traite de Crystallographie*).

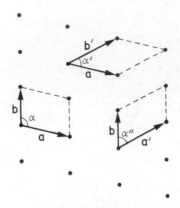

(*a*)   General Bravais lattice in two dimensions,
showing three possible primitive unit cells.

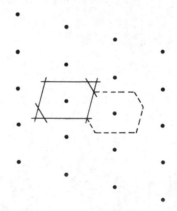

(*b*)   Coordination polygon (Wigner - Seitz cell)
for a two-dimensional lattice.

Fig. 1.11.

Fig. 1.12(*b*). This example is interesting; it shares certain symmetry with the rectangular lattice and can conveniently be referred to a rectangular unit cell defined by **a′**, **b′**. But this unit cell is *not* primitive, because there is a lattice point at the centre of the cell as well as the corners; we have a **centred rectangular lattice.** We shall meet the three dimensional analogue of this situation in section 1.6. Finally, in addition to $|\mathbf{a}| = |\mathbf{b}|$, we can have a special value of α. Two cases arise: $\alpha = 60°$ or $120°$ gives the **triangular lattice** of Fig. 1.12(*c*), with each lattice point surrounded by six neighbours at the corners of a regular hexagon; and $\alpha = 90°$ gives the **square lattice** of Fig. 1.12(*d*).

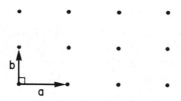

(*a*) Rectangular, $\alpha = 90°$.

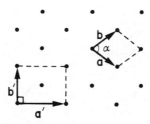

(*b*) Rhombic, $|\mathbf{a}| = |\mathbf{b}|$, equivalent to
centred rectangular.

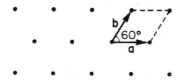

(*c*) Triangular, $|\mathbf{a}| = |\mathbf{b}|$, $\alpha = 60°$.

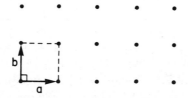

(*d*) Square, $|\mathbf{a}| = |\mathbf{b}|$, $\alpha = 90°$.

Fig. 1.12. The four special Bravais
lattices in two dimensions.

1.5.2   Crystal planes and directions

Crystal planes are built up by taking regular stepwise arrangements of unit cells as illustrated in Fig. 1.10. Usually, the faces that develop on a given crystal are those which contain fairly closely packed arrangements of atoms. Two examples of sets of parallel 'crystal planes' for a two dimensional lattice are illustrated in Fig. 1.13.* The usual notation for identifying such a plane

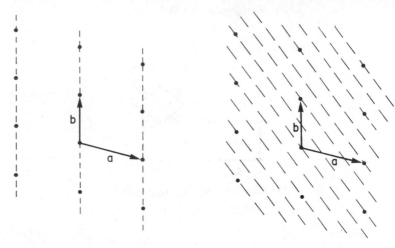

(a) The set of planes (10) in a   (b) The set of planes (32) in a
       two-dimensional Bravais lattice.   two-dimensional Bravais lattice.

Fig. 1.13.

is by means of **Miller indices.** These are derived from the intercepts on the crystal axes of the plane nearest the origin. Thus, in Fig. 1.13(b) the nearest plane to the origin has intercepts $\mathbf{a}/3$, $\mathbf{b}/2$ and is said to have Miller indices (32); in general, in three dimensions, a plane with intercepts $\mathbf{a}/h$, $\mathbf{b}/k$, $\mathbf{c}/l$ has Miller indices (hkl). Fig. 1.13(a) illustrates a special case in which one intercept is infinite so that the corresponding index is zero; the plane (10) is parallel to the $\mathbf{b}$ axis.

Some three dimensional examples are illustrated in Fig. 1.14. The plane (100) is parallel to both y and z axes, and hence to the yz plane. Negative intercepts are indicated by a bar over the corresponding index, as in ($1\bar{1}1$), ($2\bar{1}0$). For crystals of high symmetry certain planes may be related by symmetry and thus be equivalent from an atomic point of view. Thus, for crystals of cubic symmetry in which the unit cell sides $\mathbf{a}$, $\mathbf{b}$, $\mathbf{c}$, are equal in magnitude and mutually perpendicular, any permutation and changing of

---

* These planes, or rows of lattice points, may readily be picked out in a large orchard or military cemetery.

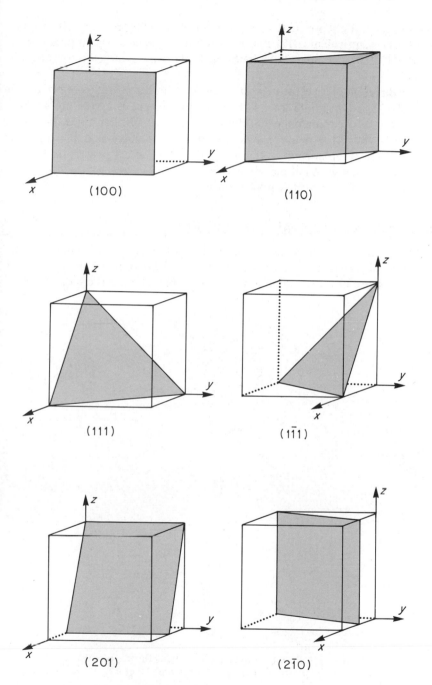

Fig. 1.14. Some crystal planes inscribed in a unit cell, with their Miller indices.

signs of the indices produces an equivalent plane: (111) is equivalent to ($1\bar{1}1$) and (201) is equivalent to ($2\bar{1}0$). It is customary to enclose the indices in curly brackets {210}, to mean 'all planes equivalent by symmetry to the given plane'.

The **direction** of a vector $\mathbf{r} = u\mathbf{a} + v\mathbf{b} + w\mathbf{c}$ in the crystal is referred to in a similar shorthand notation as 'the direction [uvw]', using *square* brackets. It is important to remember that this symbol is *not* a Miller index. For cubic crystals, however, it is a consequence of the symmetry that the direction [uvw] is that of the normal to the plane of Miller indices (uvw).* In this case therefore, which we shall mainly use for practical examples, the distinction between the two types of symbol is unimportant.

1.6 TYPICAL CRYSTAL STRUCTURES

The simplest crystal structures are those obtained by packing together identical spheres as closely as possible; such structures result from interatomic forces that are not directional and not saturable, and are therefore adopted

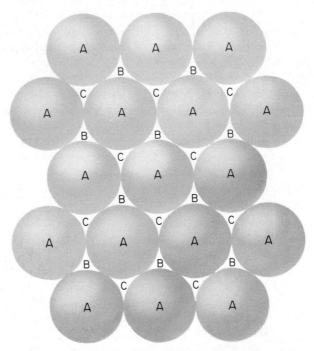

Fig. 1.15. A close-packed layer, A, showing the two possible
positions, B and C, for the next layer.

* The reason for this will become apparent in Chapter 6.

by metals and rare gas solids. A close-packed layer of spheres is illustrated by the spheres A in Fig. 1.15; their positions form a triangular Bravais lattice in two dimensions, Fig. 1.12(c). This packing can be extended to three dimensions if we place a second close packed layer over the first, with its spheres centred over interstices in the first layer. In this way each sphere in the second layer will touch three spheres in the first layer and the packing of layers will be as close as possible. Fig. 1.15 shows that such packing may be done in various ways, for the spheres of the second layer can occupy either the interstices B or the interstices C of the first layer. This leads to a variety of possible regular stacking sequences. The sequence ABCABC... gives a structure known as **cubic close-packed** (ccp) or **face-centred cubic** (fcc). A cubic unit cell of this structure is shown in Fig. 1.16; there are atoms at the corners of the unit cell and at the centre of each face. To make clear the relation to Fig. 1.15 a close-packed layer of atoms is shaded in Fig. 1.16(a); it is a (111) plane, normal to a body diagonal of the cube, and from symmetry close-packed layers of atoms can be found in all planes of the form {111}. In this structure the environment of each atom is identical in both form and orientation. The atomic positions therefore form a Bravais lattice, and the rhombo-hedral primitive unit cell of this lattice is shown in Fig. 1.16(b). But it is usual to use the larger cubic unit cell shown in Fig. 1.16 because this shows the full cubic symmetry, analogously to the use of a centred rectangular cell for the rhombic lattice in two dimensions, Fig. 1.12(b).

The environment of an atom is best visualized by means of the atomic coordination polyhedron, formed from planes which are the perpendicular bisectors of lines joining an atom to its neighbours. Clearly, the structure can be built up by stacking such polyhedra to fill space. Suppose you built a model of the structure out of plasticine spheres and then compressed it; if you then picked it apart you would find the spheres had deformed into coordination polyhedra.* The coordination polyhedron thus represents the 'sphere of influence' of an atom. The coordinations polyhedron of fcc is shown in Fig. 1.17, together with the positions of the nearest neighbour atoms. Fig. 1.17 shows the same unit cell as Fig. 1.16, but the origin is shifted by half a cell side along a cube axis, the [100] direction. This polyhedron is called the rhombic dodecahedron; it has twelve faces corresponding to contact with 12 nearest neighbors—a **coordination number** of 12. On this polyhedron it is fairly easy to identify the symmetries characteristic of a cubic structure. The rhombic dodecahedron has four three-fold axes of symmetry through opposite pairs of corners A ([111] directions); it looks the same after rotation through $2\pi/3$ about one of these axes. There are also three four-fold axes through opposite pairs of corners B ([100] directions) and six two-fold axes through the centres of opposite faces ([110] directions). Fig. 1.18 shows several rhombic dodecahedra stacked together in a pyramid.

* This method has actually been used by Bernal to study disordered (liquid) structures.

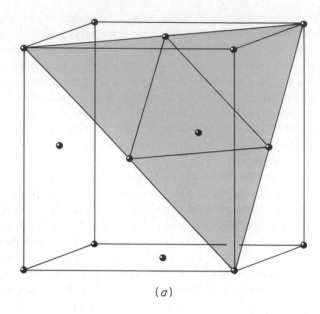

(a)

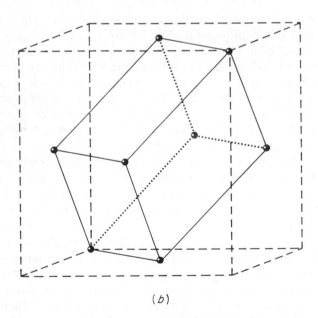

(b)

Fig. 1.16. The close-packed (face-centred) cubic structure.
(a) Conventional cubic unit cell, showing a close-packed (111) plane.
(b) Primitive rhombohedral unit cell.

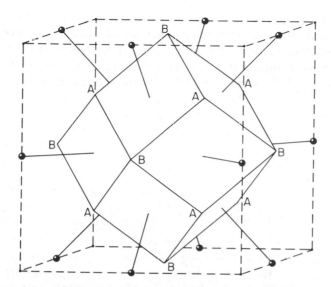

Fig. 1.17. Coordination polyhedron of the cubic close-packed structure: the rhombic dodecahedron. Bonds to nearest neighbour atoms are also shown.

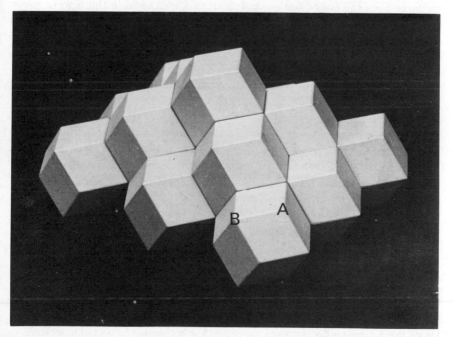

Fig. 1.18. Coordination polyhedra of fcc stacked with a (110) plane horizontal. The vertices A and B are as in Fig. 1.17.

On the right a close-packed (111) plane composed of vertices of type A can be identified, and on the left a (100) plane composed of vertices of type B; the base of the pyramid is a plane of the form {110}. It is manifest from Fig. 1.18 that the rhombic dodecahedra all stack together in the same orientation, confirming that the atomic positions in fcc constitute a Bravais lattice, and thus the atomic coordination polyhedron is a Wigner–Seitz unit cell.*

An alternative stacking of the close-packed layers in Fig. 1.15 is ABABAB .... This gives the **hexagonal close-packed** (hcp) structure illustrated in Fig. 1.19. The unit cell shown is in fact a primitive one, although it has a basis of two atoms at (000) and $(\frac{2}{3} \frac{1}{3} \frac{1}{2})$, because, as we shall see, the atomic positions do not form a Bravais lattice. The relation between the hexagonal and cubic close-packed structures is best understood by means of the atomic coordination polyhedra. Fig. 1.20(a) shows the coordination polyhedron of the cubic structure, just as in Fig. 1.17, except that the neighbouring atomic positions are linked by dotted lines to emphasize the close-packed layer structure, with six neighbours in the same layer, three in the layer below, and three in the layer above. Note that the triangles of atoms in the layers above and below are oppositely oriented. Fig. 1.20(b) is the corresponding picture for the hexagonal close-packed structure; the triangles of atoms in the layers above and below are now similarly oriented, with the result that the faces F of the coordination polyhedron appear where there were edges before and the faces G have become trapeziums instead of rhombuses. The stacking of the two types of polyhedron are compared in Fig. 1.21; rhombic dodecahedra are shown at (a) and hcp polyhedra at (b). In both cases a close-packed (111) plane is horizontal. The important thing to notice is that all polyhedra are similarly oriented in (a), but the polyhedra in alternate layers are differently oriented in (b). Thus, although all atoms have similar environment in the hcp structure the orientations of this environment differ in alternate layers, and therefore the atomic positions do *not* form a Bravais lattice and the atomic coordination polyhedron is *not* a Wigner–Seitz cell. We may also note that the coordination polyhedron now has only one threefold symmetry axis, with the result that the spacing of close-packed layers is no longer determined by symmetry. Since atoms are not really hard spheres the unit cell axial ratio (c/a) (Fig. 1.19) often differs a little from the ideal value of 1.633.

Other, more complicated, stacking sequences of close packed layers are sometimes found, particularly in the rare earth metals, but they do not concern us here; they are obtained by stacking mixtures of the two types of polyhedron shown in Figs. 1.20 and 1.21.

---

* The reader familiar with the more regular pentagonal dodecahedron may like to convince himself that it cannot occur in crystals because such objects cannot be stacked to fill space, on account of their five-fold symmetry axes. The symmetry of the pentagonal dodecahedron is, however, common in the structure of the so-called spherical viruses.

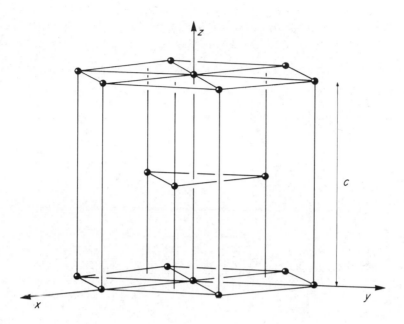

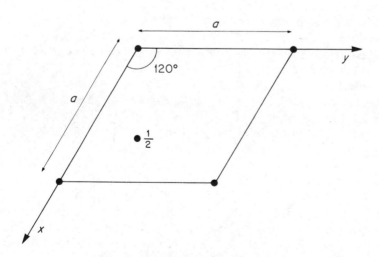

Fig. 1.19. Hexagonal close-packed structure, and plan of unit cell. Basis is $(0, 0, 0)$ and $(\frac{2}{3}, \frac{1}{3}, \frac{1}{2})$.

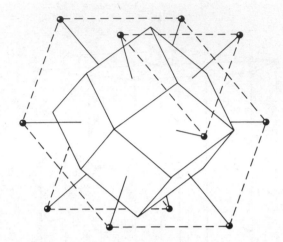

(*a*)   Coordination polyhedron of cubic close-packed structure, showing close-packed planes of nearest neighbours.

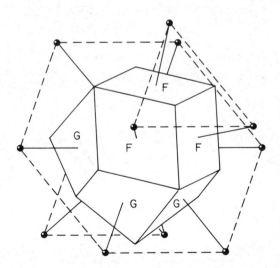

(*b*)   Coordination polyhedron of hexagonal close-packed structure, showing close-packed planes of nearest neighbours.

Fig. 1.20.

(a)   Cubic close-packed (fcc).

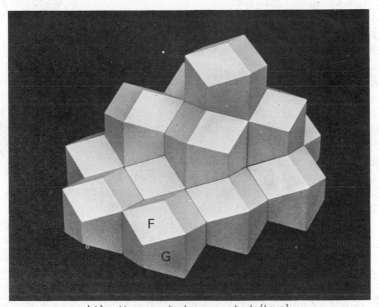

(b)   Hexagonal close-packed (hcp).

Fig. 1.21. Stacking of coordination polyhedra for close packed structures.
Faces F and G as in Fig. 1.20(b).

Another cubic structure, only slightly less closely packed than fcc is **body-centred cubic** (bcc), a cubic unit cell of which is shown in Fig. 1.22 with the central atom surrounded by its coordination polyhedron. The coordination polyhedra of all atoms in the structure are identical and similarly oriented, so the atomic positions do form a Bravais lattice, and, as with fcc, the cubic unit cell is primitive. The faces of the coordination polyhedron are eight regular hexagons and six squares. There are therefore eight nearest neighbours, giving a coordination number of 8 rather than 12, but there are also six second neighbours not much further away. The relationship between the two cubic structures can be seen with the aid of Fig. 1.23, which is a plan of four unit cells of bcc showing how a larger face centred cell can be constructed with x and y axes rotated 45°; this face-centred cell becomes cubic if the structure is stretched by a factor of $\sqrt{2}$ in the z-direction (or alternatively compressed by a factor of $\sqrt{2}$ in the x- and y-directions). In terms of the coordination polyhedra this distortion shrinks the lines BB in Fig. 1.22 and the top and bottom faces C to points which become the vertices B in Fig. 1.17; the vertices A in Fig. 1.22 become the vertices A in Fig. 1.17.

Packing considerations also determine the structure of ionic materials, but the presence of at least two types of ion makes the situation more complicated than for metals and rare gas solids. The electrostatic energy is minimized by each ion having as many neighbours of the opposite charge as possible, and the actual packing is determined partly by the charge ratio (and hence relative numbers) of the positive and negative ions, and partly by the ratio of their sizes. A single example of a common structure, that of a sodium chloride, NaCl, will serve to exhibit these principles. This structure is illustrated in Fig. 1.24; the ions occupy points of a simple cubic lattice, with Na^+ and Cl^- alternating so as to form a three-dimensional chessboard pattern. In this way each ion has six nearest neighbours of the other kind. Note that ions of a single kind occupy points of an fcc lattice, the Na^+ and Cl^- lattices being displaced from each other by half a unit cell in a [100] direction. It can also be seen from Fig. 1.24 that alternate planes of the form $\{111\}$ are occupied by Na^+ and Cl^- ions; this fact was crucial in the elucidation of the structure, the first to be determined by x-ray diffraction (see section 1.7). When this structure was first discovered the fact that there is no identifiable NaCl molecule caused some discomfort to chemists. We now know that the absence of an identifiable finite molecule is very general in inorganic crystals, and it no longer appears so shocking; we have become used to the idea of a crystal as a single giant molecule.

An example of a crystal that is a single giant covalent molecule is diamond. This structure is illustrated in Fig. 1.25. It is dominated by the tetrahedral sp^3 hybridized bonds characteristic of carbon in organic chemistry; Fig. 1.25 makes it clear that each atom is tetrahedrally bonded to four nearest neighbours. The atoms are shown as of two kinds in Fig. 1.25 to bring out the fact

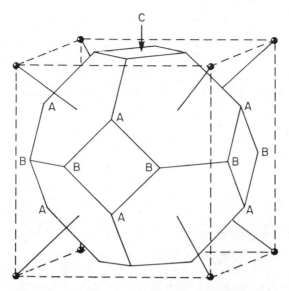

Fig. 1.22. Coordination polyhedron of body-centred cubic structure inscribed in the conventional cubic unit cell. The hexagonal faces represent contact with 8 nearest neighbours; the square faces represent contact with 6 second neighbours.

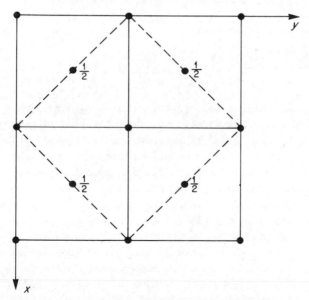

Fig. 1.23. Plan of four cubic unit cells of body-centred cubic structure showing by broken lines an alternative face-centred cell which becomes cubic by stretching along the z-axis.

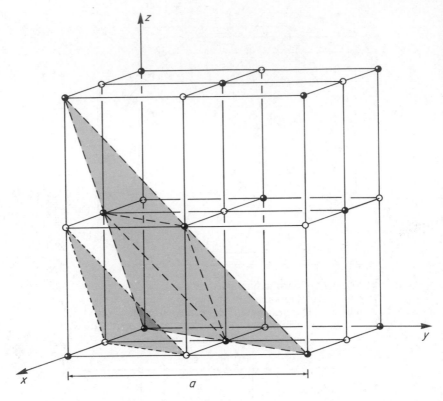

Fig. 1.24. NaCl structure, with ($\bar{1}11$) planes of Na$^+$ and Cl$^-$ ions indicated by broken lines.

that the atomic positions do not form a Bravais lattice; although all atoms have the same environment, this environment is differently oriented for the light and dark atoms. Fig. 1.25(*a*) shows a cubic unit cell of the structure, from which it can be seen that the dark atoms form an fcc Bravais lattice; the light atoms form a second fcc lattice displaced by $\frac{1}{4}$ of a unit cell in a [111] direction. The diamond structure is important for us because it is also the structure of the semiconducting elements silicon and germanium, which come below carbon in group IV of the periodic table.

A closely related structure is that of zincblende, ZnS, in which the two types of atom in Fig. 1.25 are zinc and sulphur; this is also the structure of group III–group V semiconducting compounds, such as gallium arsenide, GaAs. With two different types of atom the structure need not be purely covalent, and the bonds in these structures have some ionic character; there is some positive charge on the Zn or Ga atoms balanced by negative charge on S or As.

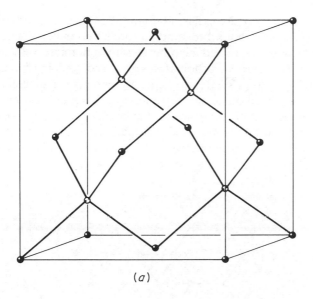

(a)

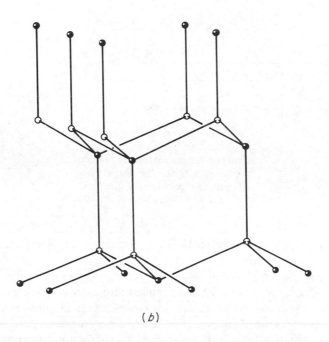

(b)

Fig. 1.25. Two views of the diamond and zincblende
structures.

A single crystal structure can exhibit more than one type of binding; thus, organic molecules are bound together in solids by van der Waals forces. An interesting example of mixed binding in a monatomic crystal is graphite. This is a layer structure of carbon atoms arranged in planar hexagonal nets, as shown in Fig. 1.26. A single layer constitutes a giant molecule, covalently bound by sp^2 hybrid bonds. The remaining p_z electrons on each atom form something like a metallic bond within the layer, so that graphite is a sort of two-dimensional metal, with good conductivity only in the plane of the layers. The residual bonding between layers is quite weak, hence the ready cleavage that gives graphite its lubricating properties.

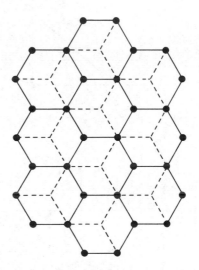

Fig. 1.26. Plan of the structure of graphite. The basic unit is a layer of carbon atoms covalently bonded in a hexagonal net; the dotted lines indicate the position of the next layer.

★ **1.7 X-RAY CRYSTALLOGRAPHY**

The x-ray spectral lines of atoms have a wavelength of order 1 Å ($= 10^{-10}$ m), rather smaller than typical interatomic spacings in a solid, which are of order 2–3 Å. Therefore, for x-rays, a crystal behaves as a three-dimensional diffraction grating. For an ordinary optical grating we can deduce the spacing from the separation of diffraction maxima, and by measuring the relative intensities of different orders we can find out about the structure of the lines of the grating. In an exactly similar way, by measuring the separation of x-ray diffraction maxima from a crystal we can find the size of the unit cell, and by measuring intensities we can find out about the arrangement of atoms within the unit cell.

The general laws of diffraction by crystals as formulated by von Laue we shall consider in Chapter 6. For the present, the more physically obvious formulation discovered by Bragg and used in his first crystal structure determinations will suffice. Consider two successive planes of atoms in the structure separated by a distance d, as in Fig. 1.27. X-rays from atoms in a given plane will be scattered in phase with each other if a specular reflection condition is satisfied, angle of incidence equal to the angle of reflection; in x-ray crystallography it is usual to use the glancing angle θ rather than the angle of incidence, so that the x-ray beam is deviated through an angle 2θ. But coherent reflection from a single plane is not sufficient; it is also necessary that successive planes should scatter in phase. From Fig. 1.27 we can see that this is so if

$$2d \sin \theta = n\lambda, \tag{1.17}$$

where n is an integer; this is **Bragg's law.**

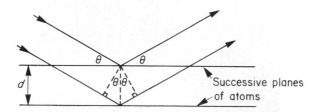

Fig. 1.27. Diagram illustrating the derivation of Bragg's law, $2d \sin \theta = n\lambda$.

To illustrate the principles of x-ray structure analysis without going into the gory details of present day refinements, we consider the first structure ever to be analysed, that of NaCl and KCl, which we have already illustrated in Fig. 1.24. This was solved by Bragg (*Proc. Roy. Soc. A,* **89,** 248 (1914)) in the same series of experiments in which he established the existence of x-ray spectral lines. He used an arrangement like an ordinary spectrometer to measure the intensity of specular reflection from a cleavage face of a crystal, and found six values of θ for which a sharp peak in intensity occurred, corresponding to three characteristic wavelengths (K, L and M x-rays) in first and second order. Then, by repeating the experiment with a different crystal face he could use Eq. (1.17) to find, for example, the ratio of the (100) and (111) plane spacings. KCl is particularly simple because both ions have the argon electron shell structure and hence scatter x-rays almost equally. He found a

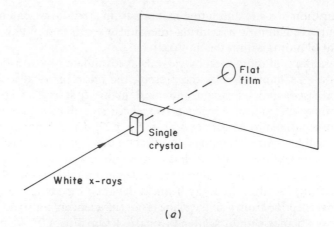

(a)

(b)

Fig. 1.28. (a) Geometry for a Laue photograph.
(b) A Laue photograph of Si with a (111) face
normal to the x-rays; note the threefold symmetry.
{after *Elements of X-ray Crystallography* by
Leonid V. Azaroff. © 68 McGraw-Hill Book
Company Inc. Used with permission}.

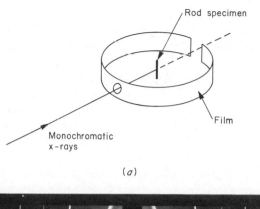

Rod specimen

Film

Monochromatic
x-rays

(*a*)

(*b*)

Fig. 1.29. (*a*) Geometry for a powder photograph.
(*b*) A powder photograph of molybdenum, taken with Co Kα radiation. The
x-rays enter the camera through the hole in the centre of the film and
leave between the ends of the film. Note that the Kα₁–Kα₂ x-ray doublet
is resolved for the back-scattered radiation near the entrance hole.
{Courtesy of H. Lipson}.

(111) plane spacing $1/\sqrt{3}$ of the (100) spacing, showing that the ions must
occupy a simple cubic lattice. Given that half the ions are positive and half
negative the structure in Fig. 1.24 is almost the only possibility, but a study
of NaCl, in which the two ions have different scattering powers, provided
convincing proof. Fig. 1.24 shows that {111} planes are alternately Na^+ and
Cl^-, so that with a phase difference of only π between successive planes there
will not be complete cancellation, but instead a weak scattered amplitude
proportional to the *difference* of the ionic scattering amplitudes. This is
just what was found for NaCl: a weak reflection at half the $\sin \theta$ value for
the main (111) reflection. On the other hand, no peculiarity was found for
(100) or (110) reflections, because these planes contain both types of atom.

Many types of x-ray camera have been invented to sort out the reflections
from different crystal planes; we shall describe only two very simple types of
x-ray photograph that are widely used for the simple substances that we
study in this book.

For a **Laue photograph,** historically the first type, a single crystal is illumin-
ated with a collinated beam of 'white' (i.e. continuous spectrum) x-radiation
as in Fig. 1.28(*a*). Each crystal plane will satisfy the Bragg condition, Eq.
(1.17), for some wavelength and a whole pattern of spots will appear (Fig.

1.28(b)), with characteristic symmetry about directions of crystal symmetry. The obvious symmetry is the main value of the method; it is used to determine the orientation of single crystals without developed external faces (e.g. metals).

The second method, called a **powder pattern,** determines values of the **Bragg angle** θ for a fixed wavelength λ; monochromatic radiation (e.g. Cu Kα, $\lambda = 1.54$ Å) is used in the arrangement of Fig. 1.29(a). The sample is either a polycrystalline wire or a rod made from grains of powder glued together; many small crystals are necessary in order that some should have each plane oriented at the Bragg angle θ to the incident x-rays. The x-rays are scattered along cones of semi-angle 2θ with the incident beam as axis, so that the photograph appears as in Fig. 1.29(b). From values of θ and rough relative intensities one can determine the unit cell and some structural information; the method is widely used as a 'fingerprint' of different crystals. Another application comes from the high resolution for radiation that is almost backscattered, as evidenced by the resolution of the Kα doublet in Fig. 1.29(b); when θ is almost 90°, Eq. (1.17) shows that it is very sensitive to the precise value of d. Therefore, very accurate unit cell dimensions can be obtained from radiation that is almost back-scattered, and this gives a valuable method of determining thermal expansion.

PROBLEMS 1

1.1 A set of normalized and mutually orthogonal p-state wavefunctions for an atom can be written in the form:

$$p_x = xf(r); \qquad p_y = yf(r); \qquad p_z = zf(r).$$

Consider the linear combination of p wavefunction

$$\psi = a_x p_x + a_y p_y + a_z p_z.$$

Find four sets of coefficients (a_x, a_y, a_z) that give normalized p-state wavefunctions with positive lobes pointing towards the corners of a regular tetrahedron. (Remember that four of the corners of a cube are corners of an inscribed regular tetrahedron.)

1.2 Consider the linear combination

$$\phi = bs + c\psi$$

where ψ is any one of the four wavefunctions calculated in Problem 1.1, and s is an s-state wavefunction, normalized and orthogonal to p_x, p_y and p_z. Find values of b and c which make the four resulting ϕ wavefunctions orthogonal to each other and normalized. Write out these four ϕ wavefunctions in terms of p_x, p_y, p_z and s. (These are sp^3 hybrid wavefunctions.)

1.3 Consider the sp^2 hybrid wavefunctions

$$\chi = \alpha s + \beta p_x + \gamma p_y.$$

Find values of α, β and γ to give normalized mutually orthogonal wavefunctions with positive lobes directed at 120° with respect to each other in the x–y plane.

1.4 Sketch a few unit cells of a simple cubic lattice and draw the following planes: (001) (101) (011) (021) (210) (211) (122).

1.5 Prove that in a crystal of cubic symmetry a direction [hkl] is perpendicular to the plane (hkl) with the same indices.

1.6 Show that the spacing d of planes (hkl) in a simple cubic lattice of side a is

$$d = a/(h^2 + k^2 + l^2)^{\frac{1}{2}}.$$

(Remember that the sum of the squares of the direction cosines of the normal to a plane is 1.)

1.7 Consider the following pattern:

$$\begin{array}{cccccccc}
\text{qp} & \text{db} & \text{qp} & \text{db} & \text{qp} & \text{db} & — & — \\
\text{db} & \text{qp} & \text{db} & \text{qp} & \text{db} & \text{qp} & — & — \\
\text{qp} & \text{db} & \text{qp} & \text{db} & \text{qp} & \text{db} & — & — \\
— & — & — & — & — & — & — & —
\end{array}$$

Indicate:
(a) a rectangular unit cell;
(b) a primitive unit cell;
(c) the basis of letters forming the contents of each type of unit cell.

1.8 Consider the face-centred cubic, body-centred cubic, hexagonal close-packed, and diamond structures.
(a) Draw plans of these structures indicating the height of atoms as a fraction of a unit cell height, and marking the unit cell sides.
(b) Give the coordinates of atoms forming the basis of the structures as fractions of a unit cell side.
(c) If the structures are formed by spheres in contact calculate the fraction of space occupied by the spheres.

1.9 A crystal has a basis of one atom per lattice point and a set of primitive translation vectors are (in Å):

$$\mathbf{a} = 3\mathbf{i}; \qquad \mathbf{b} = 3\mathbf{j}; \qquad \mathbf{c} = 1.5(\mathbf{i} + \mathbf{j} + \mathbf{k});$$

where $\mathbf{i}, \mathbf{j},$ and $\mathbf{k}$ are unit vectors in the x, y and z directions of a cartesian coordinate system. What is the Bravais lattice type of this crystal, and what are the Miller indices of the set of planes most densely populated with atoms? Calculate the volumes of the primitive unit cell and the conventional unit cell.

1.10 For the fcc and bcc structures it is possible to choose a primitive unit cell which is a rhomb, i.e. the primitive translation vectors $\mathbf{a}, \mathbf{b}, \mathbf{c}$ are all equal in magnitude and the angle between any two of them is α.
Draw diagrams showing $\mathbf{a}, \mathbf{b}, \mathbf{c}$ for each structure and calculate α in each case.

CHAPTER

2

Lattice vibrations

2.1 ELASTICITY AND ATOMIC FORCE CONSTANTS

For all the types of binding considered in the previous chapter there is some form of attractive force between atoms at large separations, but at smaller separations repulsion of the atomic electron clouds due to the exclusion principle sets in rapidly. The balance between these opposing forces determines the equilibrium separation, and near equilibrium the net force between a pair of atoms is proportional to the departure of their separation from the equilibrium value. It is therefore useful for orientation and order of magnitude calculations to consider the simple model of a crystal shown in Fig. 2.1. This consists of a simple cubic lattice of masses M joined by springs of force constant K (some cross-bracing would also be necessary to give the structure rigidity, but this does not matter for the present purpose); the lattice spacing is a. We may guess an order of magnitude for K by remembering (see, for example, Fig. 1.2) that the potential energy of a pair of atoms is raised by about 1 eV when their separation is changed from the equilibrium value by about 1 Å. If δa is the change in separation from the equilibrium value we can write the increase in potential energy as

$$\Delta V = \tfrac{1}{2}K(\delta a)^2$$

or

$$1.6 \times 10^{-19} = \tfrac{1}{2}K \times 10^{-20} \text{ joule}$$

$$K \approx 30 \text{ N m}^{-1}.$$

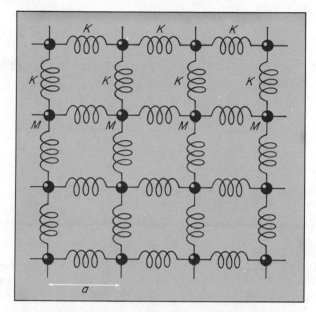

Fig. 2.1. A simple cubic lattice of masses M held at separation a by springs of force constant K.

This is readily related to the macroscopic elastic modulus E for extension
parallel to one of the principal directions of the cubic lattice. We have

$$E = \frac{\text{force/unit area}}{\text{extension/unit length}} = \frac{K\delta a/a^2}{\delta a/a} = \frac{K}{a};$$

since a typical interatomic separation is 3 Å, we have

$$E \approx \frac{30}{3 \times 10^{-10}} = 10^{11} \text{ N m}^{-2} (= 10^{12} \text{ dyne cm}^{-2})$$

which is indeed the order of magnitude of Young's modulus.*

We can see that because each atom is coupled to several others the vibrations of the atoms are not independent of each other. The modes of vibration
of such a coupled system we shall consider shortly, but for the moment we
simply note that a typical order of magnitude of vibration frequency is that
of a single mass on a single spring:

---

* The modulus E is not quite the same as Young's modulus, because it refers to extension in
which lateral contraction is prevented, whereas Young's modulus is for no lateral constraint.
E is the relevant modulus for longitudinal elastic waves in a solid, because the rest of the material
prevents lateral contraction.

$$\omega \sim \sqrt{\frac{K}{M}} \approx \left(\frac{30}{3 \times 10^{-26}}\right)^{1/2} \approx 3 \times 10^{13}\,\text{s}^{-1},$$

for atoms of atomic weight about 20. This is angular frequency; the frequency in cycles per second is

$$v = \frac{\omega}{2\pi} \approx 5 \times 10^{12}\,\text{Hz}.$$

This frequency corresponds to light in the infrared region of the spectrum; the corresponding wavelength is

$$\lambda = \frac{3 \times 10^8}{5 \times 10^{12}}\,\text{m}$$

$$= 60\,\mu\text{m},$$

a typical wavelength for infrared absorption due to molecular vibrations.

We can also calculate the velocity of sound in our lattice, by treating it, for long wavelengths, as a continuum. The velocity of sound is then

$$v_s = \sqrt{\frac{E}{\rho}}, \tag{2.1}$$

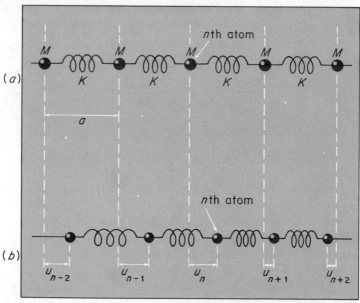

Fig. 2.2. A chain of masses M connected by springs.
(a) at equilibrium positions $x_n^0 = na$,
(b) at displaced positions $x_n = na + u_n$.

where ρ, the density, is M/a^3. Therefore

$$v_s = \left(\frac{K}{a} \times \frac{a^3}{M}\right)^{1/2} = a\sqrt{\frac{K}{M}}$$

$$\approx 3 \times 10^{-10} \times 3 \times 10^{13} \approx 10^4 \, \text{m s}^{-1}.$$

This is also correct in order of magnitude; most solids have sound velocities in the range $1\text{--}10 \, \text{km s}^{-1}$. In the next section we shall obtain this result, in the long wavelength limit, from the coupled equations of motion of the atoms, without appeal to Eq. (2.1).

2.2   DYNAMICS OF A CHAIN OF ATOMS

We can study longitudinal waves parallel to a cube edge [100] of the lattice of Fig. 2.1 by considering instead the simpler problem of the chain of atoms (one-dimensional lattice) shown in Fig. 2.2, because for a longitudinal motion the transverse springs in Fig. 2.1 have no effect. This one-dimensional model will serve to exhibit the most important qualitative features of lattice vibrations. We shall suppose that the chain consists of a very large number N of atoms, and that the last is joined to the first so as to form a ring. This last assumption is simply a device to make the chain endless so that all atoms have identical environment, and has no important effect on the problem for a long chain, where end effects are unimportant anyway. If the displacements of the atoms from their equilibrium positions are u_n, as shown in Fig. 2.2, the force on the nth atom consists of:

(i) $K(u_n - u_{n-1})$ to the left, from the spring on its left,
(ii) $K(u_{n+1} - u_n)$ to the right, from the spring on its right.

Equating this total force to the product of mass and acceleration we have

$$M\ddot{u}_n = K(u_{n+1} - 2u_n + u_{n-1}); \tag{2.2}$$

the equations of motion of *all* atoms are of this form, only the value of n varies. To solve Eqs. (2.2) we try a wavelike solution*

$$u_n = A \exp i(kx_n^0 - \omega t)$$

where x_n^0 is the *undisplaced* position of the nth atom. Substitution of this trial solution gives

$$-\omega^2 M \, e^{i(kna - \omega t)} = K\{e^{i[k(n+1)a - \omega t]} - 2e^{i(kna - \omega t)} + e^{i[k(n-1)a - \omega t]}\},$$

---

* As usual when solving vibration problems by means of complex exponentials, it is the real part of u_n that we interpret physically as atomic displacement.

or

$$-\omega^2 M = K(e^{ika} - 2 + e^{-ika})$$
$$= 2K(\cos ka - 1);$$

hence

$$\omega^2 M = 4K \sin^2 \tfrac{1}{2}ka. \tag{2.3}$$

The maximum frequency $2\sqrt{(K/M)}$ given by Eq. (2.3) is a natural cut-off frequency for the lattice; for frequencies above this value k must be imaginary to satisfy Eq. (2.3), so that we have an exponentially decaying motion that does not propagate into the lattice.

We notice that n has cancelled out in Eq. (2.3), so that the equations of motion of *all* atoms lead to the same algebraic relation between ω and k. This shows that our trial function for u_n is indeed a solution of Eqs. (2.2). It is also important to notice that we started from the equations of N *coupled* harmonic oscillators (Eqs. (2.2)); if one atom starts vibrating it does not continue with constant amplitude, but transfers energy to the others in a complicated way; the vibrations of individual atoms are not simple harmonic because of this energy exchange among them. Our solutions, on the other hand, are *uncoupled* oscillations called **normal modes**; each k has a definite ω given by Eq. (2.3) and oscillates independently of the other modes. We should however expect the number of modes to be the same as the number, N, that we started with; let us see whether this is the case.

Eq. (2.3) is plotted in Fig. 2.3; it is a periodic function, giving several k values for each ω less than the maximum value, $2\sqrt{(K/M)}$. Not all values are allowed, however, for the nth atom is identical with the $(N + n)$th, as the chain is joined on itself. Therefore

$$u_n = u_{N+n},$$

which requires that there should be an integral number of wavelengths in the length of our ring of atoms,

$$Na = p\lambda,$$

or

$$k = \frac{2\pi}{\lambda} = \frac{2\pi p}{Na}, \tag{2.4}$$

where p is an integer. There are thus N possible k values (with associated ω) in a range $2\pi/a$ of k, say the range

$$-\frac{\pi}{a} < k \leqslant \frac{\pi}{a}.$$

Fig. 2.3 shows that this restricted range of k does indeed include all possible values of the frequency ω and the group velocity ($d\omega/dk$), as well as giving the N modes of vibration we expect for N atoms. What, if anything, is the physical significance of wavenumbers outside this range?

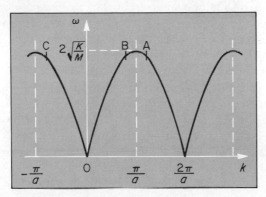

Fig. 2.3. Normal mode frequencies for a chain of atoms. Note that the wavenumbers at A, B, C all have the same frequency and correspond to the same instantaneous atomic displacements (see Fig. 2.4). Point B represents a wave moving to the right, points A and C a wave moving to the left.

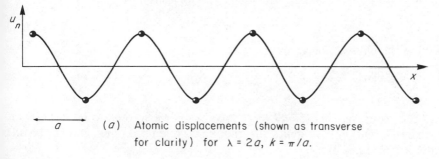

(a)   Atomic displacements (shown as transverse for clarity) for $\lambda = 2a$, $k = \pi/a$.

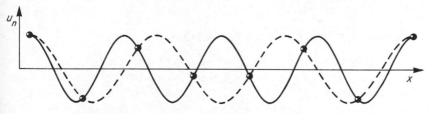

(b)   Atomic displacements for $\lambda = \frac{7}{4}a$ ($k = 8\pi/7a$, full curve), showing the equivalent description by a wave with $\lambda = \frac{7}{3}a$ ($k = 6\pi/7a$, broken curve).

Fig. 2.4.

To understand this, consider the instantaneous atomic displacements shown in Fig. 2.4; we are really considering longitudinal waves, but the displacements are shown as transverse in Fig. 2.4 because this makes their wavelike nature easier to visualize. Fig. 2.4(a) shows the displacements for $k = \pi/a$, which gives the maximum frequency; alternate atoms oscillate in antiphase, so that the midpoint of each spring is at rest. The masses therefore behave as if held by two springs each of spring constant $2K$, giving the frequency $\omega = (4K/M)^{1/2}$ that we have calculated.

Now consider the displacements for a slightly larger $k = (8\pi/7a)$, shown by the solid curve in Fig. 2.4(b) and corresponding to the point A in Fig. 2.3. The displacements can also be represented by the longer wave, shown dotted in Fig. 2.4(b), for which $k = (6\pi/7a)$; this corresponds to points B or C in Fig. 2.3. Thus, points A, B, and C correspond to the same instantaneous atomic displacements as well as the same frequency. At B $d\omega/dk > 0$, so we have a wave travelling to the right; A and C both represent a wave travelling to the left and are thus completely equivalent. We therefore conclude that adding any multiple of $(2\pi/a)$ to k does not alter the atomic displacements we are describing, or the group velocity, and is without physical significance. We need only consider the range $-\pi/a < k \leqslant \pi/a$, which contains just the N modes of vibration we expected.

We also note that for $ka \ll 1$ Eq. (2.3) reduces to

$$\omega^2 M = Kk^2 a^2,$$

so that the phase velocity (ω/k) is given by

$$\frac{\omega}{k} = a\sqrt{\frac{K}{M}},$$

and the group velocity ($d\omega/dk$) also has this value, so there is no dispersion for small k. This limit is identical with long wavelength sound, and our result agrees with Eq. (2.1).

2.3   CHAIN OF TWO TYPES OF ATOM

Let us now consider the dynamics of a chain of two types of atom, masses M and m, as shown in Fig. 2.5. This is the one-dimensional analogue of the NaCl structure shown in Fig. 1.24. To emphasize the more complicated type of motion that is now possible, we show in Fig. 2.5(b) displaced positions in which the two types of atom are displaced in opposite directions.

The equations of motion can be written down in the same way as Eq. (2.2), but there are now two distinct types of equation: those for the masses M

$$M\ddot{u}_n = K(u_{n+1} - 2u_n + u_{n-1}); \tag{2.5}$$

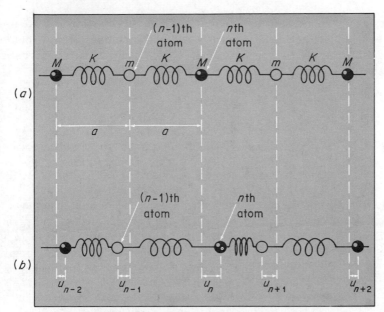

Fig. 2.5. A chain of unequal masses connected by springs.
(a) at equilibrium positions $x_n^0 = na$,
(b) at displaced positions $x_n = na + u_n$.

and those for the masses m

$$m\ddot{u}_{n-1} = K(u_n - 2u_{n-1} + u_{n-2}). \tag{2.6}$$

For the masses M we may assume as before a solution of the form

$$u_n = A \exp i(kx_n^0 - \omega t)$$

where x_n^0 is the undisplaced atomic position. There is now an extra unknown quantity, the relative amplitude and phase of the vibrations of the two types of atom; this we allow for by taking for the masses m

$$u_n = A\alpha \exp i(kx_n^0 - \omega t),$$

where α is a complex number giving the relative amplitude and phase. Substitution in Eqs. (2.5) and (2.6) then gives

$$-\omega^2 M \, e^{i(kna - \omega t)} = K\{\alpha \, e^{i[k(n+1)a - \omega t]} - 2e^{i(kna - \omega t)} + \alpha \, e^{i[k(n-1)a - \omega t]}\}$$

and

$$-\alpha\omega^2 m \, e^{i[k(n-1)a - \omega t]} = K\{e^{i(kna - \omega t)} - 2\alpha \, e^{i[k(n-1)a - \omega t]} + e^{i[k(n-2)a - \omega t]}\},$$

or, by cancelling out the common factors as before,

$$-\omega^2 M = 2K(\alpha \cos ka - 1)$$
$$-\alpha\omega^2 m = 2K(\cos ka - \alpha). \tag{2.7}$$

Thus, instead of a single algebraic equation for ω as a function of k, we now have a pair of algebraic equations for α and ω as functions of k. As before, the fact that n does not appear in Eqs. (2.7) indicates that our assumed solution is of the correct form. Eqs. (2.7) may be rewritten in the form

$$\alpha = \frac{2K \cos ka}{2K - \omega^2 m} = \frac{2K - \omega^2 M}{2K \cos ka}, \tag{2.8}$$

from which by cross-multiplication we obtain a quadratic equation for ω^2:

$$mM\omega^4 - 2K(M + m)\omega^2 + 4K^2 \sin^2 ka = 0. \tag{2.9}$$

The solutions to Eq. (2.9) are sketched in Fig. 2.6; we notice that they are periodic in k with period (π/a), whereas for one type of atom we had period $(2\pi/a)$. However, the length of a unit cell of the lattice in Fig. 2.5 is $2a$, so it is true in *both* cases that ω is periodic in k with period $2\pi/$(unit cell length).

The limiting solutions to Eq. (2.9) at the points O, A, B, C, in Fig. 2.6 are easily obtained. For $ka \ll 1$, $\sin ka \approx ka$ and we have

$$\omega^2 = \frac{K(M + m) \pm [K^2(M + m)^2 - 4K^2 mM k^2 a^2]^{1/2}}{mM}$$

$$= \frac{K(M + m)}{mM} \left\{ 1 \pm \left[1 - \frac{4mM}{(M + m)^2} k^2 a^2 \right]^{1/2} \right\}$$

$$\approx \frac{K(M + m)}{mM} \left\{ 1 \pm \left[1 - \frac{2mM}{(M + m)^2} k^2 a^2 \right] \right\}$$

$$\approx \frac{2K(M + m)}{mM} \quad \text{or} \quad \frac{2Kk^2 a^2}{(M + m)}.$$

By substituting these values of ω in Eq. (2.8) we find the corresponding values of α

$$\alpha \approx -\frac{M}{m} \quad \text{or} \quad 1.$$

The first solution corresponds to point A in Fig. 2.6; M and m oscillate in antiphase with their centre of mass at rest and the frequency is given by the spring constant $2K$ and the reduced mass $M^* = Mm/(M + m)$. The second solution represents long wavelength sound waves in the neighbourhood of the point O in Fig. 2.6 and the velocity of sound is

$$v_s = \frac{\omega}{k} = a\sqrt{\frac{2K}{(M + m)}}.$$

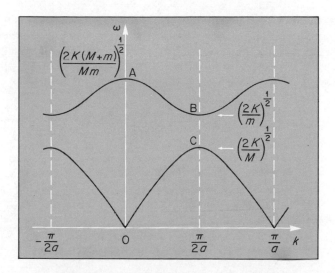

Fig. 2.6. Normal mode frequencies of a chain of two types of atom. At A the two types are oscillating in antiphase with their centre of mass at rest; at B the lighter mass m is oscillating and M is at rest; and at C, M is oscillating and m is at rest.

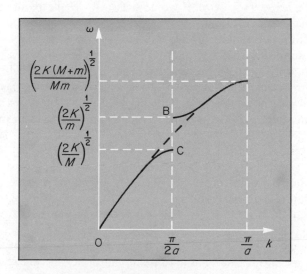

Fig. 2.7. Fig. 2.6. replotted for comparison with Fig. 2.3 in the limit $m \to M$.

Thus only the mean density matters and both types of atom oscillate with the same amplitude and phase; sound is a macroscopic phenomenon and the result is essentially the same as for a chain of one type of atom.

The other limiting solution of Eq. (2.9) is for $ka = (\pi/2)$, $\sin ka = 1$; in this case

$$\omega^2 = \frac{K(M + m) \pm [K^2(M + m)^2 - 4K^2 mM]^{1/2}}{mM}$$

$$= \frac{K(M + m) \pm K(M - m)}{mM}$$

$$= \frac{2K}{m} \quad \text{or} \quad \frac{2K}{M},$$

with, from Eq. (2.8), the corresponding amplitude ratios $\alpha = \infty$ or $\alpha = 0$ respectively. In this limit we have a half-wavelength of $2a$, the spacing between atoms of the same kind. In the first solution m oscillates and M is at rest (point B in Fig. 2.6), and the frequency therefore depends only on m; in the second solution M oscillates and m is at rest (point C in Fig. 2.6).

It is instructive to compare our present results with those of section 2.2 for a chain of one type of atom. In Fig. 2.7 we plot the lower frequency branch of Fig. 2.6 in the region $|k| < (\pi/2a)$, and the higher frequency branch in the region $(\pi/2a) < |k| < (\pi/a)$. If we now let $m \to M$ the points B and C in Fig. 2.7 come together and we recover Fig. 2.3, as we must.

There is thus a certain arbitrariness about how we assign k values to the modes of a diatomic lattice. The most direct comparison with a monatomic lattice is obtained with the assignment shown in Fig. 2.7, where there is only one ω for each k and these are N modes in the range $-(\pi/a) < k \leqslant (\pi/a)$. It is more usual, however, to assign the lowest possible k as in Fig. 2.6; there are now two branches with $N/2$ modes on each branch in the range $-(\pi/2a) < k \leqslant (\pi/2a)$. This latter convention has the useful feature that for both types of lattice the range of k values used is $2\pi/$(unit cell length).

The relation between these different assignments of k values is clarified by studying atomic displacements as in Fig. 2.4(b), but in Fig. 2.8 two types of atom alternate in the structure. This has the consequence that in addition to the two possible k values of Fig. 2.4(b), $(8\pi/7a)$ and $(6\pi/7a)$, with $\alpha = 1$, we can assign $k = (\pi/7a)$ with $\alpha = -1$; the negative value of α precludes the assignment of this last k value to the monatomic lattice. Fig. 2.8 is over-simplified in that we have chosen $|\alpha| = 1$ to emphasize the correspondence with the monatomic lattice. If $M \approx m$ we have $|\alpha| \approx 1$ except when $ka \approx (\pi/2)$,

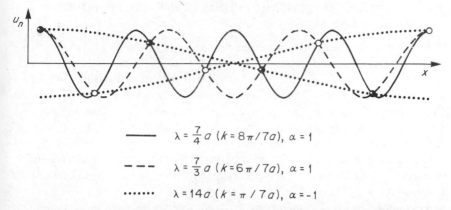

$$\text{———} \quad \lambda = \frac{7}{4}a \ (k = 8\pi/7a), \ \alpha = 1$$

$$\text{- - -} \quad \lambda = \frac{7}{3}a \ (k = 6\pi/7a), \ \alpha = 1$$

$$\cdots\cdots \quad \lambda = 14a \ (k = \pi/7a), \ \alpha = -1$$

Fig. 2.8. Atomic displacements as in Fig. 2.4(b), but with two different types of atom, showing three equivalent descriptions.

but otherwise $|\alpha|$ can be quite different; however, it is generally true that the sign of α depends on the assignment of k.

The lower branch in Fig. 2.6 is called the **acoustic** branch and the upper is called the **optical** branch. The former name stems from the fact that near $k = 0$ the lower branch represents long wavelength elastic waves with a frequency independent velocity ($\omega \propto k$), and longitudinal elastic waves are identical with sound waves.

The reason for the name 'optical' is that the upper branch for small k corresponds, as we have seen, to vibrations of the two types of atom in antiphase; in ionic crystals the resulting charge oscillations give a strong coupling to the electromagnetic field, which causes strong selective reflection at the frequency of point A in Fig. 2.6. As we saw in section 2.1 this frequency lies in the infrared region of the spectrum, and selective reflection from ionic crystals is a useful way of obtaining approximately monochromatic infrared radiation; multiple reflection can be used to improve the monochromaticity. The radiation formed in this way is known as 'Reststrahlen' (residual rays) and this phenomenon gave the first experimental indication of the magnitude of atomic vibration frequencies in solids; we shall discuss the phenomenon in more detail in the next section.

It is worth remembering at this point that Figs. 2.3 and 2.6 refer to one-dimensional models and are therefore oversimplifications. In particular we have considered only longitudinal vibrations, whereas in 3 dimensions transverse vibrations of two polarizations are also possible, giving in all 3 acoustic modes and 3 optical modes. In general all modes have different frequencies, although along directions of high symmetry in the crystal the two transverse modes are degenerate.

★ **2.4   COUPLING OF ATOMIC VIBRATIONS TO LIGHT IN IONIC CRYSTALS**

In a diatomic crystal the uncoupled dispersion relations $\omega(k)$ of the possible wave motions for small k are as shown in Fig. 2.9. If the polarizability of the crystal is neglected light will travel with velocity c (about 10^5 times greater than the velocity of the acoustic mode) and will not be coupled to the atomic vibrations. If polarizability is taken into account light becomes strongly coupled to the optical mode near point A, where they match in frequency and wavelength. We shall consider the simple model of a crystal made up of ions which are themselves unpolarizable, so that polarization is entirely due to ion motion; this will serve to exhibit the important qualitative features.

The point A in Fig. 2.9 is so near $k = 0$ that we may neglect the k dependence of the optical mode frequency, and take the equations of motion for $k = 0$. In this case all positive ions (mass M, say) have the same displacement u_+ and all negative ions have the same displacement u_-, so that Eqs. (2.5) and (2.6) become

$$Mü + 2K(u_+ - u_-) = eE$$
$$mü + 2K(u_- - u_+) = -eE, \qquad (2.10)$$

where we have added the force due to an electric field E which may be present; e is the ionic charge. If we subtract $M/(M + m)$ times the second of Eqs. (2.10) from $m/(M + m)$ times the first we obtain a single equation for the relative displacement $(u_+ - u_-)$:

$$M^*(ü_+ - ü_-) + 2K(u_+ - u_-) = eE, \qquad (2.11)$$

where we have introduced the reduced mass $M^* = mM/(M + m)$. Our one-dimensional model is no longer adequate for considering the interaction with electromagnetic radiation, so we replace Eq. (2.11) by the three-dimensional equivalent

$$M^*\ddot{\mathbf{w}} + 2K\mathbf{w} = eE, \qquad (2.12)$$

where $\mathbf{w}$ is the vector relative displacement of the ions. Eq. (2.12) implies an assumption of isotropy, which is true for cubic crystals sufficiently near $\mathbf{k} = 0$.

The coupling to light arises from the fact that ion motion gives rise to regions of net charge, and hence electric fields. Thus, consider the positive charge contained in the volume $\delta x \, \delta y \, \delta z$ when the ions are undisplaced; after displacement by $\mathbf{u}_+$ with components (u_x, u_y, u_z) the same charge is

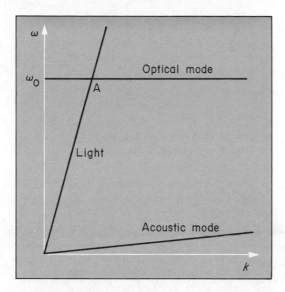

Fig. 2.9. $\omega(k)$ for uncoupled atomic vibrations and
light in a diatomic crystal.

contained in a volume

$$(\delta x + \delta u_x)(\delta y + \delta u_y)(\delta z + \delta u_z) =$$

$$\left(\delta x + \frac{\mathrm{d}u_x}{\mathrm{d}x}\delta x\right)\left(\delta y + \frac{\mathrm{d}u_y}{\mathrm{d}y}\delta y\right)\left(\delta z + \frac{\mathrm{d}u_z}{\mathrm{d}z}\delta z\right). \qquad (2.13)$$

Since for long wavelengths the fractional changes in atomic displacement in
an atomic spacing are very small, all quantities like $(\mathrm{d}u_x/\mathrm{d}x)$ are very much
less than 1, and the volume (2.13) may therefore be written as

$$\delta x\,\delta y\,\delta z(1 + \operatorname{div}\mathbf{u}_+),$$

with

$$\operatorname{div}\mathbf{u}_+ \ll 1.$$

This dilation of the volume occupied by the charge implies that the positive
charge density, ρ_+, is related to the value for undisplaced ions, ρ_0, by

$$\rho_+ = \rho_0(1 - \operatorname{div}\mathbf{u}_+);$$

similarly, the negative charge density is given by

$$\rho_- = -\rho_0(1 - \operatorname{div}\mathbf{u}_-),$$

so that the net charge density ρ is given by

$$\rho = \rho_+ + \rho_- = -\rho_0 \operatorname{div} \mathbf{w}. \tag{2.14}$$

We also have

$$\operatorname{div} \mathbf{E} = \rho/\varepsilon_0,$$

so that Eq. (2.14) gives

$$\operatorname{div}(\mathbf{E} + \mathbf{W}) = 0, \tag{2.15}$$

where

$$\mathbf{W} = \rho_0 \mathbf{w}/\varepsilon_0.$$

In fact, the polarization per unit volume $\mathbf{P}$ is $\varepsilon_0 \mathbf{W}$, so that Eq. (2.15) is just div $\mathbf{D} = 0$, but we prefer to use $\mathbf{W}$ both to remind ourselves of the connection with ionic displacement and to save writing ε_0 all over the place. Replacement of $\mathbf{w}$ by $\mathbf{W}$ enables Eq. (2.12) to be written in the form

$$\ddot{\mathbf{W}} + \omega_0^2 \mathbf{W} - \omega_0^2 \chi \mathbf{E}/\varepsilon_0 = 0, \tag{2.16}$$

where ω_0 is the optical mode frequency and the static electric susceptibility χ^* is obtained from

$$\chi \mathbf{E} = \mathbf{P} = \rho_0 \mathbf{w} = \frac{e\rho_0}{2K} \mathbf{E},$$

using Eq. (2.12). Thus

$$\chi = e^2/2KV$$

where V is the volume per ion pair.

Since we are considering the motion of all charges explicitly, from a microscopic point of view, it is Maxwell's equations *in vacuo* that must be solved in conjunction with Eq. (2.16); these are*

$$\operatorname{curl} \mathbf{E} = -\frac{\partial \mathbf{B}}{\partial t} \qquad \operatorname{curl} \mathbf{B} = \frac{1}{c^2}\frac{\partial \mathbf{E}}{\partial t} + \mu_0 \mathbf{j}. \tag{2.17}$$

---

* For the definition of χ see Appendix E.

* If the second equation looks unfamiliar, remember that from a *macroscopic* point of view we usually write $\mathbf{j}$ as the sum of free currents, microscopic currents causing magnetization, and microscopic charge movements changing polarization:

$$\mathbf{j} = \mathbf{j}_0 + \operatorname{curl} \mathbf{M} + \frac{\partial \mathbf{P}}{\partial t}.$$

If we substitute this expression in Eq. (2.17) and bring together the terms in curl and $\partial/\partial t$, we obtain the usual equation relating the vectors $\mathbf{H}$ and $\mathbf{D}$. By taking the microscopic view we avoid introducing these additional vectors.

Elimination of **B** from Eqs. (2.17) gives

$$-\operatorname{curl}\operatorname{curl}\mathbf{E} = \frac{1}{c^2}\frac{\partial^2\mathbf{E}}{\partial t^2} + \mu_0\frac{\partial\mathbf{j}}{\partial t};$$

but $\mathbf{j} = \rho_0\dot{\mathbf{w}} = \varepsilon_0\dot{\mathbf{W}}$, so that

$$-\operatorname{curl}\operatorname{curl}\mathbf{E} = \frac{1}{c^2}\left[\frac{\partial^2\mathbf{E}}{\partial t^2} + \frac{\partial^2\mathbf{W}}{\partial t^2}\right], \tag{2.18}$$

which is to be solved with Eq. (2.16). This is not as bad as it looks if we assume a wavelike solution $\exp i(kx - \omega t)$.

The simplest case is a purely longitudinal wave, for which $\operatorname{curl}\mathbf{E} = 0$, so that $\partial^2(\mathbf{E} + \mathbf{W})/\partial t^2 = 0$ from Eq. (2.18), and hence for an oscillatory solution $\mathbf{E} + \mathbf{W} = 0$, so that Eq. (2.16) becomes

$$\mathbf{W} + \varepsilon\omega_0^2\mathbf{W} = 0, \tag{2.19}$$

where $\varepsilon = 1 + \chi/\varepsilon_0$ is the static dielectric constant. This is simply an oscillation at frequency $\varepsilon^{1/2}\omega_0$ for all k; there is no electromagnetic wave, but the frequency of the optical mode is modified by electrostatic forces.

For a *transverse* wave $\operatorname{div}\mathbf{W} = 0$, so that from Eq. (2.15) $\operatorname{div}\mathbf{E} = 0$; we also have $\operatorname{curl}\operatorname{curl}\mathbf{E} = -\nabla^2\mathbf{E}$, so that Eq. (2.18) becomes, for a wave in the x-direction,

$$\frac{\partial^2\mathbf{E}}{\partial x^2} = \frac{1}{c^2}\frac{\partial^2}{\partial t^2}(\mathbf{E} + \mathbf{W}). \tag{2.20}$$

If we now try a solution

$$\mathbf{E} = \mathbf{E}_0\,e^{i(kx - \omega t)}$$
$$\mathbf{W} = \mathbf{W}_0\,e^{i(kx - \omega t)},$$

Eq. (2.20) gives

$$k^2c^2\mathbf{E}_0 = \omega^2(\mathbf{E}_0 + \mathbf{W}_0),$$

and Eq. (2.16) gives

$$(\omega_0^2 - \omega^2)\mathbf{W}_0 = \chi\omega_0^2\mathbf{E}_0/\varepsilon_0$$

from which

$$\frac{k^2c^2}{\omega^2} = 1 + \frac{\chi\omega_0^2/\varepsilon_0}{\omega_0^2 - \omega^2}$$

or

$$k^2 = \frac{\omega^2}{c^2}\left(\frac{\varepsilon\omega_0^2 - \omega^2}{\omega_0^2 - \omega^2}\right).$$

(2.21)

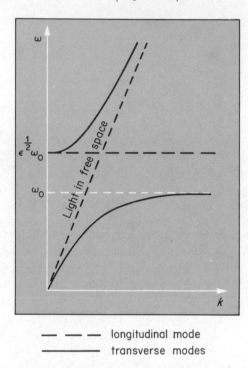

——— longitudinal mode
———— transverse modes

Fig. 2.10. Optical modes and light in an ionic
crystal.

This dispersion relation is sketched in Fig. 2.10, together with the longitudinal mode we have already discussed. The general form of the curve is easily obtained by noting the following limits:

for $\omega \ll \omega_0$, $k^2 = \varepsilon\omega^2/c^2$;
$\omega \gg \omega_0$, $k^2 = \omega^2/c^2$;
$\omega \to \varepsilon^{1/2}\omega_0$, $k^2 \to 0$;
$\omega \to \omega_0$, $k^2 \to \infty$.

It is particularly worth noting that for $\omega_0 < \omega < \varepsilon^{1/2}\omega_0$ we have $k^2 < 0$, so that k is imaginary. This means that no wave can propagate—we have an evanescent wave that decays exponentially as we go into the crystal. Therefore, in this frequency range, we have *total external reflection* of radiation incident on the crystal from outside; this is the 'Reststrahlen' phenomenon.

We note from Eq. (2.21) that the frequency dependent dielectric constant $\varepsilon(\omega)$ and refractive index $n(\omega)$ are given by

$$\varepsilon(\omega) = n^2(\omega) = \left(\frac{\varepsilon\omega_0^2 - \omega^2}{\omega_0^2 - \omega^2}\right). \tag{2.22}$$

Fig. 2.11(a) shows some actual measurements of infrared refractive index for comparison; note that they are plotted as a function of wavelength rather than frequency. Both Eq. (2.22) and Fig. 2.11(a) show the typical anomalous dispersion associated with a resonance absorption at frequency ω_0; this is made use of in constructing prisms for infrared spectroscopy. According to Eq. (2.22) the dielectric constant has the static value well below ω_0 and tends to unity above ω_0; this is only qualitatively borne out by Fig. 2.11(a), which shows an almost constant refractive index greater than unity in the region $\lambda \sim 1\ \mu m$. The discrepancy is the result of ionic polarizability which we have neglected in our model; the further rise in refractive index at the left of Fig. 2.11(a) heralds the approach of an ionic absorption frequency. We also note that our model contains no damping, and hence we do not obtain the resonance absorption near ω_0 associated with anomalous dispersion. This absorption is a quantum process in which a photon is absorbed and its energy transferred to lattice vibrations.

Although resonance absorption is largely in the vicinity of the transverse optical mode frequency ω_0, we have already noted that total external reflection is to be expected over the whole frequency range, $\omega_0 < \omega < \varepsilon^{1/2}\omega_0$, between the transverse and longitudinal optical modes; in this region $\varepsilon(\omega)$ is negative and $n(\omega)$ is purely imaginary. Fig. 2.11(b) shows that measurements of reflection coefficient bear this out qualitatively, but the reflection is never total. This is another consequence of our neglect of damping; there is no transmission through a crystal more than a few wavelengths thick in this frequency range, but some of the energy is absorbed rather than reflected.

2.5 PHONONS

So far we have considered the mechanics of lattice vibrations in a completely classical way. To the extent that the normal modes of vibration we have found are truly harmonic and independent, the transition to quantum mechanics is easily made, for we know that the energy levels of a harmonic oscillator of angular frequency ω are

$$\varepsilon_n = (n + \tfrac{1}{2})\hbar\omega. \tag{2.23}$$

Since Eq. (2.23) represents a set of uniformly spaced energy levels, it is possible to regard the state ε_n as constructed by adding n 'excitation quanta', each of energy $\hbar\omega$, to the ground state. You will have already met this view-

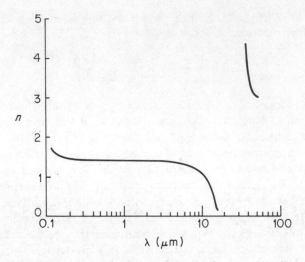

(*a*) Experimental refractive index of LiF as a function of
wavelength (plotted logarithmically).

[after H.W. Kohls, *Ann. Physik*, **29**, 433 (1937)].

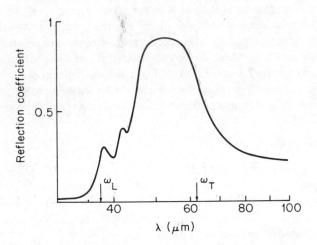

(*b*) Reflection coefficient of a thick NaCl crystal.

[after A. Mitsuishi *et al.*, *J. Opt. Soc. Amer.*, **52**, 14 (1962)].

Fig. 2.11.

point in the context of electromagnetic radiation of angular frequency ω; there we say that the state ε_n corresponds to the presence of n **photons** each of energy $\hbar\omega$. The reality of such energy carrying particles is shown, for example, in the photoelectric effect.

It is often convenient to treat lattice vibrations in an analogous way, and introduce the idea of **phonons** of energy $\hbar\omega$ as quanta of excitation of the lattice. Our normal modes are plane waves extending throughout the crystal lattice, and correspondingly the phonons are not localized particles; the uncertainty principle demands that the position cannot be determined because the momentum* $\hbar k$ is exact. However, just as with photons or electrons, one can construct a fairly localized wavepacket by combining modes of slightly different frequency and wavelength. Thus, if we take waves with a spread in **k** of order $(\pi/10a)$ we can make a wavepacket localized to within about 10 atomic spacings. Such a wavepacket represents a fairly localized phonon moving with group velocity $(d\omega/d\mathbf{k})$. We can therefore treat phonons as localized particles within the limits of the uncertainty principle. The $\omega(\mathbf{k})$ curve for lattice vibrations can be interpreted, if both axes are multiplied by $\hbar$, as a relation between energy and crystal momentum for phonons $(E = \hbar\omega, \mathbf{p} = \hbar\mathbf{k})$.

Like photons, phonons are bosons and are not conserved; they can be created or destroyed in collisions. These properties follow from Eq. (2.23); phonons are bosons because n can take any value, and they are not conserved because n can change with time. We shall meet examples of the usefulness of the idea of phonons later. In section 7.3 we shall see that phonons can be created or absorbed when neutrons are scattered from a solid, leading to a direct experimental measurement of $\omega(\mathbf{k})$; and in Chapter 8 we shall see that the thermal conductivity of insulators can be understood by applying kinetic theory to the gas of phonons present in a crystal at finite temperature.

2.6   SPECIFIC HEAT FROM LATTICE VIBRATIONS

In most solids the energy given to lattice vibrations is the dominant contribution to the specific heat; in non-magnetic insulators it is the only contribution. Other contributions arise in metals, from the conduction electrons, and in magnetic materials from magnetic ordering.

We have seen in our examples of one-dimensional lattices that the coupling together of atomic vibrations leads to a band of normal mode frequencies from zero up to some maximum value. Calculation of the lattice energy

---

* Actually, in a harmonic crystal phonons do not carry momentum, because each atom vibrates symmetrically about its mean position. The quantity $\hbar k$, which we shall see later is in some respects analogous to momentum, we shall call **crystal momentum**.

and specific heat of a solid therefore falls into two parts: the evaluation of the contribution of a single oscillator; and the summation over the frequency distribution of oscillators.

2.6.1   Energy and specific heat of a harmonic oscillator

We have already noted that the energy levels of a quantum harmonic oscillator are

$$\varepsilon_n = (n + \tfrac{1}{2})\hbar\omega, \tag{2.23}$$

so that the average energy $\bar{\varepsilon}$ in thermal equilibrium at temperature T is given by

$$\bar{\varepsilon} = \sum_n p_n \varepsilon_n.$$

Here p_n is the probability of the oscillator being in level ε_n; this is given by a Boltzmann factor so that

$$\bar{\varepsilon} = \frac{\displaystyle\sum_{n=0}^{\infty} (n + \tfrac{1}{2})\hbar\omega \exp[-(n + \tfrac{1}{2})\hbar\omega/k_B T]}{\displaystyle\sum_{n=0}^{\infty} \exp[-(n + \tfrac{1}{2})\hbar\omega/k_B T]}$$

$$= \tfrac{1}{2}\hbar\omega + \frac{\displaystyle\sum_{n=0}^{\infty} n\hbar\omega \exp(-n\hbar\omega/k_B T)}{\displaystyle\sum_{n=0}^{\infty} \exp(-n\hbar\omega/k_B T)} \tag{2.24}$$

$$= \tfrac{1}{2}\hbar\omega + k_B T^2 \frac{\partial}{\partial T}\left[\ln \sum_{n=0}^{\infty} \exp(-n\hbar\omega/k_B T)\right].$$

But

$$\sum_{n=0}^{\infty} e^{-n\hbar\omega/k_B T} = 1 + e^{-\hbar\omega/k_B T} + e^{-2\hbar\omega/k_B T} + \cdots$$

$$= (1 - e^{-\hbar\omega/k_B T})^{-1},$$

so that

$$\bar{\varepsilon} = \tfrac{1}{2}\hbar\omega + \frac{\hbar\omega}{e^{\hbar\omega/k_B T} - 1}. \tag{2.25}$$

This mean energy is readily interpreted in terms of phonons. For bosons of energy $\hbar\omega$, which are not conserved, the average number present in thermal equilibrium at temperature T is

$$n(\omega) = \frac{1}{e^{\hbar\omega/k_{\mathrm{B}}T} - 1},\tag{2.26}$$

as in the case of photons in black-body radiation (Mandl,[2] Chapter 10). Multiplication of Eq. (2.26) by $\hbar\omega$ gives the second term in Eq. (2.25) as the contributions of the phonons to the energy. The first term in Eq. (2.25), $\frac{1}{2}\hbar\omega$, is the zero-point energy, which cannot be frozen out because of the uncertainty principle. Fig. 2.12(a) shows that the mean energy tends to this value in the low temperature limit $k_{\mathrm{B}}T \ll \hbar\omega$. At high temperatures $k_{\mathrm{B}}T \gg \hbar\omega$ we can expand the exponential to obtain

$$\begin{aligned}
\bar{\varepsilon} &= \tfrac{1}{2}\hbar\omega + \hbar\omega \Bigg/ \left[\frac{\hbar\omega}{k_{\mathrm{B}}T} + \frac{1}{2}\left(\frac{\hbar\omega}{k_{\mathrm{B}}T}\right)^2 + \cdots \right] \\
&= \tfrac{1}{2}\hbar\omega + k_{\mathrm{B}}T \left[1 - \frac{1}{2}\left(\frac{\hbar\omega}{k_{\mathrm{B}}T}\right) + \cdots \right] \\
&= k_{\mathrm{B}}T \left[1 + O\left(\frac{\hbar\omega}{k_{\mathrm{B}}T}\right)^2 \right],
\end{aligned}$$

so that the classical equipartition value is obtained in this limit; note that a one-dimensional harmonic oscillator gives $k_{\mathrm{B}}T$, not $\frac{1}{2}k_{\mathrm{B}}T$, because the potential energy, as well as the kinetic energy, is a quadratic contribution to the total energy. The specific heat C is found by differentiating Eq. (2.25) with respect to temperature:

$$C = \frac{\mathrm{d}\bar{\varepsilon}}{\mathrm{d}T} = k_{\mathrm{B}} \left(\frac{\Theta}{T}\right)^2 \frac{e^{\Theta/T}}{(e^{\Theta/T} - 1)^2},\tag{2.27}$$

where $\Theta = \hbar\omega/k_{\mathrm{B}}$. This is shown in Fig. 2.12(b); the specific heat vanishes exponentially at low temperatures and tends to the classical value k_{B} at high temperatures.

The general features of Fig. 2.12 are common to all quantum systems: the energy tends to the zero-point energy at low temperatures and to the classical equipartition value at high temperatures; the specific heat tends to zero at absolute zero and rises to the classical value in such a way that the 'missing area' under the classical specific heat curve (shaded in Fig. 2.12(b)) is equal to the zero point energy.

The first quantum theory of the specific heat of solids was due to Einstein. He made the extreme simplifying assumption that all $3N$ vibrational modes of a three-dimensional solid of N atoms had the same frequency, so that the whole solid had a heat capacity $3N$ times Eq. (2.27). Even this very crude model gave the correct limit at high temperatures, an atomic heat capacity

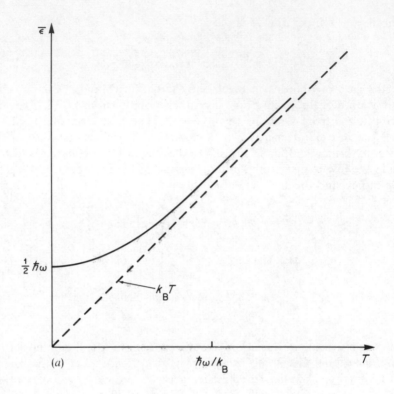

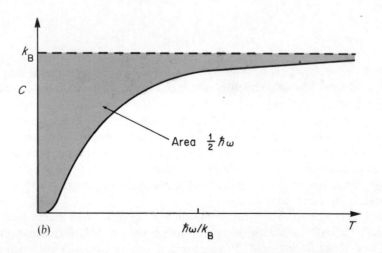

Fig. 2.12. Mean energy and specific heat of a harmonic oscillator.

of $3R$, as first found empirically by Dulong and Petit; this result depends only on the classical theorem of the equipartition of energy. Einstein's model also gave, correctly, a specific heat tending to zero at absolute zero, but the temperature dependence near $T = 0$ did not agree with experiment. We shall now account for this discrepancy by considering the actual distribution of vibration frequencies in a solid, starting from our one dimensional model of section 2.2.

2.6.2 The density of states

We saw in section 2.2 that for a chain of atoms joined to make a ring the allowed wavenumbers are given by

$$k = \frac{2\pi}{\lambda} = \frac{2\pi p}{Na}, \qquad (2.4)$$

where p is an integer, so that the allowed states are uniformly distributed in k at a density $\rho(k)$ such that the number of states in the range $k \to k + dk$ is given by

$$\rho(k)\,dk = \frac{L}{2\pi}\,dk, \qquad (2.28)$$

where $L = Na$ is the length of the chain. These allowed k values correspond to running waves, so both positive and negative k values are significant. It is often convenient to consider instead a chain with *fixed* ends, in which case the normal modes are standing waves and we have an integral number of half-wavelengths in the chain, so that $L = n\lambda/2$ and

$$\rho(k)\,dk = \frac{L}{\pi}\,dk. \qquad (2.29)$$

However, for standing waves only positive k has physical significance, so we obtain the number of states with $|k| < k_0$ by integrating Eq. (2.28) from $-k_0$ to k_0 or Eq. (2.29) from 0 to k_0; in either case the answer is Lk_0/π, independently of the boundary conditions we have used.

To calculate an energy or specific heat by summing over normal mode oscillators we need the density of states per unit frequency range $\rho(\omega)$; if we pick a certain number dn of states covering a range dk of k and a range $d\omega$ of ω we clearly have

$$dn = \rho(k)\,dk = \rho(\omega)\,d\omega$$

so that

$$\rho(\omega) = \rho(k)\frac{dk}{d\omega}. \qquad (2.30)$$

This may be applied to calculate the density of states for a chain of atoms by using Eq. (2.3) for $\omega(k)$ to give

$$\frac{d\omega}{dk} = a\sqrt{\frac{K}{M}}\cos \tfrac{1}{2}ka,$$

so that

$$
\begin{aligned}
\rho(\omega) &= \frac{L}{\pi a}\sqrt{\frac{M}{K}}\sec \tfrac{1}{2}ka \\
&= \frac{N}{\pi}\sqrt{\frac{M}{K}}\frac{2\sqrt{K/M}}{[(4K/M) - \omega^2]^{1/2}} \\
&= \frac{2N}{\pi}\left(\frac{4K}{M} - \omega^2\right)^{-1/2}.
\end{aligned}
\tag{2.31}
$$

This density of states is plotted in Fig. 2.13; we notice that it tends to infinity as the cutoff frequency $2(K/M)^{1/2}$ is approached from below, because the group velocity $d\omega/dk$ tends to zero here. If we had ignored dispersion of sound at short wavelengths we would have obtained the constant density of

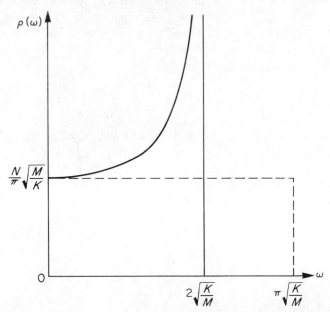

Fig. 2.13. Density of states for a chain of atoms: full curve. Density of states if dispersion of sound is ignored: broken curve.

states shown by the broken line in Fig. 2.13, for which the total number of states does not reach the value N until frequency $\omega = \pi(K/M)^{1/2}$.

We shall not proceed further with our one-dimensional example, since it cannot be compared with experiment. Instead, we shall now generalize Eq. (2.29) to three dimensions. First, consider the two-dimensional standing-wave pattern in a box of side L shown in Fig. 2.14(a); one of the running waves that go to make up this pattern is shown in Fig. 2.14(b). From the diagram we see that

$$\lambda = \lambda_x \cos \theta = \lambda_y \sin \theta,$$

so that λ does not behave as a vector; but if we take reciprocals $k = 2\pi/\lambda$, $k_x = 2\pi/\lambda_x$, etc., we get

$$k_x = k \cos \theta$$
$$k_y = k \sin \theta,$$

which is the relation for a vector, so that $\mathbf{k}$ can be represented by a point in $\mathbf{k}$-space as in Fig. 2.14(c). The boundary conditions for standing waves require $L = n_x \lambda_x / 2$, $L = n_y \lambda_y / 2$, so that

$$k_x = n_x \pi / L$$
$$k_y = n_y \pi / L .$$

(2.32)

This result can be extended easily to three dimensions; for a cubical box of side L the allowed states are represented by a simple cubic lattice in $\mathbf{k}$-space of side π/L, so that the density of states in $\mathbf{k}$-space is given by

$$\rho(\mathbf{k})\, d^3\mathbf{k} = \left(\frac{L}{\pi}\right)^3 d^3\mathbf{k}$$

$$= \frac{V}{\pi^3} d^3\mathbf{k}$$

(2.33)

where $d^3\mathbf{k}$ is an element of volume in $\mathbf{k}$-space and V is the volume of the box; for a large box the result (2.33) is in fact independent of the shape of the box.

If things depend only on the magnitude of $\mathbf{k}$, not its direction, (this is not true in a crystal, but it is often used as an approximation) it is convenient to write

$$d^3\mathbf{k} = \tfrac{1}{8} \times 4\pi k^2\, dk,$$

(2.34)

as $\tfrac{1}{8}$th of the volume of a spherical shell in $\mathbf{k}$-space; the factor $\tfrac{1}{8}$ arises because for standing waves only the octant in which k_x, k_y, k_z are all positive is

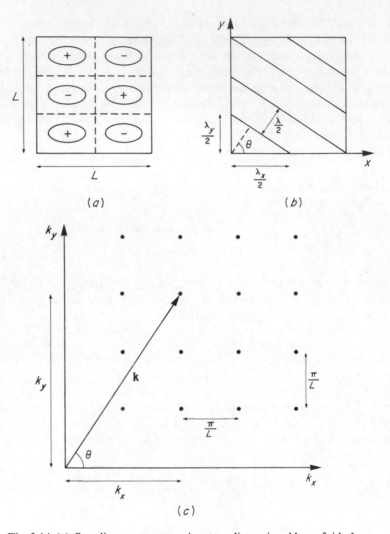

Fig. 2.14. (a) Standing wave pattern in a two dimensional box of side L.
(b) One of the running waves that make up this standing wave pattern.
(c) Diagram in **k**-space of allowed modes.

meaningful. Therefore

$$\rho(\mathbf{k})\, d^3\mathbf{k} = \frac{Vk^2}{2\pi^2}\, dk$$

$$= g(k)\, dk, \tag{2.35}$$

defining a new density of states function $g(k)$ per unit magnitude of k. As with our one-dimensional result, Eq. (2.35) could also have been obtained by considering running waves in a box with periodic boundary conditions;* Eq. (2.33) would then contain $(L/2\pi)^3$, but the factor $\frac{1}{8}$ would be missing from Eq. (2.34) because negative k values would now be significant.

If ω is a function of the magnitude of k we obtain from Eq. (2.35) a result analogous to Eq. (2.30):

$$g(\omega) = g(k)\frac{dk}{d\omega} = \frac{Vk^2}{2\pi^2}\frac{dk}{d\omega}, \tag{2.36}$$

in which, of course, k has to be expressed as a function of ω.

2.6.3   The Debye approximation

The contribution of lattice vibrations to the internal energy E of a solid can be obtained by integrating the mean energy of a simple harmonic oscillator given by Eq. (2.25) over frequency, weighted according to the density of states $g(\omega)$:

$$E = \int_0^\infty \left[\tfrac{1}{2}\hbar\omega + \frac{\hbar\omega}{e^{\hbar\omega/k_B T} - 1}\right]g(\omega)\,d\omega. \tag{2.37}$$

Strictly, $\omega(\mathbf{k})$ should be calculated from interatomic force constants and combined with Eq. (2.33) to give $(g(\omega))$; but in three dimensions this is a heavy calculation even on a computer. Debye obtained a very good approximation to the answer by neglecting acoustic dispersion, i.e. by assuming a density of states curve analogous to the broken line in Fig. 2.13 instead of the full line.

With an isotropic velocity of sound v_s Eq. (2.36) gives

$$g(\omega) = \frac{Vk^2}{2\pi^2}\frac{dk}{d\omega} = \frac{V\omega^2}{2\pi^2 v_s^3};$$

in fact there is one longitudinal mode with $v_s = v_L$ and two transverse modes with $v_s = v_T$, so that altogether

$$g(\omega) = \frac{V\omega^2}{2\pi^2}\left(\frac{1}{v_L^3} + \frac{2}{v_T^3}\right). \tag{2.38}$$

As in Fig. 2.13, we have to impose a cutoff frequency ω_c such that

$$\int_0^{\omega_c} g(\omega)\,d\omega = 3N,$$

* This is a convenient fiction in three dimensions, since a three-dimensional object cannot be deformed so as to join up on itself in all three directions at once. The fiction is nevertheless sometimes useful since it enables running waves to be considered as normal modes.

in order to ensure that we have the correct number of normal modes for N atoms in three dimensions; by integrating Eq. (2.38) up to ω_c we have

$$\frac{V\omega_c^3}{6\pi^2}\left(\frac{1}{v_L^3} + \frac{2}{v_T^3}\right) = 3N. \tag{2.39}$$

Substitution of Eqs. (2.38) and (2.39) in Eq. (2.37) then gives

$$E = \frac{9N}{\omega_c^3}\int_0^{\omega_c}\left[\tfrac{1}{2}\hbar\omega + \frac{\hbar\omega}{e^{\hbar\omega/k_BT} - 1}\right]\omega^2\,d\omega$$

$$= \tfrac{9}{8}N\hbar\omega_c + \frac{9N}{\omega_c^3}\int_0^{\omega_c}\frac{\hbar\omega^3\,d\omega}{e^{\hbar\omega/k_BT} - 1}. \tag{2.40}$$

The specific heat C_v (constant volume because this keeps the frequencies constant) may then be obtained by differentiating Eq. (2.40) with respect to temperature, or by integrating Eq. (2.27) over the frequency distribution, as

$$C_v = 9R\left(\frac{T}{\Theta_D}\right)^3\int_0^{\Theta_D/T}\frac{x^4\,e^x\,dx}{(e^x - 1)^2}, \tag{2.41}$$

where we have introduced the variable $x = \hbar\omega/k_BT$ and defined the Debye temperature Θ_D as $\hbar\omega_c/k_B$; we have also taken N as Avogadro's number, so that $Nk_B = R$ and C_v is an atomic heat. For $T \gg \Theta_D$, x is always small, and by expanding the exponential the integrand reduces to x^2 so that

$$C_v \approx 3R, \tag{2.42}$$

the classical equipartition value; this limit is independent of our assumption about the density of states, and depends only on our assumption of harmonic forces. Anharmonic effects, though vital for thermal expansion (see Flowers and Mendoza,[3] section 8.4) usually have only a small effect on the specific heat.

At low temperatures $T \ll \Theta_D$ the limit of the integral is essentially infinite, and it is just a pure number, $4\pi^4/15$ in fact, so that

$$C_v = \frac{12R\pi^4}{5}\left(\frac{T}{\Theta_D}\right)^3, \tag{2.43}$$

the Debye T^3 law. Fig. 2.15 illustrates the excellent agreement of Eq. (2.43) with experiment for a non-magnetic insulator; we shall see later that there are extra contributions to the low temperature specific heat of other substances. The specific heat vanishes much more slowly than the exponential behaviour for a single harmonic oscillator because the vibration spectrum extends down to zero frequency. In the low temperature limit the integrand is only appreciable for small ω, where our assumption about the density of

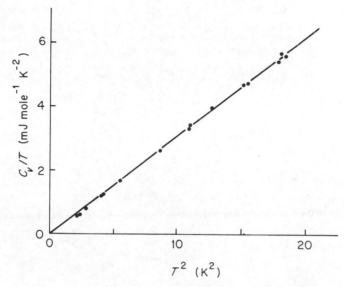

Fig. 2.15. Low temperature heat capacity of KCl, plotted so as
to demonstrate the T^3 law.
{after P. H. Keesom and N. Perlman, *Phys. Rev.*, **91**, 1354 (1954)}.

states curve is correct, so this limit is also exact. The Debye approximation
therefore serves as a convenient interpolation between the exact limits of
Eqs. (2.42) and (2.43).

It is interesting to compare the above results with the theory of black-
body radiation (Mandl,[2] Chapter 10). Eq. (2.43) implies an internal energy
proportional to T^4, as for black-body radiation. However, there is no maximum allowed frequency for photons, because vacuum does not have an
atomic structure, and consequently the energy density is proportional to
T^4 at all temperatures for black-body radiation. The other difference is
that there are three allowed polarizations for phonons, one longitudinal and
two transverse, but only the two transverse polarizations are allowed for
photons.

Because it is exact in both the high and low temperature limits the Debye
formula, Eq. (2.41), gives quite a good representation of the specific heat of
most solids, even though the actual phonon density of states curve may differ
appreciably from the Debye assumption; this is illustrated for copper in
Fig. 2.16. The true density of states, deduced from $\omega(k)$ curves determined by
neutron scattering (section 7.3), is compared with the Debye assumption
in Fig. 2.16(a). The difference arises from two main effects: dispersion of
sound lowers the cutoff frequency and causes a rise in the density of states

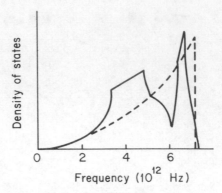

(*a*) Density of states for copper deduced from
neutron-scattering experiments (solid curve); Debye
density of states, fitted to elastic constants, is shown
by broken curve.

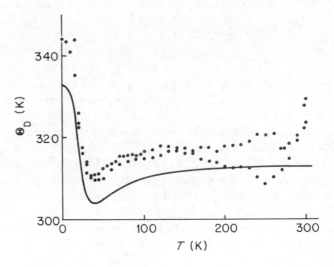

(*b*) Temperature dependence of Θ_D for copper deduced from
the density of states curve above (solid line), compared
with various experimental points, deduced from specific
heat measurements.
[after E.C. Svensson, B.N. Brockhouse and J.M. Rowe,
Phys. Rev., **155**, 619-32 (1967)].

Fig. 2.16.

just below it, as in our one dimensional example (Fig. 2.13); and the variation of cutoff wavenumber with crystallographic orientation blurs the sharp cutoff in the density of states so that the actual maximum frequency is raised somewhat. The general result, as in Fig. 2.16(a), is that the maximum frequency can be quite close to the Debye value, but the centre of gravity of the frequency distribution is lowered. This means that the main rise in the specific heat comes at a somewhat lower temperature than one would expect from Eq. (2.41).

Departures from the Debye theory are best investigated by using Eq. (2.41) backwards to calculate $\Theta_D(T)$ at each temperature from the measured* $C_v(T)$; a non-constant Θ_D indicates departures from the Debye law. In Fig. 2.16(b) $\Theta_D(T)$ deduced from measured specific heats is compared with $\Theta_D(T)$ calculated from the true density of states curve in Fig. 2.16(a). It can be seen that the general trend is very similar; the small systematic difference is attributable to the fact that the neutron experiments measured $\omega(k)$ at room temperature, not at the temperature of the heat capacity measurements. We have entirely ignored the anharmonic forces which give rise to a temperature dependence of $\omega(k)$, as well as to thermal expansion (Flowers and Mendoza,[3] section 8.4).

The result shown in Fig. 2.16(a), that the maximum vibration frequency is of the order of the Debye cutoff frequency, also applies to diatomic crystals. This serves to explain the correlation between Debye temperature and Reststrahl frequency, shown in Table 2.1, which was in fact noticed before the atomic mechanism of Reststrahlen was understood.

Table 2.1. Comparison of Debye Θ_D and Reststrahl frequency ω_R.

Salt	Θ_D	$\hbar\omega_R/k$
KCl	235	217
KBr	174	179
KI	132	153

PROBLEMS 2

2.1 Show that the total momentum of the vibrating linear chain considered in section 2.2 is given by

$$P(k) = -i\omega M A \, e^{-i\omega t} \sum_{n=1}^{N} e^{ikna},$$

* Actually, of course, C_p is measured and C_v must be deduced by thermodynamics (Mandl,[2] section 5.3); also, for metals, the contribution of conduction electrons to the total specific heat (section 4.21) must be deducted.

and hence that with periodic boundary conditions $u_{N+n} = u_n$, $P(k) \equiv 0$ for $k \neq 0$, so that a phonon carries no momentum. For $k = 0$, $P \neq 0$ if $Lt_{\omega \to 0}(\omega A) \neq 0$; what motion of the linear chain does this represent?

2.2 Obtain expressions for the specific heat due to longitudinal vibrations of a chain of identical atoms:

(a) in the Debye approximation;

(b) using the exact density of states (Eq. (2.31)).

With same constants K and M, which expression gives the greater specific heat and why?

Show that at low temperatures both expressions give the same specific heat, proportional to T.

2.3 We may make a model of the stretching vibrations of a polyethylene chain $-CH=CH-CH=CH- \ldots$ by considering a linear chain of identical masses M with alternating force constants K_1 and K_2. Show that the characteristic frequencies of such a chain are given by

$$\omega^2 = \frac{K_1 + K_2}{M}\left[1 \pm \left\{1 - \frac{4K_1 K_2 \sin^2 \frac{1}{2}ka}{(K_1 + K_2)^2}\right\}^{1/2}\right]$$

where a is the repeat distance of the chain (note that the relative lengths of single and double bonds are irrelevant; why?).

By obtaining values for ω as $k \to 0$ and $k \to \pm \pi/a$, sketch the dispersion curves for the optical and acoustic branches of the phonon spectrum.

2.4 The unit cell side of NaCl is 5.6 Å, and Young's modulus in a [100] direction is $5 \times 10^{10} \text{N m}^{-2}$. Estimate the wavelength at which electromagnetic radiation is strongly reflected by a sodium chloride crystal, explaining any assumptions you make.

(Atomic weights: Na = 23; Cl = 37.)

2.5 The relation between frequency v and wavelength λ for surface tension waves on a liquid of density ρ and surface tension σ is

$$v^2 = \frac{2\pi\sigma}{\rho\lambda^3}.$$

Use this result to construct a 'Debye theory' of the surface contribution to the internal energy of a liquid. Obtain the analogue of the Debye T^3 law for the surface contribution to the specific heat of liquid helium very near absolute zero.

Given that σ is the surface free energy ($F = E - TS$), how does σ vary with temperature near absolute zero?

2.6 Use Eq. (2.26) to show that in thermal equilibrium at temperature T the average energy of a sufficiently long wavelength mode is $k_B T$.

At temperatures much less than the Debye temperature Θ_D, approximately how many modes will be excited?

Use your answer to show that for $T \ll \Theta_D$ the specific heat due to atomic vibrations is of order $Nk(T/\Theta_D)^3$, where N is the number of atoms in the solid.

2.7 Diamond (atomic weight of carbon = 12) has Young's modulus of 10^{12}N m^{-2} and a density of 3.5 g cm^{-3}. Draw a sketch with numbered axes of the specific heat of diamond as a function of temperature, and explain the major features shown on your sketch. (You may ignore crystalline anisotropy and the difference between longitudinal and shear elastic moduli.)

CHAPTER

Mobile electrons

3.1 MOBILE ELECTRONS IN COVALENT CRYSTALS

One of the most obvious and important properties of solids is that many of them are electrically conducting. In such substances, metals or semi-conductors, there must clearly be some electrons that are not localized in atoms or covalent bonds, but are more or less free to move through the crystal. Our main purpose in this chapter is to understand how such mobile electron states arise. We shall consider semiconductors first, because in these substances the number of mobile electrons is small (i.e. several orders of magnitude smaller than the number of atoms), so that we may with some confidence apply an independent particle model and ignore the interactions between mobile electrons.

The important semiconducting materials, silicon and germanium, form covalent crystals with the diamond structure (Fig. 1.25), in which each atom forms four tetrahedrally directed sp^3 covalent bonds; semiconducting compounds such as GaAs form the analogous zincblende structure. These substances are all, like diamond, insulators at absolute zero. To understand their conductivity at finite temperature it is useful to consider first what happens when an extra electron is added to a structure of saturated covalent bonds; this we do, for a one-dimensional model, in section 3.2. In section 3.3 we shall consider how such electrons are actually provided in semi-conductors at finite temperature.

3.2 DYNAMICS OF AN EXTRA ELECTRON ON A CHAIN OF ATOMS

In section 1.2.2 we applied our fundamental equation for coupled quantum-mechanical probability amplitudes c_n,

$$i\hbar \frac{dc_n}{dt} = \sum_m E_{nm} c_m, \tag{1.7}$$

to the problem of the H_2^+ ion, one electron in the field of a pair of protons. We could clearly apply the same method to an electron in the field of a whole chain of protons. In fact we wish to consider an extra electron on a long chain of covalently bound atoms. Since the covalent binding electrons fill an atomic shell, the extra electron must go in the next shell of one of the atoms; if we ignore interaction of this extra electron with the closed shells we have a hydrogen-like situation.

We simplified the H_2^+ problem by considering only two probability amplitudes: c_a, the probability amplitude for finding the electron in the 1s state of atom A; and c_b, the probability amplitude for finding the electron in the 1s state of atom B. We shall, for simplicity, make an analogous assumption for the present problem. We consider only the probability amplitudes c_n for the electron to be found in the next unoccupied state of the nth atom; as with H_2^+, we suppose that the probability amplitudes for higher atomic states are negligible. With this assumption, Eqs. (1.7) give

$$i\hbar \frac{dc_n}{dt} = Ec_n - Ac_{n-1} - Ac_{n+1}, \tag{3.1}$$

(compare Eqs. (1.9)), if we further assume that the probability amplitude c_n is coupled only to the probability amplitudes on its nearest neighbours. You will notice that Eq. (3.1) is very similar in form to Eq. (2.2) for the lattice vibrations of a chain of atoms; the only essential difference is that we have $i(d/dt)$ instead of (d^2/dt^2), because our uncoupled 'oscillator' is a Schrödinger equation, not a classical harmonic oscillator.

As in section 2.2 we assume for convenience that the chain is joined to itself to form a ring* and try a solution of the form

$$c_n \propto \exp i(kx_n - \omega t),$$

where $x_n = na$ is the position of the nth atom. Substitution in Eq. (3.1) gives

$$\hbar\omega \, e^{i(kna - \omega t)} = E \, e^{i(kna - \omega t)} - A \, e^{i[k(n-1)a - \omega t]} - A \, e^{i[k(n+1)a - \omega t]},$$

---

* This makes the stationary states running waves rather than standing waves. It is a convenient way to approach the problem of an infinite chain, because it has the essential property that the chain has no ends.

i.e.

$$\hbar\omega = E - A\,e^{-ika} - A\,e^{ika}$$

$$= E - 2A\cos ka. \tag{3.2}$$

This is plotted in Fig. 3.1; we see that ω is a periodic function of k with period $(2\pi/a)$, as in the case of lattice vibrations. It can also be seen that our assumption of coupling only to the nearest neighbours is not essential to this periodicity; for if there were terms in c_{n+2} and c_{n-2} in Eq. (3.1) these would give rise to a term in $\cos 2ka$ in Eq. (3.2) and similarly for more distant neighbours. In other words, more complicated coupling can only add harmonics to the curve of $\omega(k)$; the fundamental periodicity remains at $(2\pi/a)$, depending only on the lattice spacing.

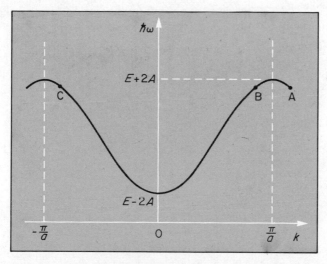

Fig. 3.1. Energy of an electron on a chain of atoms.

It is also true that the wavenumber range $-(\pi/a) < k \leqslant (\pi/a)$ describes all possible physical situations, by an exactly similar argument to that used for lattice waves. We may now take the ordinate of Fig. 2.4 as representing c_n, so that Fig. 2.4(b) shows how c_n values given by a wave with $|k| > (\pi/a)$ (point A in Fig. 3.1) can equivalently be represented by a wave with $|k| < (\pi/a)$ (points B or C in Fig. 3.1). Points A and C in Fig. (3.1) represent states with a negative group velocity $(d\omega/dk)$, and thus represent electron wave packets moving to the left. A and C are thus completely equivalent; point B represents an otherwise similar wavepacket moving to the right.

If we take the energy levels shown in Fig. 3.1 as referring to an electron on a finite chain of N atoms, we find, as in the case of lattice vibrations (section

2.2) that there are just N allowed states in the range $-(\pi/a) < k \leqslant (\pi/a)$. Thus, since the chain is joined on itself to form a ring, we require an integral number of wavelengths in the total length $L = Na$, so that

$$Na = p\lambda = 2\pi p/k,$$

or

$$k = \frac{p}{N} \cdot \frac{2\pi}{a},$$

giving N states in a range $(2\pi/a)$ of k.

In section 1.2.2 we argued that A was positive, which would imply a minimum energy at $k = 0$, as shown in Fig. 3.1. However, that argument depended on using atomic $1s$ wavefunctions, which are positive everywhere, as base states, and is therefore not generally applicable. The minimum could be at $k = (\pi/a)$, or even, with coupling to more distant neighbours, at some intermediate value. But for our present purposes it is the existence of an energy minimum that is important, independently of the k value at which it occurs; we shall therefore, for convenience, continue to discuss the $\omega(k)$ relation given by Eq. (3.2) and Fig. 3.1. If we have a number of extra electrons present at temperature T such that $k_B T \ll A$, they will all have an energy close to the minimum value; it is then convenient to define an excitation energy $\varepsilon = \hbar\omega - (E - 2A)$ and expand Eq. (3.2) for small k to give

$$\varepsilon = Ak^2a^2$$

$$= \frac{\hbar^2 k^2}{2m^*} \tag{3.3}$$

where the **effective mass** m^* is defined as $(\hbar^2/2Aa^2)$. This may be compared with the result for electrons in free space $\varepsilon = p^2/2m = \hbar^2k^2/2m$. Eq. (3.3) thus indicates that *extra electrons travel freely through the crystal like particles with mass m^**. This result is so important that it is worth giving a little thought to what it depends on. A well-defined $\omega(k)$ curve exists because we were able to find a plane wave solution $\exp i(kx_n - \omega t)$ to Eqs. (3.1); it is because these undamped plane waves can be combined to form unattenuated groups travelling with group velocity $(d\omega/dk)$ that we can speak of particles travelling freely through the lattice. Undamped plane wave solutions occur because we have a regular lattice, so that all N of Eqs. (3.1) are *identical in form*. Our result, therefore, is that *a perfect crystal is transparent to electrons*; it is not quite a vacuum, in that the electrons have an effective mass m^* rather than their real mass, analogous to a refractive index for light. This is the fundamental fact that has enabled the transistor to supersede the vacuum tube in electronics and semiconductor counters largely to replace proportional counters in nuclear instrumentation.

Although we arrived at Eqs. (3.1) by considering an extra electron in our linear lattice, you may have noticed that the arguments leading to these equations are really more general than that; all that is actually necessary is that c_n should be the probability amplitude for the nth atom to be in a state that differs in some way from that of the others, and that these 'different' states should form our base states. We considered the case where the 'different' state consists in the presence of an extra electron, but there are several other possibilities, some of which we shall encounter later, all leading to different types of **excitation** travelling freely through the crystal lattice.

A possibility of equal importance in semiconductors to that we have already considered is when the 'different' state consists in the *absence* of an electron, so that c_n is the probability amplitude for an electron to be missing from the nth atom.[†] In this case the quantity that has excitation energy $\varepsilon = (\hbar^2 k^2/2m^*)$ and moves through the crystal with group velocity $(\hbar k/m^*)$ is the lack of an electron, or a **hole**; since it represents the absence of charge $-e$ from the region of the wave packet, it behaves as a particle of charge $+e$.

3.3   METHODS OF PROVIDING ELECTRONS AND HOLES

3.3.1   Donor and acceptor impurities

If an atom from group V of the periodic table (such as phosphorus or arsenic) is added to molten silicon or germanium it crystallizes, when the melt is cooled, in a *substitutional* position in the lattice; the impurity takes the place of a Si or Ge atom in the diamond structure (Fig. 1.25). It is important that the impurity takes up a substitutional, rather than an interstitial, position, because this means that after forming the four covalent bonds demanded by the structure there is an extra electron left over which can occupy one of the mobile states we have just discussed, and hence carry electric current. It is for this reason that the band of mobile states from $(E - 2A)$ to $(E + 2A)$ in Fig. 3.1, is known as the **conduction band.**

Of course such a mobile electron will leave the P or As impurity with a positive charge, which will tend to bind the mobile electron to the impurity, just as a proton will bind an electron to form a hydrogen atom. We can estimate the strength of this binding by treating the extra electron as a particle of mass m^*, in accordance with Eq. (3.3), and replacing the covalent crystal lattice by a medium of dielectric constant ε_r. Ordinary hydrogen atom theory then gives the bound energy levels

$$E_n = -\frac{m^* e^4}{2\varepsilon_r^2 \hbar^2 n^2} \cdot \frac{1}{(4\pi\varepsilon_0)^2}$$

$$= \frac{m^* c^2 \alpha^2}{2\varepsilon_r^2 n^2} \tag{3.4}$$

[†] Or, equivalently, the nth covalent bond between atoms.

where $\alpha = (e^2/4\pi\varepsilon_0\hbar c) \approx (1/137)$ is the fine structure constant; the spatial extent of the corresponding wavefunctions is indicated by the corresponding Bohr radii,

$$r_n = \frac{\varepsilon_r n^2 \hbar^2}{m^* e^2} 4\pi\varepsilon_0 = \frac{\varepsilon_r n^2 \hbar}{m^* c\alpha} \tag{3.5}$$

The effective mass varies considerably with direction of travel in the crystal, but a typical value for Si or Ge is of order 0.1 electron masses; the dielectric constant of germanium is 15.8 and of silicon is 11.7. We can therefore estimate the ground state binding energy of the extra electron by taking $m/m^* = 10$ and $\varepsilon_r = 10$ in Eq. (3.4), so that

$$E_1 = -\left(\frac{m^*}{m\varepsilon_r^2}\right) 13.6 \text{ eV} \approx -0.01 \text{ eV},$$

and the extent of the ground state wavefunction is given by Eq. (3.5) as

$$r_1 = \left(\frac{\varepsilon_r m}{m^*}\right) 0.53 \text{ Å} \approx 50 \text{ Å}.$$

Thus, the combination of small effective mass and large dielectric constant gives very weak binding of the extra electron to the impurity and a very extended wavefunction for the bound state. Since the bound state wavefunction extends over many atomic diameters, our approximation of using a macroscopic dielectric constant to describe the lattice polarizability should not be too bad. Note that our estimate of binding energy, 0.01 eV, is rather less than $k_B T$ at room temperature, 0.026 eV, so we would expect plenty of the impurity atoms to be ionized, with their extra electrons free to move through the crystal lattice; the degree of ionization will be discussed more fully in the next section.

Eq. (3.4) gives an infinite series of bound states, but this applies only to the idealized case of a single impurity in an infinite crystal. In practice we can expect Eq. (3.4) to apply only if the mean separation between impurities is large compared with the size of the bound state wavefunction.† Since the size of the wavefunction, according to Eq. (3.5), goes as the square of the quantum number n, the highest impurity concentration for which we expect Eq. (3.4) to apply is proportional to $(1/n^6)$. The limiting concentration is thus of order $10^{18}/\text{cm}^3$ for $n = 1$, $10^{16}/\text{cm}^3$ for $n = 2$, and $10^{15}/\text{cm}^3$ for $n = 3$. In practice, therefore, the notion of hydrogenic bound states applies only to the lowest few levels. The important parameter in calculating electrical properties is the energy difference between the lowest bound state and the lowest mobile state, the bottom of the conduction band. The presence of a few more bound states much closer to the bottom of the conduction band

† What happens when this condition is not satisfied is discussed in section 4.1.2.

has very little effect and is usually ignored; we count the electron as ionized into the conduction band if it is excited out of the lowest bound state. In other words, we count all states other than the lowest bound state as part of the conduction band, although really a few states at the bottom of this band are localized near impurities rather than mobile.

Thus, the region near the minimum of Fig. 3.1 becomes as in Fig. 3.2; the bound state is called a **donor impurity level** because it is capable of giving an electron to the conduction band. Because the bound state is somewhat localized in space it requires several waves of different wavelength to make a suitable wavepacket, and there is therefore not a definite **k** associated with it. But as the wavepacket is rather large (about 10 atomic diameters) only a rather small range of **k** around **k** $= 0$, about $(\pi/10a)$, is required to construct it; the horizontal line for the donor level in Fig. 3.2 is intended as a reminder of this range of **k** values. At absolute zero the extra electron will occupy the bound impurity level, but relatively little energy is required to ionize it into the conduction band.

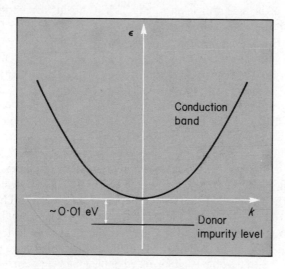

Fig. 3.2. Mobile electron states in the conduction band of a semiconductor crystal, with a donor impurity level below the band. The impurity state does not have definite k because it is localized in space.

Everything we have said about group V impurities and electrons applies equally to group III impurities and holes. A group III element, such as boron or aluminium, has one too few electrons to form the four covalent bonds demanded by the diamond structure of Si or Ge. This missing electron will form a mobile hole, as discussed at the end of section 3.2, except for a tendency to be bound to the B^- or Al^- ion, just as we have discussed for electrons and P^+ or As^+ ions. We thus have an exactly similar diagram for hole energies, as in Fig. 3.3(a). However, in order to consider electrons and holes together it is convenient to refer both to a diagram of *electron* energy; this can be done by considering the band of energy levels which is fully occupied at absolute zero by the electrons forming the covalent bonds. This band of levels is called the **valence band** and is shown in Fig. 3.3(b); we shall now show that holes may be equivalently represented on the hole energy diagram of Fig. 3.3(a) or the electron energy diagram of Fig. 3.3(b), provided that these two diagrams are related by changing the sign of both ε and k.

Thus, at absolute zero the impurity level at $-E_A$ in Fig. 3.3(a) is occupied by a hole and all other levels are vacant; in Fig. 3.3(b) the whole valence band is occupied by electrons and the **acceptor impurity level** at E_A is vacant.* If we now supply energy E_A (by means of a photon, for example) the hole is raised to the state $\varepsilon = 0$, $k = 0$ in Fig. 3.3(a) or an electron is promoted from $\varepsilon = 0$, $k = 0$ to the acceptor level in Fig. 3.3(b). Addition of a further energy $\varepsilon(k)$ will now move the hole to the point illustrated in Fig. 3.3(a). Note that this state has positive group velocity $\hbar^{-1}(\partial\varepsilon/\partial k)$, so that a wavepacket formed from this and neighbouring states represents a region of positive charge moving to the right. In order that Fig. 3.3(b) should represent the same thing we must ensure that the unoccupied state in the valence band has a positive group velocity; we transfer the unoccupied state to $-k$, as illustrated. Thus, we may say that Figs. 3.3(a) and (b) represent the same state of affairs because in both cases we have added to the ground state at $T = 0$ energy $(E_A + \varepsilon(k))$ and momentum $\hbar k$.†

A word is in order here about the validity of the independent particle model, for at first sight Fig. 3.3(b) seems to imply that we are treating the electrons in the covalent bonds as independent particles. This is not so, however. We obtained our mobile states by considering a single electron or hole and argued, reasonably, that for *small* numbers of electrons or holes these should behave independently. Thus a few occupied hole states in Fig. 3.3(a) will behave almost independently of each other, and correspondingly a few empty states in Fig. 3.3(b) will behave almost independently of each other. But it does *not* follow that the many *occupied* states in Fig. 3.3(b) behave independently of each other, and in general they do not. The problem

---

* Adding an impurity does not alter the total number of electron energy levels. Rather, levels are detached from the conduction and valence bands to form the donor and acceptor impurity levels. Therefore, although a crystal containing an acceptor atom has one too few electrons to form all the covalent bands, the valence band has lost an energy level to form the localized acceptor level. There are thus just enough electrons to fill the valence band at $T = 0$.

† However, as with phonons, $\hbar k$ is not really a momentum, it is only similar in some respects. We call it **crystal momentum**; the relation to the true momentum of an electron is discussed in section 9.2.

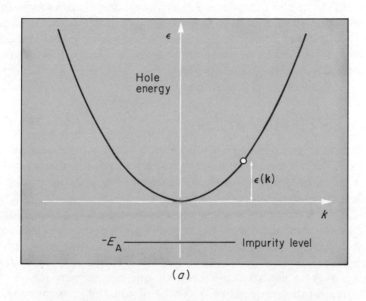

(a)

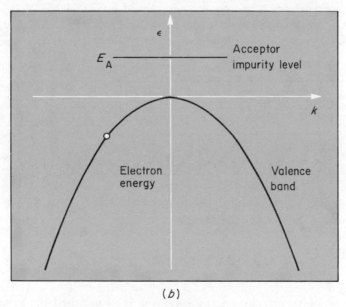

(b)

Fig. 3.3. Hole states and acceptor impurity level:
(a) hole energy levels,
(b) equivalent electron energy levels.

of the validity of the independent particle model when many states are occupied is one we shall have to face when we consider metals in Chapter 4 and particularly in Chapter 9. For the moment we note that Fig. 3.3(b) is just a way of describing the motion of independent holes: they behave *as if* the valence band in Fig. 3.3(b) were almost full of *independent* electrons but no actual independence of electrons is implied.

3.3.2 Thermal excitation of carriers

In order to calculate the thermal excitation of electrons in the conduction band and holes in the valence band we must first put together Figs. 3.2 and 3.3(b) so as to obtain the complete electron energy diagram Fig. 3.4. The zero of energy is conventionally taken at the top of the valence band, and the energy E_G is the minimum energy required to create an electron–hole pair; this can be determined, for example, by the onset of photon absorption in the infrared. Fig. 3.4 is a simplification of the actual situation in Si and Ge in that for these materials the conduction band minimum is not at $\mathbf{k} = 0$, but this makes no essential difference for our present purposes. Table 3.1 shows values of the energies in Fig. 3.4 for silicon and germanium. The fact that E_G corresponds to a frequency in the infrared makes silicon and germanium opaque in the visible, and their high dielectric constant gives them a high reflection coefficient and consequent metallic lustre.

Table 3.1. Characteristic energies in Si and Ge.

	E_G (eV)	E_D (eV) for P	E_D (eV) for As	E_A (eV) for B	E_A (eV) for Al
Si	1.08	0.045	0.049	0.045	0.057
Ge	0.66	0.012	0.013	0.010	0.010

To calculate the thermal occupation of the energy levels in Fig. 3.4 we need to know the Fermi distribution function,

$$f(\varepsilon) = \frac{1}{e^{(\varepsilon - \mu)/k_B T} + 1}, \tag{3.6}$$

in which the chemical potential μ is to be adjusted to obtain the right total number of particles. The energy level $\varepsilon = \mu(T)$ is called the **Fermi level**. We shall follow the usual convention in statistical mechanics and define the **Fermi energy** ε_F as $\varepsilon_F = \mu(0)$.* We also require the density of states per unit energy range, which may be obtained from our general density of

* But note that in semiconductor physics it is common practice to define $\varepsilon_F(T) \equiv \mu(T)$.

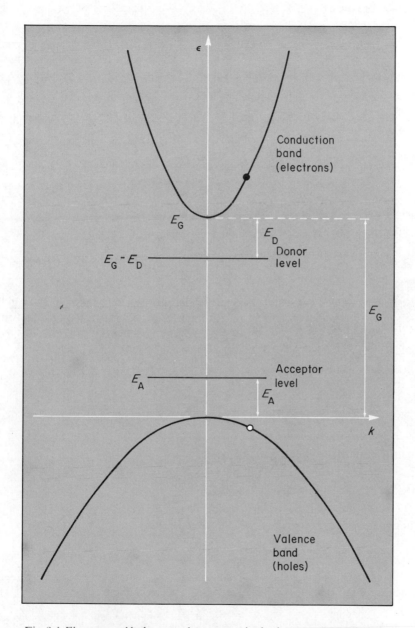

Fig. 3.4. Electron and hole states shown on a single electron energy diagram; E_G is the minimum energy to create an electron–hole pair.

states expression†

$$g(k)\,dk = \frac{Vk^2}{2\pi^2}\,dk \qquad (2.28)$$

by setting $g(\varepsilon)\,d\varepsilon = g(k)\,dk$ (as for Eq. (2.29)); thus

$$g(\varepsilon) = \frac{Vk^2}{2\pi^2}\frac{dk}{d\varepsilon}. \qquad (3.7)$$

In this we have, for a particle of mass m^*,

$$\varepsilon = \frac{\hbar^2 k^2}{2m^*}, \qquad (3.3)$$

so that

$$\frac{d\varepsilon}{dk} = \frac{\hbar^2 k}{m^*} = \frac{\hbar}{m^*}(2m^*\varepsilon)^{1/2}.$$

Also, each state of translational motion represented by Eq. (3.7) can be occupied by two electrons of opposite spin, so that, introducing a factor 2 to allow for this, our final density of states per unit energy range is

$$g(\varepsilon) = 2 \cdot \frac{V}{2\pi^2} \cdot \frac{2m^*\varepsilon}{\hbar^2} \cdot \frac{m^*}{\hbar(2m^*\varepsilon)^{1/2}}$$

$$= \frac{V}{2\pi^2\hbar^3}(2m^*)^{3/2}\varepsilon^{1/2}. \qquad (3.8)$$

For the conduction band Eq. (3.3) is replaced by

$$\varepsilon = E_G + \frac{\hbar^2 k^2}{2m_e},$$

where m_e is the effective mass of electrons, so that the density of states is

$$g(\varepsilon) = \frac{V}{2\pi^2\hbar^3}(2m_e)^{3/2}(\varepsilon - E_G)^{1/2}; \qquad (3.8a)$$

correspondingly, for the valence band the electron energy is

$$\varepsilon = -\frac{\hbar^2 k^2}{2m_h}, \qquad (3.3b)$$

† The reader unfamiliar with this result should consult section 2.6.2 or Mandl,[2] Appendix B, at this stage.

where m_h is the effective mass of holes, so that the density of states is

$$g(\varepsilon) = \frac{V}{2\pi^2\hbar^3}(2m_h)^{3/2}(-\varepsilon)^{1/2}, \tag{3.8b}$$

The density of states given by Eqs. (3.8a) and (3.8b) for the energy levels shown in Fig. 3.4 is sketched in Fig. 3.5; in addition to the parabolic densities of states in conduction and valence bands there are δ-functions at energies E_A and $(E_G - E_D)$ corresponding to the N_A acceptor impurities and N_D donor impurities respectively. The Fermi function, Eq. (3.6), is also shown

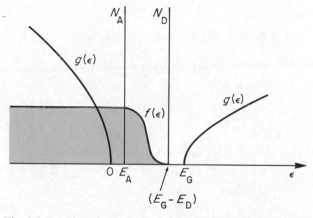

Fig. 3.5. Density of states $g(\varepsilon)$ and Fermi functions $f(\varepsilon)$ for a semiconductor with the level scheme of Fig. 3.4.

in Fig. 3.5; we note that at ordinary temperatures $k_B T \ll E_G$, so that provided the Fermi level comes somewhere in the band gap (as illustrated; this is almost always the case) both conduction and valence bands come in the 'tails' of the Fermi function, which simplifies our calculations. Thus, for electrons in the conduction band,

$$f(\varepsilon) \approx e^{(\mu-\varepsilon)/k_B T},$$

so that the total number of electrons per unit volume is given by

$$n = \frac{1}{V}\int_{E_G}^{\infty} f(\varepsilon)g(\varepsilon)\,d\varepsilon$$

$$= \frac{1}{2\pi^2\hbar^3}\int_{E_G}^{\infty}(2m_e)^{3/2}(\varepsilon - E_G)^{1/2}\,e^{(\mu-\varepsilon)/k_B T}\,d\varepsilon$$

$$= \frac{(2m_e)^{3/2}}{2\pi^2\hbar^3}e^{(\mu-E_G)/k_B T}\int_0^{\infty}(\varepsilon - E_G)^{1/2}\,e^{-(\varepsilon-E_G)/k_B T}\,d(\varepsilon - E_G)$$

$$= N_c\,e^{(\mu-E_G)/k_B T}, \tag{3.9}$$

where

$$N_c = 2\left(\frac{2\pi m_e k_B T}{h^2}\right)^{3/2}.$$
(3.10)

N_c is the effective number of levels per unit volume in the conduction band if we imagine them concentrated at the bottom of the band. It is, of course, temperature dependent, because the levels are not so concentrated. The number of holes per unit volume may be calculated similarly. The occupation number for holes is

$$1 - f(\varepsilon) \approx e^{(\varepsilon - \mu)/k_B T}$$

so that the total number of holes in unit volume is*

$$p = \frac{1}{V}\int_{-\infty}^{0} [1 - f(\varepsilon)]g(\varepsilon)\,d\varepsilon$$

$$= \frac{(2m_h)^{3/2}}{2\pi^2\hbar^3}\int_{-\infty}^{0} (-\varepsilon)^{1/2}\,e^{(\varepsilon - \mu)/k_B T}\,d\varepsilon$$

$$= \frac{(2m_h)^{3/2}}{2\pi^2\hbar^3}\,e^{-\mu/k_B T}\int_{0}^{\infty} \varepsilon^{1/2}\,e^{-\varepsilon/k_B T}\,d\varepsilon$$

$$= N_v\,e^{-\mu/k_B T},$$
(3.11)

where

$$N_v = 2\left(\frac{2\pi m_h k_B T}{h^2}\right)^{3/2}$$
(3.12)

is the effective number of states per unit volume in the valence band.

The effect of doping a pure semiconductor with donor or acceptor impurities is to shift the chemical potential, as we shall discuss shortly, and hence alter the electron and hole concentrations. But first let us note that an important result *independent of μ* can be obtained by multiplying together Eqs. (3.9) and (3.11):

$$np = N_c N_v\,e^{-E_G/k_B T}.$$
(3.13)

This result, that the product of electron and hole concentrations, in a given semiconductor, is a function only of the temperature and is, for example, independent of impurity concentration, is an example of the **law of mass action** in chemistry; it is analogous to the constant (at a given temperature) product of hydrogen and hydroxyl ion concentrations in different aqueous solutions. The 'chemical reaction' concerned is the equilibrium between

---

* We use the symbol p for positive carrier concentration and n for negative carrier concentration.

electron hole pairs and thermal energy, in the form of lattice vibrations or black body radiation.

In a pure semiconductor electrical neutrality requires that $n = p$, so that

$$e^{(2\mu - E_G)/k_B T} = N_v/N_c,$$

or

$$\mu = \tfrac{1}{2}E_G + \tfrac{3}{4}k_B T \ln (m_h/m_e); \tag{3.14}$$

since $k_B T \ll E_G$ the second term is small and the Fermi level $\varepsilon = \mu$ is essentially in the middle of the gap. Also from Eq. (3.13)

$$n = p = (N_c N_v)^{1/2} e^{-E_G/2k_B T}. \tag{3.15}$$

The carrier concentrations given by Eq. (3.15) are called **intrinsic carrier concentrations**, because they are an intrinsic property of the pure semiconductor; the electrical conductivity that they give rise to is likewise called **intrinsic conductivity**.

To find out in what circumstances we may expect to observe experimentally the intrinsic behaviour characteristic of pure material, let us consider what happens when a small amount of donor impurity is added. At room temperature small amounts of impurity alter the chemical potential μ only slightly, so that the donor level remains in the high energy tail of the Fermi function. Consequently, the donor level is almost unoccupied, so that most donor atoms are ionized. The number of free charge carriers will therefore be seriously affected unless the number of donors N_D is small compared with the number of intrinsic carriers given by Eq. (3.15). Thus, taking equal effective masses for simplicity, the material will show intrinsic behaviour if

$$N_D \ll N_v e^{-E_G/2k_B T}.$$

With $m/m_e \sim 10$ as before we find from Eq. (3.10) that at room temperature $N_v \sim 10^{18}/cm^3$ (remember the density of *atoms* is of order $10^{22}/cm^3$). Also at room temperature $e^{-E_G/2k_B T} \sim 10^{-4}$, so that the condition for intrinsic behaviour becomes

$$N_D \ll 10^{14}/cm^3,$$

or an impurity content of less than 1 in 10^8. The technical problem of making crystals of such extraordinarily high purity is one reason why semiconductors were not widely used earlier; the breakthrough that made modern developments possible was the discovery of **zone-refining**. This depends on the fact that impurities are more soluble in the liquid than in the solid, so that if a molten zone is moved along a crystal the impurities will be drawn along with it. In practice the crystal is held just below the melting point in a furnace and a small zone is melted by an auxiliary induction heater; by pulling the sample through the induction heater the molten zone is moved along it.

By repeatedly passing the molten zone along the sample in the same direction the donor and acceptor concentrations can be made as low as 1 part in 10^{10}.

Intrinsic semiconductors produced in this way are the starting point of semiconductor technology, but for most purposes the purified material is subsequently doped with controlled amounts of donors or acceptors to give n-type material (majority carriers electrons) or p-type material (majority carriers holes).

In general the electron and hole concentrations can be calculated by combining with the law of mass action, Eq. (3.13), the requirement of electrical neutrality

$$n + N_A^- = p + N_D^+, \tag{3.16}$$

in which N_D^+ and N_A^- are the concentrations of ionized donors and acceptors respectively; these are given in terms of the Fermi distribution function, Eq. (3.6), by

$$N_D^+ = N_D[1 - f(E_G - E_D)],$$

and $$\tag{3.17}$$

$$N_A^- = N_A f(E_A).$$

In other words, the total number of holes in valence band and donor level is equal to the total number of electrons in conduction band and acceptor level.

The commonest situation is that in which some impurities of both types are present. Then, at absolute zero, donors will be ionized in filling up the acceptor levels until this is no longer possible; the minority impurity will then be fully ionized and the majority impurity will be partly ionized. Thus, if there are more donors than acceptors, the acceptor levels will be fully occupied by electrons and the donor levels will be partly occupied; since only the Fermi level can be partly occupied at $T = 0$ we have $\mu = (E_G - E_D)$. Moreover, provided the fraction of ionized donors is neither very large nor very small, the Fermi level will remain close to this value as the temperature is raised, so that Eq. (3.9) gives

$$n \approx N_c \, e^{-E_D/k_B T}. \tag{3.18}$$

Note the very much smaller exponent in Eq. (3.18) than in Eq. (3.15) for intrinsic material, and consequently the larger number of carriers. Material of this type, with more electrons than holes, is known as **n-type**. In an exactly similar way excess acceptors give **p-type** material, with more holes than electrons, and by a similar argument

$$p \approx N_v \, e^{-E_A/k_B T}.$$

The situation is slightly different if the impurities are entirely of one type (say donors). Then for any appreciable amount of impurity Eq. (3.16) reduces to

$$n \approx N_D^+. \tag{3.19}$$

At first the Fermi level remains below the donor level and we have $N_D^+ \approx N_D$ so that

$$n \approx N_D;$$

as the donor concentration is raised we reach a situation where the Fermi level is between the donor level and the conduction band. Provided $k_B T < E_D$ (in practice this means below room temperature) the situation is now similar to that for intrinsic material except that the donor level takes the place of the valence band. Eq. (3.17) gives

$$N_D^+ = N_D \, e^{(E_G - E_D - \mu)/k_B T},$$

which with

$$n = N_c \, e^{(\mu - E_G)/k_B T} \tag{3.9}$$

and Eq. (3.19) gives

$$n \approx (N_c N_D)^{1/2} \, e^{-E_D/2k_B T}, \tag{3.20}$$

analogously to Eq. (3.15). Note that the exponents of Eqs. (3.18) and (3.20) differ by a factor of two, so appreciable amounts of minority impurity have quite a noticeable effect.

3.4   TRANSPORT PROPERTIES

3.4.1   Conductivity

In the absence of collisions electrons or holes in a crystal will obey the acceleration equation

$$m^* \frac{d\mathbf{v}}{dt} = e\mathbf{E} \tag{3.21}$$

for the drift velocity $\mathbf{v}$ in an electric field $\mathbf{E}$; e is the algebraic charge on the carriers, positive for holes and negative for electrons.† We saw in section 3.2 that electrons or holes travel freely through a perfect crystal, but in order to create free carriers we normally add impurities to the crystal, as discussed in section 3.3, and these impurities then act as collision centres. In addition,

† At this stage we may regard Eq. (3.21) as justified by the analogy between real mass and effective mass defined by Eq. (3.3). A formal justification of Eq. (3.21) will be given in section 9.2.

the thermal vibrations of the crystal lattice mean that its spacing is no longer quite uniform, and this provides an additional, thermally activated, mechanism for scattering electrons and holes. All these mechanisms tend to remove drift momentum from the carriers so that in the absence of an electric field the drift velocity decays to zero. We allow for this effect by modifying Eq. (3.21) to

$$m^* \left(\frac{d\mathbf{v}}{dt} + \frac{\mathbf{v}}{\tau} \right) = e\mathbf{E}, \tag{3.22}$$

where the relaxation time τ is a measure of the collision rate. If we suppose an electron loses all its drift momentum at each collision, then τ is the mean interval between collisions; in any case it is a quantity of the same order of magnitude. Note that we have assumed for simplicity that τ is independent of the thermal group velocity v_G of the carriers.

In a steady state Eq. (3.22) gives for a single type of carrier

$$\mathbf{v} = \frac{e\tau}{m^*} \mathbf{E}.$$

The absolute value of the constant of proportionality between drift velocity and electric field is called the **mobility** μ of the carriers, and is given by

$$\mu = |\mathbf{v}|/|\mathbf{E}| = |e|\tau/m^*. \tag{3.23}$$

The electric current $\mathbf{j}$ is $ne\mathbf{v}$ where n is the number of carriers per unit volume, so that Ohm's law $\mathbf{j} = \sigma\mathbf{E}$ is obeyed and from Eq. (3.23) the conductivity σ is given by

$$\sigma = \frac{ne^2\tau}{m^*} = n|e|\mu. \tag{3.24}$$

When both types of carrier are present the conductivity is a sum of electron and hole contributions so that Eq. (3.24) becomes

$$\sigma = \frac{ne^2\tau_e}{m_e} + \frac{pe^2\tau_h}{m_h} = |e|(n\mu_e + p\mu_h). \tag{3.24a}$$

Some experimental conductivities for arsenic-doped germanium are plotted logarithmically against $(1/T)$ in Fig. 3.6. The steep line on the left represents intrinsic behaviour, only reached by the purest sample. The lines of smaller slope on the right[†] correspond to a carrier concentration given approximately by Eq. (3.20), with the donors partly ionized. In the intermediate region the donors are fully ionized, so that the number of electrons is equal to the number of donors, and the decrease in conductivity

---

[†] For impurity concentrations of $10^{13}/cm^3$ and $10^{16}/cm^3$. The different behaviour at concentrations $\sim 10^{18}/cm^3$ will be discussed in section 4.1.2.

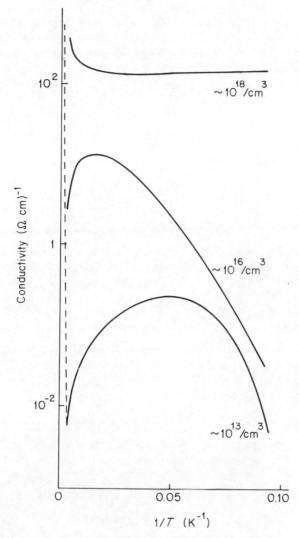

Fig. 3.6. Temperature dependence of the conductivity of
three samples of n-type germanium, of the approximate
impurity concentration marked.
{after P. P. Debye and E. M. Conwell, *Phys. Rev.*, **93**, 693
(1954)}.

with rising temperature is due to increased scattering by thermally-excited
lattice vibrations.

Let us see if we can account roughly for the order of magnitude of conductivity shown in Fig. 3.6. If we take a moderately low temperature such as
that of liquid N_2 (77 K) the scattering will be largely due to impurities

except in the purest samples. A typical conductivity, from Fig. 3.6, is $\sigma \sim 10\,\Omega^{-1}\,\text{cm}^{-1}\,(=10^3\,\Omega^{-1}\,\text{m}^{-1})$ for an impurity concentration of $10^{16}/\text{cm}^3\,(=10^{22}/\text{m}^3)$; from Eq. (3.24), with $e \approx 10^{-19}$ coulomb a typical mobility is thus

$$\mu = \frac{\sigma}{n|e|} \approx \frac{10^3}{10^{22} \times 10^{-19}}$$

$$= 1\,\text{m}^2\,\text{V}^{-1}\,\text{s}^{-1} = 10^4\,\text{cm}^2\,\text{V}^{-1}\,\text{s}^{-1},$$

and from Eq. (3.23) with $m^* \approx 10^{-28}$ g a typical collision time is

$$\tau = \frac{\mu m^*}{|e|} \approx \frac{1 \times 10^{-31}}{10^{-19}}$$

$$= 10^{-12}\,\text{s}.$$

We can relate this to the collision cross section of an impurity by elementary kinetic theory (Flowers and Mendoza,[3] section 6.4.1). The mean free path l is given by

$$l = \frac{1}{N_i X}$$

where N_i is the concentration of collision centres (taken as the concentration of impurities), and X is the collision cross-section (we use X instead of the customary σ to avoid confusion with the conductivity); this in turn is related to the collision time by

$$\tau = l/v_G$$

where v_G, the thermal group velocity of carriers, is of order $(k_B T/m^*)^{1/2}$. We thus have, at liquid nitrogen temperature $(k_B T \approx 10^{-14}\,\text{erg} = 10^{-21}$ joule),

$$\tau = \frac{1}{N_i X}\left(\frac{m^*}{k_B T}\right)^{1/2}$$

or

$$X \approx \frac{1}{N_i \tau}\left(\frac{m^*}{k_B T}\right)^{1/2} \approx \frac{1}{10^{22} \times 10^{-12}}\left(\frac{10^{-31}}{10^{-21}}\right)^{1/2}$$

$$= 10^{-15}\,\text{m}^2 = 10^{-11}\,\text{cm}^2.$$

An impurity thus seems to be a very large object, with a collision diameter of about 3×10^{-8} m, or 300 Å. This is of the same order as the size of the bound state wavefunction for a neutral impurity, but under our conditions $(k_B T \sim E_D, N_D \ll N_v)$ the donors will be mostly ionized, so that it is Coulomb scattering by an ionized donor that is important. This can be calculated from the Rutherford scattering formula; for the present purpose we make a rough estimate by saying that the radius inside which scattering is appreciable is that at which the Coulomb potential energy $(e^2/4\pi\varepsilon_0\varepsilon_r r)$ is equal to the kinetic energy $(\sim k_B T)$ of the incident electron,

$$r \approx \frac{e^2}{4\pi\varepsilon_0\varepsilon_r kT} \approx \frac{10^{-38}}{10^{-10} \times 10 \times 10^{-21}}$$

$$= 10^{-8}\,\mathrm{m} = 100\,\text{Å},$$

so that the effective collision diameter is about 200 Å, in agreement with our rough estimate from the conductivity.

3.4.2   Hall effect

When a metal or semiconductor is placed in a magnetic field **B** and current density **j** passed through it a transverse electric field $\mathbf{E_H}$ is set up given by

$$\mathbf{E_H} = R_H\mathbf{B} \times \mathbf{j}, \tag{3.25}$$

which is the defining equation for the **Hall coefficient** R_H. The geometry of the experiment is shown in Fig. 3.7. The origin of the effect is the Lorentz force $e\mathbf{v} \times \mathbf{B}$ on a moving charge in a magnetic field. Fig. 3.7 shows the direction of motion of electrons and holes corresponding to a current j_x, and the

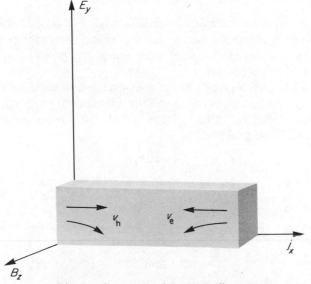

Fig. 3.7. Geometry of the Hall effect.

curved arrows show the direction in which a magnetic field will tend to deflect the carriers. This deflection rapidly builds up a space charge, and consequent electric field E_H in the y-direction, so that the current continues to flow in the x-direction, as it must for a long rod with electrical connections at the ends. The situation may be analysed, for a single type of carrier, by generalizing Eq. (3.22) to include magnetic fields:

$$\frac{d\mathbf{v}}{dt} + \frac{\mathbf{v}}{\tau} = \frac{e}{m^*}(\mathbf{E} + \mathbf{v} \times \mathbf{B}).\qquad(3.26)$$

In a steady state for the geometry of Fig. 3.7 this becomes in component form, with $v_y = 0$,

$$\frac{v_x}{\tau} = \frac{e}{m^*}E_x,$$

$$0 = \frac{e}{m^*}(E_y - v_xB),$$

so that from the first equation the resistivity is unchanged, and from the second

$$E_y = v_xB = \frac{j_xB}{ne},\qquad(3.27)$$

so that

$$R_H = 1/ne.$$

This result relies on our assumption of a single relaxation time τ, independent of electron thermal velocity; failure of this assumption may modify Eq. (3.27) by a numerical factor of order unity, and give rise to a change of resistance with magnetic field.

The Hall effect thus gives a direct measurement of carrier concentration, and moreover, its sign gives the sign of the charge carriers (positive for holes, negative for electrons). Also, by combining with the conductivity (Eqs. (3.23) and (3.24)) we find

$$\mu = |R_H|\sigma,\qquad(3.28)$$

so that mobilities can be determined.

Note that for the example we considered in section 3.4.1

$$R_H = \frac{1}{ne} = \frac{1}{10^{22} \times 10^{-19}}$$

$$= 10^{-3}\,\Omega\,\text{m}\,\text{T}^{-1} = 10^{-5}\,\Omega\,\text{cm}\,\text{gauss}^{-1}$$

compared with a resistivity $(1/\sigma) = 10^{-1}\,\Omega\,\text{cm}\,(10^{-3}\,\Omega\,\text{m})$; the ohmic and

Hall electric fields will therefore be equal in a field of 1 tesla. At this magnetic field the electric field is at 45° to the current flow, or the **Hall angle** is 45°. This magnetic field, the Hall field B_0, is a useful measure of the strength of the Hall effect; from the above calculation we see that it is given by

$$B_0 = \frac{1}{|R_H|\sigma} = 1/\mu. \tag{3.29}$$

It is thus the small number of carriers and high mobility in semiconductors that gives them a small Hall field and a large Hall angle, and hence makes them useful for the construction of probes to measure magnetic fields. Fairly lightly doped material is most suitable for this purpose, so that the carrier concentration is equal to the impurity concentration, independent of temperature in the operating range.

If more than one type of carrier is present the situation is more complicated, in that the terms of Eq. (3.26) contain contributions from both electrons and holes so that in a steady state it becomes

$$\mathbf{j} = n\mu_e e(\mathbf{E} + \mathbf{v}_e \times \mathbf{B}) + p\mu_h e(\mathbf{E} + \mathbf{v}_h \times \mathbf{B}), \tag{3.30}$$

where e is now a positive quantity, and we are taking account of negative signs for electrons explicitly. Although our boundary conditions require $j_y = 0$, there may now be cancelling electron and hole currents in the y-direction. These currents will however be proportional to B, so that their contribution to the x component of Eq. (3.30), through the $\mathbf{v} \times \mathbf{B}$ terms will be of order B^2; we ignore this for simplicity and write Eq. (3.30) in component form as

$$j_x = eE_x(n\mu_e + p\mu_h)$$
$$0 = eE_y(n\mu_e + p\mu_h) - eB_z(n\mu_e v_{ex} + p\mu_h v_{hx})$$
$$= eE_y(n\mu_e + p\mu_h) + eB_z E_x(n\mu_e^2 - p\mu_h^2),$$

where the signs of v_e and v_h are taken from Fig. 3.7. Elimination of E_x between these two equations gives

$$E_y = -\frac{j_x B_z(n\mu_e^2 - p\mu_h^2)}{e(n\mu_e + p\mu_h)^2},$$

or

$$R_H = \frac{p\mu_h^2 - n\mu_e^2}{e(p\mu_h + n\mu_e)^2}. \tag{3.31}$$

Thus a minority carrier can determine the sign of the Hall coefficient if its mobility is high enough. It is interesting to work out also the y-component of electron current j_{ey} which is balanced by an equal and opposite hole current j_{hy}. We have

$$
\begin{aligned}
j_{ey} &= n\mu_e e(E_y + \mu_e E_x B_z) \\
&= n\mu_e e E_x B_z \left[\frac{p\mu_h^2 - n\mu_e^2}{p\mu_h + n\mu_e} + \mu_e \right] \\
&= \frac{j_x B_z \, n\mu_e p\mu_h (\mu_e + \mu_h)}{(n\mu_e + p\mu_h)^2} \\
&= j_x B_z (\mu_e + \mu_h) \frac{\sigma_e \sigma_h}{(\sigma_e + \sigma_h)^2},
\end{aligned}
\tag{3.32}
$$

where σ_e and σ_h are the electron and hole contribution to the conductivity. Eq. (3.32) shows that we have a steady flow of electrons and holes in the negative y-direction, which is largest when the electron and hole conductivities are comparable. This requires creation of electron–hole pairs at one side of the sample, absorbing energy, and their mutual annihilation at the other side, releasing energy. Consequently, a transverse temperature gradient, known as the Ettinghausen effect, will develop; such an effect is indicative of the presence of two types of carrier.

3.4.3 Cyclotron resonance

In a cyclotron use is made of the fact that in a magnetic field non-relativistic charged particles of mass m^* move in circular orbits at an angular frequency $\omega_c = eB/m^*$ independent of their energy, and can thus absorb energy from a suitably phased radiofrequency electric field at this frequency. In semiconductors the same principle can be used to determine the effective mass m^*; for a single group of carriers we analyse the situation by writing Eq. (3.26) in component form for B in the z-direction and E in the (x, y) plane:

$$
\frac{dv_x}{dt} + \frac{v_x}{\tau} = \frac{e}{m^*}(E_x + v_y B),
$$

$$
\frac{dv_y}{dt} + \frac{v_y}{\tau} = \frac{e}{m^*}(E_y - v_x B).
$$

The symmetry of these two equations enables us to reduce them to one by adding i times the second to the first and writing $v_x + iv_y = u$, $E_x + iE_y = \mathscr{E}$. This gives†

$$
\frac{du}{dt} + \frac{u}{\tau} = \frac{e}{m^*}(\mathscr{E} - iuB).
\tag{3.33}
$$

---

† This equation is an example of the general rule that two dimensional vectors can be represented by complex numbers on an Argand diagram.

The obvious form of solution to try in Eq. (3.33) is $u = u_0 e^{-i\omega t}$, $\mathscr{E} = \mathscr{E}_0 e^{-i\omega t}$, where the amplitudes u_0 and $\mathscr{E}_0$ are complex numbers. The physical meaning of this type of solution is just that $v_x = u_0 \cos \omega t$ and $v_y = u_0 \sin \omega t$ (because $e^{i\theta} = \cos \theta + i \sin \theta$), so that it corresponds to motion in a circle. Similarly, $\mathscr{E}$ is the electric field of circularly polarized radiation; the magnetic field of the radiation (which we have not written down) is negligible at the frequencies that concern us. With this trial solution Eq. (3.33) reduces to the algebraic equation

$$\left(\frac{1}{\tau} - i\omega\right)u_0 = \frac{e}{m^*}(\mathscr{E}_0 - iu_0 B)$$

from which we obtain the specific impedance as

$$Z = \frac{\mathscr{E}_0}{neu_0} = \frac{m^*}{ne^2}\left(\frac{1}{\tau} - i\omega + i\frac{eB}{m^*}\right). \tag{3.34}$$

This is like the impedance of a series L–C–R circuit; the reactance vanishes, giving a minimum impedance and maximum power absorption from a constant voltage source, at the resonant frequency $\omega_c = eB/m^*$. For a sharp resonance we need $\omega_c \tau \gg 1$, i.e. the electron should complete several orbits between one collision and the next; in a field of 1 tesla our example of section 3.4.1 gives

$$\omega_c = \frac{eB}{m^*} \approx \frac{10^{-19} \times 1}{10^{-31}} = 10^{12} \text{ s}^{-1}$$

so that

$$\omega_c \tau \approx 1.$$

The observation of cyclotron resonance is therefore not trivial; it requires quite high purity and liquid helium or hydrogen temperatures to reduce scattering by lattice vibrations. The conductivity must not be too high, however, or the radiofrequency field will not penetrate the sample because of the skin effect. Moreover, the angular frequency we calculated above is inconveniently high; it corresponds to $\nu \approx 2 \times 10^{11}$ Hz = 200 GHz, or a wavelength of about 1 mm. A wavelength of about 1 cm in a field of about 0.1 tesla is easier, if τ can be made sufficiently long.

Equation (3.34) is correct for holes as it stands; if e is negative ω must also be made negative to give a resonance, i.e. the opposite circular polarization must be used; a little thought will verify that the senses of rotation are those indicated by the curved arrows in Fig. 3.7. If several types of carrier are present they behave as impedances in parallel, and separate resonances for each can be picked out if $\omega_c \tau$ is large enough; plane polarized radiation can also be used, as it can be considered as a coherent sum of left and right circular polarizations, but if this is done electrons and holes cannot be distinguished. Some typical results are shown in Fig. 3.8.

It is of interest to identify the various terms in the impedance, Eq. (3.34). The first is the D.C. resistance; the second is the inductance from electron inertia; and the third is the Hall voltage, which appears capacitative or inductive according to the sense of circular polarization.

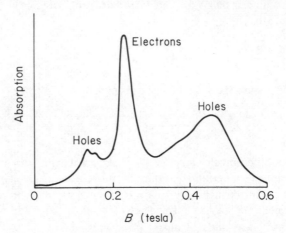

Fig. 3.8. Cyclotron resonance in Si at 4 K, using plane-polarized radiation at 23 GHz; the carriers were optically excited.
{after R. N. Dexter, H. J. Zieger and B. Lan, *Phys. Rev.*, **104**, 637 (1956)}.

★ **3.5 SEMICONDUCTOR DEVICES**

3.5.1 Junction diode

The fundamental semiconductor device is a junction between p-type and n-type material (a p–n junction). An energy level diagram for such a junction in thermal equilibrium is shown in Fig. 3.9; in this diagram energy levels are shown as a function of position in the crystal, and no distinction is made between different **k** values. Electrons in the conduction bands are indicated schematically by solid circles and holes in the valence bands by open circles.

The factor controlling the relative positions of the levels on the two sides of the junction is the necessity of a **uniform chemical potential** μ in thermal equilibrium; the Fermi level μ is therefore the same on the two sides. If we imagine the p- and n-type materials placed in contact, this equilibrium is achieved by a small transfer of electrons from n-type to p-type, where they annihilate with holes, leaving a region with no free carriers near the junction called the **depletion layer** The ionized donors and acceptors remaining in

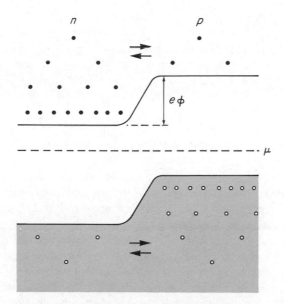

Fig. 3.9. A *p-n* junction in equilibrium, with equal
and opposite flows of electrons (●) and of holes (○),
as indicated by the arrows.

the depletion layer leave the *n*-type positively charged and the *p*-type
negatively charged. This dipolar layer at the interface produces the potential
energy gradient at the bottom of the conduction band and the top of the
valence band shown in Fig. 3.9; remember that Fig. 3.9 is a diagram of
electron energy, and therefore a region of low energy is a region of high
electrostatic potential. It is important to remember that this potential
difference between *p*-type and *n*-type in thermal equilibrium is in the nature
of a contact potential ϕ; in any complete circuit there are compensating
contact potentials at other junctions so that no current flows in thermal
equilibrium.

We may estimate the thickness d of the depletion layer as follows. Without
solving the electrostatic problem completely, we can say that the potential
difference across the junction is related to the charge Q per unit area of the
dipole layer by

$$\frac{Q}{\varepsilon_r \varepsilon_0} \approx \frac{\phi}{d}. \tag{3.35}$$

For comparable concentrations of donors and acceptors on the two sides of
the junction we also have

$$Q \approx N_D e d$$

which with Eq. (3.35) yields

$$d \approx \left(\frac{\varepsilon_r \varepsilon_0 \phi}{N_D e}\right)^{1/2}$$

$$\approx \left(\frac{10 \times 10^{-11} \times 1}{10^{22} \times 10^{-19}}\right)^{1/2} \approx 3 \times 10^{-7} \text{ m} = 0.3 \text{ } \mu\text{m},$$

(3.36)

if we assume $\varepsilon_r \approx 10$, $\phi \approx 1$ V, $N_D \approx 10^{16}/\text{cm}^3$ $(10^{22}/\text{m}^3)$.

In thermal equilibrium electrons pass both ways through the junction at constant *total* energy, but with a kinetic energy greater by $e\phi$ on the n side. If, for the moment, we ignore electron collisions, we can easily calculate the flow of electrons from p-type to n-type by multiplying the density of states in the p-type conduction band by the occupation number and component of velocity towards the junction, and integrating over energy from the bottom of the p-type conduction band upwards. At first sight, the balancing flow from n-type to p-type can be similarly calculated by taking the corresponding quantities for the n-type conduction band and integrating over allowed energies i.e. from the bottom of the p-type conduction band upwards. But at each energy the n-type density of states is greater than that in the p-type, and also the n-type electron velocity is greater, so it seems that there must be a net flow from n to p, in contradiction with thermal equilibrium! The fallacy in this argument may be exposed by considering Fig. 3.10. Electrons going from p-type to n-type are accelerated down the potential gradient in the junction and emerge in a narrow cone (Fig. 3.10(a)). Correspondingly, electrons incident from the n-type side of the junction must have trajectories within this cone to be transmitted (Fig. 3.10(b)); if they are incident at too large an angle they are reflected; although they have enough energy to surmount the barrier, they do not have enough momentum normal to the barrier. Thus only a fraction of the density of states is available for transitions from n-type to p-type, and this reduces the flow from n to p to a value equal to that from p to n. The electron flow in both directions in equilibrium is thus what one would naïvely calculate from the side of the junction where electrons are *minority* carriers, and is proportional to the minority carrier concentration n_p. Similarly, the hole current in both directions in equilibrium is proportional to the minority carrier concentration p_n.

Now consider the situation when a potential difference V is applied across the junction, with the p-side positive (Fig. 3.11; remember that a positive potential lowers electron energy). Since there are hardly any carriers in the junction region the potential drop will appear there, as shown in Fig. 3.11. It can be seen from the figure that the flow of electrons from p to n is unaltered; but at a given energy level the occupation number on the n-side is a factor $\exp(eV/kT)$ greater than on the p-side, so that the flow from n to p is greater by this factor. Similarly, the flow of holes from n to p is as in equilibrium, but

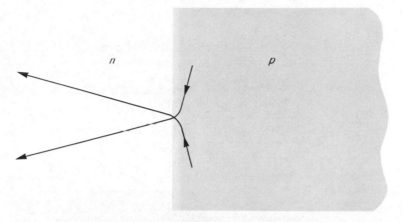

(*a*) Electrons from the *p* side are accelerated in the junction, and emerge in a narrow cone of directions.

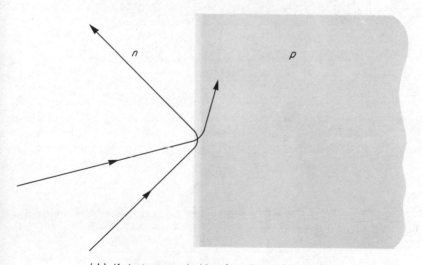

(*b*) If electrons are incident from the *n* side outside this narrow cone they are reflected, not transmitted.

Fig. 3.10.

the flow of holes from p to n is increased by a factor $\exp(eV/k_BT)$. The net current from p to n can therefore be written as

$$I = I_0(e^{eV/k_BT} - 1), \tag{3.37}$$

where I_0 is the total current in thermal equilibrium in each direction. Similarly, if a voltage is applied in the reverse direction (Fig. 3.12) the electron

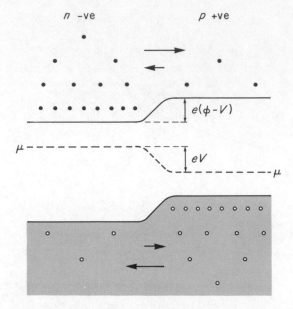

Fig. 3.11. A *p-n* junction with forward bias. Electron and hole flows indicated by arrows. There is a large net current from *p* to *n*.

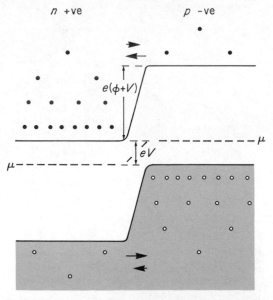

Fig. 3.12. A *p-n* junction with reverse bias. Electron and hole flows indicated by arrows. There is a small net current from *n* to *p*.

flow from n to p is *reduced* by a factor $\exp(eV/k_B T)$, and Eq. (3.37) still applies, with the sign of V reversed.

Equation (3.37) gives an exponentially increasing current for forward bias and rapid approach to a saturation current I_0 for reverse bias. It is a rectifier characteristic, and Eq. (3.37) is in fact the best rectifier characteristic that can be obtained using carriers of charge e. For an interesting discussion showing that the mechanical analogue of such a rectifier (a ratchet) cannot violate the second law of thermodynamics, consult Feynman, *Lectures on Physics*, Vol. I, Chapter 46 (Addison–Wesley, 1963).

If our assumption of no collisions were correct the saturation current I_0 would, by elementary kinetic theory, be given by

$$I_0 = \tfrac{1}{4}e(n_p \bar{c}_e + p_n \bar{c}_h)A, \qquad \text{(WRONG)} \qquad (3.38)$$

where A is the junction area, n_p and p_n are the minority carrier concentrations, as before, and $\bar{c}_e$ and $\bar{c}_h$ are the electron and hole thermal velocities. But in fact the carrier mean free-path is typically an order of magnitude smaller than the thickness of the depletion layer, so that passage through the junction is diffusive rather than collision free, and Eq. (3.38) is *not* correct.

It can be seen from Figs. 3.11 and 3.12 that the essential point in our argument for the rectifier characteristic, Eq. (3.37), was that carriers should pass right through the junction while retaining the thermal population appropriate to the Fermi level on the side from which they originated. This means that excess carriers must have a long lifetime for recombination of electron hole pairs, since this is the mechanism by which the thermal equilibrium concentration is restored. In fact, the recombination lifetime τ_R is typically 10^{-6} to 10^{-3} s, compared with a collision interval τ_c of order 10^{-12} s. The question we have to ask is, can a carrier diffuse right through a junction in a recombination lifetime? Since diffusion is a random walk process the **diffusion length** L which a carrier moves on average during a lifetime is given by

$$L^2 \sim D\tau_R,$$

where D is the diffusion coefficient (see Flowers and Mendoza,[3] section 6.2) which is of order

$$D \sim \bar{c}^2 \tau_c,$$

so that

$$L \sim \bar{c}(\tau_c \tau_R)^{1/2}. \qquad (3.39)$$

With $\bar{c} \sim 10^5$ m s^{-1}, $\tau_c \sim 10^{-12}$ s, $\tau_R \sim 10^{-4}$ s, Eq. (3.39) gives $L \sim 1$ mm, which is comfortably larger than the thickness of the depletion layer. We therefore expect the theory leading to Eq. (3.37) to apply. But because of collisions the effective carrier velocity is not $\bar{c}$ but the mean diffusive velocity

L/τ_R ; Eq. (3.38) is therefore replaced by

$$I_0 \sim \frac{e}{\tau_R}(n_p L_e + p_n L_h)A. \tag{3.40}$$

Thus since

$$\frac{L}{\tau_R} \sim \bar{c}\left(\frac{\tau_c}{\tau_R}\right)^{1/2}$$

the current is reduced by a factor $(\tau_c/\tau_R)^{1/2}$ by the presence of collisions. The most important feature of Eq. (3.40), however, is that I_0 is still proportional to the minority carrier concentrations, and therefore quite small. Moreover, because of Eq. (3.13), I_0 varies with temperature roughly as $\exp(-E_G/k_B T)$.

3.5.2   Counter

Semiconductor counters make use of the fact that Li, which is an *interstitial* impurity, also acts as a donor in Si and Ge. It gives its single valence electron to the conduction band and forms no covalent bands; consequently it diffuses quite readily through the crystal lattice. If a *p*-type crystal is heavily doped with Li on one face, a *p-n* junction is formed, which broadens as the Li diffuses. When a potential difference is applied the electric field appears across the intrinsic region at the *p-n* junction and causes a drift motion of the diffusing Li$^+$ ions. This results in the development of a large intrinsic region in which the concentration of Li extremely accurately balances the acceptor concentration. With such a large intrinsic region no current normally flows under reverse bias, but if electron hole pairs are created by the passage of ionizing radiation, the electrons are collected in the *n* region and the holes in the *p* region. Because of the small energy required to create an electron–hole pair many pairs are created and the statistical fluctuations in pulse height for totally absorbed particles are small; consequently the resolution is extremely good, a few keV at 1 MeV.

3.5.3   Junction transistor

An *npn* transistor consists of a thin ($\sim 10\ \mu$m) layer of *p*-type material in the middle of a slab of *n*-type. It is essentially a symmetrical device, but in operation one *p-n* junction, known as emitter-base, is forward biased, and the other, known as collector-base, is reverse biased. The energy levels under operating conditions are shown in Fig. 3.13. There is a large electron flow from emitter to base across the forward-biased junction; because the base

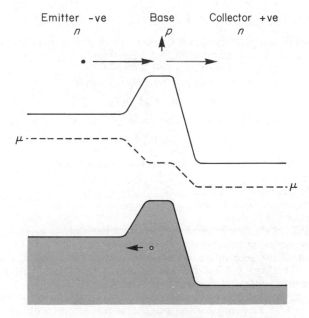

Fig. 3.13. An npn transistor under operating bias. The current, mainly electrons, is controlled by the forward biased emitter-base junction; because the base is thin, most of the current goes straight through to the collector.

(p-layer) is so thin the electron concentration does *not* equilibrate in the base region, and most of these electrons diffuse to the second junction and are accelerated into the collector region– only a small fraction flow to the base electrode.

Because the collector-base junction is reverse biased, the collected current hardly depends on the collector voltage, so the device has a high output impedance, which is conducive to high voltage gain, with a suitable load resistor. On the other hand, the collected current is almost the same as the emitter current which is exponentially sensitive to the base-emitter voltage through Eq. (3.35).

To avoid too low an input impedance it is desirable to keep the total base current as small as possible. Some base current of electrons is inevitable, but hole conduction makes no contribution to the amplification, and is therefore to be avoided as far as possible. The p region is therefore much more lightly doped than the n region, to keep the hole concentration small.

pnp transistors are exactly similar in action with the roles of electrons and holes reversed.

PROBLEMS 3

3.1 A sample of silicon is purified until it contains only 10^{12} donors/cm³. Below what temperature will it cease to show intrinsic behaviour?
(Take $E_G = 1$ eV, $E_D = 0.05$ eV, $m_e = m_h = 0.2m$)

3.2 The electron energy near a valence band edge is given by

$$E = -10^{-37}k^2 \text{ joules}$$

for a state of wavenumber $\mathbf{k}$ m⁻¹. An electron is removed from the state

$$\mathbf{k} = 10^9 \hat{\mathbf{k}}_x \text{ m}^{-1},$$

where $\hat{\mathbf{k}}_x$ is a unit vector in the $\mathbf{k}_x$ direction, and all other states are occupied. Calculate:
(a) the effective mass of the hole;
(b) the wavevector of the hole;
(c) the velocity of the hole;
(d) the energy of the hole referred to the valence band edge.
(Each answer must include the sign (or direction).)

3.3 Three samples of germanium doped with arsenic have donor concentrations N_D of 5.5×10^{16} cm⁻³, 1.7×10^{15} cm⁻³, and 1.4×10^{14} cm⁻³. All samples have acceptor concentrations N_A less than 10^{13} cm⁻³.

Sketch the behaviour of carrier concentration n as a function of temperature T (plot $\ln n$ vs. $1/T$). Give physical reasons for the main features of your sketch.

($E_G = 0.67$ eV for Ge, $E_D = 0.0127$ eV for As, intrinsic electron density $n_i = 2.5 \times 10^{13}$ cm⁻³ at room temperature.)

3.4 Indium antimonide has dielectric constant $\varepsilon_r = 17$ and electron effective mass $m_e = 0.014m$.
Calculate:
(a) the donor ionization energy (why is the acceptor ionization energy different?);
(b) the radius of the ground state orbit;
(c) the donor concentration at which orbits around adjacent impurities begin to overlap. What effects occur at about this concentration, and why?
(Ionization potential of hydrogen = 13.6 eV, Bohr radius = 0.53 Å.)

3.5 A sample of germanium is doped with a single type of impurity. Outline the measurements you would make to determine the sign and concentration of carriers and their mobility and effective mass.

If there are 10^{14} donors/cm³, what are the conditions necessary for satisfactory observation of cyclotron resonance.

(Collision diameter of donor = 300 Å, effective mass of electron = 10^{-31} kg.)

3.6 5 µA flows through a simple p-n junction diode at room temperature when it is reverse biased with 0.15 V. Calculate the current flow when it is forward-biased with the same voltage.

3.7 In a *tunnel diode* the n-side is so heavily doped that the Fermi level is in the conduction band, and the p-side is so heavily doped that the Fermi level is in the valence band. Draw an energy level diagram for such a diode with zero bias. Sketch and explain the current-voltage characteristic of this device.

4

Metals

4.1 EVIDENCE FOR INDEPENDENT MOBILE ELECTRONS

We saw in section 3.2 that the stationary states for an extra electron in a regular crystal lattice are waves extending throughout the crystal, from which mobile wavepackets can be constructed. At small concentrations these mobile electrons are approximately independent of each other, but as the concentration is increased the independent particle approximation becomes less obviously valid, and it is no longer clear that the electronic states are mobile. This point was brought out in our discussion of the H_2 molecule in sections 1.3.1 and 1.4 where we found that a wavefunction constructed according to the independent particle approximation gives an appreciable probability of finding both electrons near the same proton. At large internuclear separations this wavefunction tends to a linear combination of states $(H + H)$, corresponding to two neutral atoms, and $(H^+ + H^-)$, corresponding to two ions. These states have quite different energies and the linear combination of them given by the independent particle approximation is therefore not a stationary state; the lowest stationary state corresponds to one electron near each proton.

Analogously to our development from the H_2^+ ion to the H_2 molecule in Chapter 1, we can first apply the result of section 3.2 to a single extra electron on a chain of protons or Na^+ ions. Suppose now we go on adding electrons until we have one per ion, i.e. a chain of H or Na atoms. Then, according to the independent particle approximation, we have a band of N mobile states in which to put them, for a chain of N atoms. The exclusion principle allows one spin up and one spin down electron in each state, so we only need to use the states in the lower half of the band (Fig. 4.1(a); c.f. Fig. 3.1). Since not all states are occupied we can also produce an asymmetrical distribution as in Fig. 4.1(b); this corresponds to a flow of electric current, since there are more electrons with positive group velocity than negative group velocity.

However, as in the case of the hydrogen molecule, the independent particle approximation implies an appreciable probability that there will be two electrons near some ions and none near others, and we know that when the atoms are far apart such states do not occur, because they require an atomic ionization energy ~ 10 eV to produce them. Instead, we have one electron on each atom and the electrons are *not* mobile; whether this remains true as the atomic separation is decreased to the usual equilibrium value is not clear.

Thus, the mobility of electrons in high concentrations is far less obvious theoretically than for the low concentrations encountered in semiconductors. We shall therefore begin by examining the *experimental* evidence for the occurrence of independent mobile electrons.

4.1.1   Solid and liquid metals

Metals exist. This basic experimental fact is probably as important for solid state physics as the equally simple observation that the sky is dark at night is for cosmology. Historically, the idea of mobile electrons in solids originated with the free electron model of a metal which we shall discuss in section 4.2. At room temperature the resistivity of a typical metal is of order 10^{-8} Ω m (10^{-6} Ω cm), much lower than that of a semiconductor. Moreover, the resistance *decreases* approximately linearly as the temperature falls below room temperature, levelling off to a constant value at low temperatures (Fig. 4.2) which is highly dependent on specimen purity, and can be of order 10^{-4} of the room temperature resistivity for highly purified annealed single crystals. The extra resistance of an impure sample is approximately independent of temperature over the whole temperature range; this fact constitutes **Matthiesen's rule.** This extra resistance is presumably due to electron scattering by impurities, and its temperature independence therefore suggests that the mobile electron concentration is independent of temperature, in contrast to the exponential temperature dependence in a semiconductor. The temperature dependent part of the resistance (which is independent of impurities) can then reasonably be attributed to temperature dependent scattering of electrons by lattice vibrations; this we shall discuss further in Chapter 9.

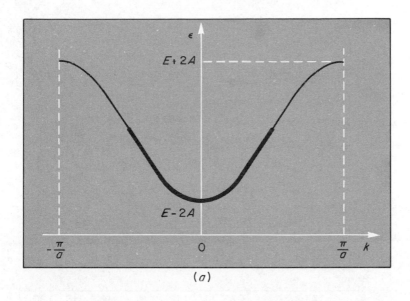

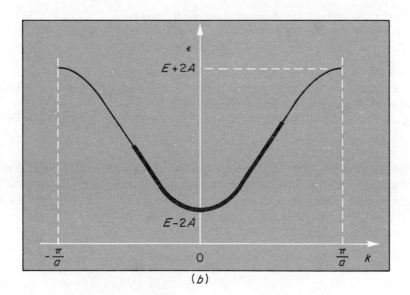

Fig. 4.1. Occupied independent electron states for a chain of H or Na atoms indicated by heavy line.
(a) Thermal equilibrium at $T = 0$,
(b) A current carrying state.

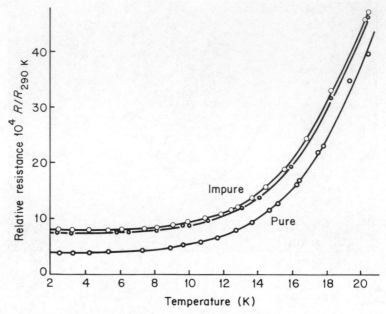

Fig. 4.2. Resistivity–temperature curves for sodium.
{after D. K. C. Macdonald and K. Mendelssohn, *Proc. Roy. Soc. A*, **202**, 103 (1950)}.

We now estimate some typical orders of magnitude for a metal, using our result of Chapter 3, for the electrical conductivity σ:

$$\sigma = \frac{ne^2\tau}{m^*} = n|e|\mu. \tag{3.24}$$

For this purpose we shall assume a number of electrons equal to the number of atoms ($\sim 10^{29}/m^3$ at metallic density $\sim 10\ g/cm^3$), and see whether this assumption leads to reasonable answers; we take m^* as the mass of a free electron. Then at room temperature the mobility is

$$\mu = \frac{\sigma}{n|e|} \sim \frac{10^8}{10^{29} \times 10^{-19}}$$

$$= 10^{-2}\ m^2\ V^{-1}\ s^{-1} = 10^2\ cm^2\ V^{-1}\ s^{-1};$$

near absolute zero for a pure sample σ rises to about $10^{12}\ \Omega^{-1}\ m^{-1}$ so that

$$\mu \sim 10^2\ m^2\ V^{-1}\ s^{-1} = 10^6\ cm^2\ V^{-1}\ s^{-1},$$

which is the order of mobility attainable in the purest semiconductor samples at low temperatures, so that these mobility estimates do indeed seem reasonable. It is also of interest to estimate the relaxation time:

$$\tau = \mu m^*/|e|$$

$$\sim \frac{10^{-2} \times 10^{-30}}{10^{-19}} = 10^{-13} \text{ s at room temperature}$$

$$\sim \frac{10^2 \times 10^{-30}}{10^{-19}} = 10^{-9} \text{ s near } T = 0.$$

We can obtain a more direct indication of mobile electron concentration by measuring the Hall effect, since

$$R_H = 1/ne \tag{3.27}$$

for a single type of carrier with a single relaxation time. Thus, if N is the number of *atoms* in unit volume, the quantity $1/(R_H Ne)$ gives an estimate of the number of mobile electrons per atom; a selection of values of this quantity is shown in Table 4.1.

Table 4.1

Metal	Group	$1/(R_H Ne)$
Na	I	−1.1
K		−1.0
Cu	IB	−1.2
Au		−1.4
Be	II	+0.2
Mg		−1.4
Cd	IIB	+2.0
Al	III	−2.5

We see that the number of mobile electrons on this criterion is indeed of the order of one per atom, but the situation is not simple in that positive Hall coefficients, indicating that the mobile carriers are holes, sometimes occur; thus Be appears to have 0.2 holes/atom.

It is also an experimental fact that liquid metals exist, with an electrical conductivity not very different from that of solid metals near the melting point. Therefore the existence of a regular crystal lattice, which we assumed in section 3.2, cannot be a necessary condition for the existence of mobile electron states. Without a regular crystal lattice we cannot have the plane

wave solutions that we had in section 3.2, because the equations making up
Eqs. (3.1) will all have different values for the coupling coefficients A. But
it must nevertheless be possible to have wavefunctions extending throughout
the system from which moving wavepackets can be constructed. Although in
the solid the atoms vibrate about fixed lattice sites, whereas in the liquid the
atoms continually exchange positions, these two types of thermal motion
must be comparably effective in scattering electrons, in order to account for
the similar resistivities of solid and liquid.

4.1.2 Impurity bands in semiconductors

Liquid metals thus show that we can have mobile electrons in a dense
array of atoms even when the atoms are not arranged on a regular lattice;
on the other hand, we have seen that for well-separated atoms the electrons
are not mobile.

While we cannot experimentally vary the equilibrium separation of a
collection of atoms continuously (because of the transition from solid or
liquid to vapour), this can in effect be achieved by studying the behaviour
of impurities in a doped semiconductor. We have seen in section 3.3.1 that
such impurities effectively behave as rather large 'atoms' about 100 Å in
size with a rather small ionization energy of about 0.01 eV. The normal
conduction mechanism (for donor impurities) is by ionization of electrons
into the conduction band, analogous to electric conduction in a gas hot
enough to be thermally ionized. Since these impurity 'atoms' are about
100 Å diameter, they will begin to overlap bound state wavefunctions at a
concentration of order $10^{18}/cm^3$. Fig. 3.6 shows that at a concentration
of this order a remarkable change in the conductivity occurs: it rises sharply
with concentration and becomes approximately independent of temperature;
the variation of resistivity with concentration at constant temperature is
shown in Fig. 4.3. Moreover, Hall effect measurements show that above this
critical concentration the number of carriers is temperature independent over
a wide range and equal to the number of impurities. In other words the
conductivity quite suddenly becomes *metallic in character* at about the
concentration where the localized impurity wavefunctions begin to overlap
appreciably. Since the electrons now appear to be mobile *without* being
ionized we presume that the impurity levels have ceased to be localized and
have formed a band of mobile states, as in a liquid metal.

4.1.3 Soft x-ray emission spectra

The evidence we have adduced so far shows that electrons are mobile,
but does not convincingly show that they are independent of each other.
It is quite conceivable that the charge transport required for electrical
conductivity could be provided by some sort of collective state involving
many coupled electrons, and the relaxation time τ could be the lifetime of

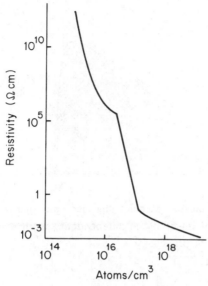

Fig. 4.3. Dependence of resistivity of germanium on impurity concentration at 2.5 K. For more than about 10^{17} impurities/cm^3 the conductivity is independent of temperature and metallic in character.
{after N. F. Mott, *Phil. Mag.* (8), **6**, 287 (1961)}.

such a collective state. To investigate the extent to which mobile electrons are independent we must see how well-defined is the energy to add an electron to the conduction band or remove one from it. Such evidence is provided by x-ray emission spectra; if an electron is ejected from an atomic K or L shell by electron bombardment, K or L x-rays are emitted when an electron falls from a higher level to replace it. In particular, if there is a partly occupied conduction band, as shown in Fig. 4.1, we expect a corresponding band of emitted x-ray photon energies. On an independent particle model we will have a roughly parabolic density of states, as for a semiconductor, which is filled up to the Fermi energy ε_F at absolute zero, giving the number of electrons per unit energy range shown in Fig. 4.4(*a*). For a monovalent metal, such as Li or Na, ε_F will be half way up the conduction band, as in Fig. 4.1; at finite temperature the cutoff at ε_F will be blurred by an amount of order $k_B T$ by the Fermi distribution function. We expect a similar shape to the x-ray emission spectrum; it need not be identical, because intrinsic transition probabilities may depend on energy, but the property of sharp cutoff at ε_F should be preserved. This is demonstrated for sodium in Fig. 4.4(*b*).

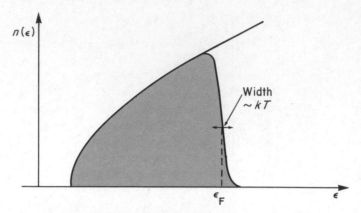

(*a*) Expected number of electrons per unit energy range for a monovalent metal, on an independent particle model.

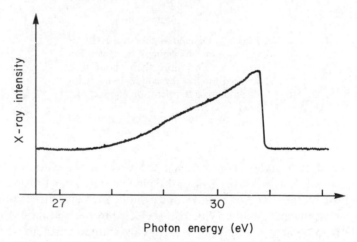

(*b*) Experimental L x-ray emission for sodium.
[after H.W.B. Skinner, *Phil. Trans.*, **239**, 95 (1940)].

Fig. 4.4.

On the other hand, if conduction electrons were *not* independent, the energy of a single electron would not be a well-defined quantity and we should expect the x-ray emission cutoff to be blurred by some short lifetime τ. We can set a lower limit on this τ by noting that the observed cutoff is as sharp as $k_B T$ at room temperature, so that

$$\tau > \hbar/k_B T \sim 10^{-34}/(5 \times 10^{-21}) = 2 \times 10^{-14}\,\text{s}.$$

4.2   THE FREE ELECTRON MODEL

Given that mobile electronic states exist in metals, the simplest model we can construct is a *free electron model*, in which we suppose that the electrons are independent with energies $\hbar^2 k^2/2m^*$. This is an even more drastic assumption than the independent particle model shown in Fig. 4.1, for we are ignoring the departure from parabolic form of the $E(\mathbf{k})$ curve of Fig. 4.1. Our assumption amounts, in fact, to supposing that the electrons move in a uniform potential rather than the true periodic potential provided by the positive ions. We shall see in section 4.3 how the periodic ion potential produces an $E(\mathbf{k})$ curve essentially of the form of Fig. 4.1, but it is instructive to see first how far we can proceed with the simpler free-electron model.

4.2.1   The equilibrium Fermi gas‡

In thermal equilibrium at temperature T the number of electrons per unit energy range, $n(\varepsilon)$, is given by

$$n(\varepsilon) = g(\varepsilon) f(\varepsilon).$$

In this expression $g(\varepsilon)$ is the density of available states per unit energy range, including a factor 2 for spin degeneracy, which we have already calculated as

$$g(\varepsilon) = \frac{V}{2\pi^2 \hbar^3}(2m^*)^{3/2}\,\varepsilon^{1/2}, \tag{3.8}$$

and $f(\varepsilon)$ is the Fermi distribution function

$$f(\varepsilon) = \frac{1}{e^{(\varepsilon - \mu)/k_B T} + 1}. \tag{3.6}$$

The function $n(\varepsilon)$ is illustrated in Fig. 4.5. At absolute zero Eq. (3.6) reduces to a step function, so that all levels below μ are occupied, and all levels above μ are empty. Because of this special significance of $\mu(0)$ it is usual to write

$$\mu(0) = \varepsilon_F = k_B T_F$$

thereby defining the **Fermi energy** ε_F and **Fermi temperature** T_F†. The value of the Fermi energy is determined by the electron density. Thus, at absolute zero we may calculate the total number of particles by integrating the density of states from zero up to the Fermi energy:

$$N = \int_0^{\varepsilon_F} g(\varepsilon)\,d\varepsilon = \frac{V}{3\pi^2 \hbar^3}(2m^*\varepsilon_F)^{3/2} = \tfrac{2}{3}\varepsilon_F g(\varepsilon_F). \tag{4.1}$$

---

‡ See also Mandl,[2] section 11.4.
† Note that the Fermi level μ varies slightly with temperature, but ε_F and T_F are constants.

Inversion of Eq. (4.1) gives the Fermi energy as

$$\varepsilon_F = \frac{\hbar^2}{2m^*}\left(\frac{3\pi^2 N}{V}\right)^{2/3}, \tag{4.2}$$

and by using $\varepsilon_F = \hbar^2 k_F^2/2m^*$ we obtain also the **Fermi wavenumber** k_F as

$$k_F = \left(\frac{3\pi^2 N}{V}\right)^{1/3}. \tag{4.3}$$

Typical orders of magnitude may be obtained by setting $(N/V) \approx 3 \times 10^{28}/m^3$ and m^* of the order of a free electron mass so that

$$\varepsilon_F \sim \frac{(10^{-34})^2}{2 \times 10^{-30}}(10^{30})^{2/3}$$

$$= 5 \times 10^{-19} \text{ joule}$$

$$\sim 5 \text{ eV},$$

and

$$T_F \sim 5 \times 10^4 \text{ K},$$

$$k_F \sim 1 \text{ Å}^{-1}.$$

Thus ε_F is of the order of atomic ionization energies and k_F is of the order of the reciprocal of an atomic spacing. Since at ordinary temperatures $T \ll T_F$, the electron gas in a metal is normally highly degenerate. This has the consequence that very few electrons are ordinarily thermally excited; in Fig. 4.5 those in the shaded area with $\varepsilon < \varepsilon_F$ are shifted to the shaded area

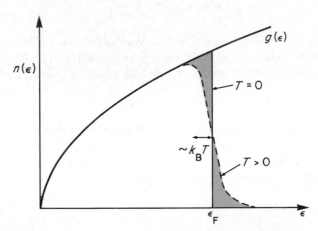

Fig. 4.5. Number of electrons in unit energy range on the free electron model. The shaded area shows the change in distribution between absolute zero and finite temperature.

with $\varepsilon > \varepsilon_F$. To evaluate the thermal energy (and hence the specific heat) exactly we should have to calculate these shaded areas by expanding $f(\varepsilon)$ about ε_F, but a rough estimate is easily obtained by inspection of Fig. 4.5. The number of electrons excited thermally is of order $\frac{1}{2}g(\varepsilon_F)k_B T$ and they are on average excited by an amount of order $2kT$, so that the thermal energy (difference in internal energy from the value at $T = 0$) is

$$U(T) - U(0) \sim g(\varepsilon_F)(k_B T)^2$$

$$= \tfrac{3}{2}\frac{N}{\varepsilon_F}(k_B T)^2,$$

where the last step follows from Eq. (4.1). Differentiating with respect to T we obtain the electronic specific heat,

$$C_v \sim 3Nk_B\left(\frac{T}{T_F}\right);$$

the exact result from a proper calculation of $U(T)$ for $T \ll T_F$ is

$$C_v = \frac{\pi^2}{2}Nk_B\left(\frac{T}{T_F}\right). \tag{4.4}$$

At room temperature, therefore, the electronic specific heat is of order 1% of the lattice specific heat, and hardly observable. The free electron model was first introduced before quantum theory, and one of its greatest difficulties was that the specific heat of a *classical* electron gas is $\frac{3}{2}Nk_B$, which is clearly not observed; quantum theory solved this problem.

It is worth noting that the smallness of the electronic specific heat is an example of a general quantum mechanical result, applicable to both fermions and bosons, and thus independent of the exclusion principle. The existence of discrete quantum states leads to a unique quantum mechanical ground state, which in turn implies that the entropy S tends to zero at absolute zero– the third law of thermodynamics (see, for example, Mandl,[2] section 4.7). At a finite temperature the entropy of any system is given by

$$S(T) = \int_0^T \frac{C_v}{T} \, dT;$$

this integral converges at the lower limit only if C_v goes to zero at least as fast as T as $T \to 0$. This is the general quantum mechanical result for specific heats at low temperatures, and we see that the limiting behaviour is actually reached by a free electron Fermi gas; C_v vanishes as slowly as possible.

We may estimate the temperature at which a quantum gas changes from a small specific heat to a classical specific heat as follows. Classical behaviour is expected if the de Broglie wavelength λ_{DB} of a particle of energy $k_B T$ is small compared with the interparticle spacing; this condition is

$$\lambda_{DB} \sim \frac{\hbar}{(m^*k_B T)^{1/2}} \ll \left(\frac{V}{N}\right)^{1/3}.$$

We thus expect a change from classical to quantum behaviour at a temperature T_0 of order

$$T_0 \sim \frac{\hbar^2}{m^*k_B}\left(\frac{N}{V}\right)^{2/3}.$$

This result is really just a dimensional one; $k_B T_0$ is the only quantity of the dimensions of energy that can be formed from $\hbar$, m^*, and the particle density. From Eq. (4.2) we see that $T_0 \sim T_F$, but our present argument shows that there is no essential connection between a characteristic temperature of this order and Fermi statistics. T_F is so large and the electronic specific heat so small because electrons have a very small mass but are packed at the same number density as atoms.

Because different contributions to internal energy are additive, we may add the electronic specific heat Eq. (4.4) to the lattice specific heat Eq. (2.36) to obtain the total specific heat of a metal at low temperatures in the form

$$C_v = aT + bT^3. \tag{4.5}$$

This functional form may be checked, and the constants a and b determined, by plotting (C_v/T) as a function of T^2; an example is given in Fig. 4.6. We see

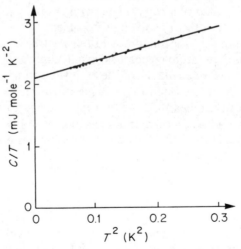

Fig. 4.6. Separation of electronic and lattice specific heats at low temperatures, for potassium.
{after W. H. Lien and N. E. Phillips, *Phys. Rev.*, **133**, A1370 (1964)}.

that the two contributions are comparable at a temperature in the range 1–10 °K; the electronic contribution dominates at the lowest temperatures because of its weaker temperature dependence.

4.2.2   Transport properties

Our expressions of Chapter 3 for the electrical conductivity

$$\sigma = \frac{ne^2\tau}{m^*},$$ (3.24)

and for the Hall coefficient

$$R_{\mathrm{H}} = \frac{1}{ne},$$ (3.27)

can be applied directly to a free electron gas of particle density $n = N/V$. The only difference is that we are now dealing with a degenerate Fermi gas, whereas in the case of semiconductors we essentially had a classical electron gas because all occupied states were far from the Fermi level. We have seen in Table 4.1 that the number of electrons deduced from Eq. (3.27) is frequently of the order of the number of valence electrons, as one would expect on a free electron model, but there are a number of notable exceptions in which the carriers appear to be holes.

It is of interest to calculate also the *thermal* conductivity due to free electrons; in our case the standard formula of the elementary kinetic theory of gases (see, for example, Flowers and Mendoza,[3] section 6.5.3) can be written as

$$K = \tfrac{1}{3}C_v v_{\mathrm{F}} l$$

where C_v is the specific heat *per unit volume* and l is the mean free path; it is appropriate to use $v_{\mathrm{F}} = \hbar k_{\mathrm{F}}/m^*$ as a thermal velocity, rather than the average velocity, since only electron states within about $k_{\mathrm{B}}T$ of ε_{F} are partly occupied, and it is therefore only the occupation numbers of these states that change during small departures from equilibrium. With $l = v_{\mathrm{F}}\tau$, $\varepsilon_{\mathrm{F}} = \tfrac{1}{2}m^* v_{\mathrm{F}}^2$ and Eq. (4.4) for the specific heat we have

$$K = \tfrac{1}{3}C_v v_{\mathrm{F}}^2 \tau = \tfrac{1}{3}\frac{\pi^2}{2}\frac{Nk_{\mathrm{B}}}{V}\left(\frac{T}{T_{\mathrm{F}}}\right)\frac{2\varepsilon_{\mathrm{F}}}{m^*}\tau$$

$$= \frac{\pi^2}{3}\frac{nk_{\mathrm{B}}^2 T\tau}{m^*}.$$ (4.6)

It is interesting to note that Eq. (4.6) is also true in order of magnitude for a *classical* gas: the specific heat is larger by a factor of order (T_{F}/T) and the square of the thermal velocity is smaller by the same factor. Note that the combination $n\tau/m^*$ occurs in both Eq. (4.6) and Eq. (3.24), so that by dividing them we obtain a result independent of the electron gas parameters:

$$\frac{K}{\sigma T} = \frac{\pi^2}{3}\left(\frac{k_B}{e}\right)^2 = 2.45 \times 10^{-8} \text{ W }\Omega\text{ K}^{-2}. \tag{4.7}$$

This is the Wiedemann–Franz law; it is found to be approximately, but not exactly, obeyed by many metals. This is not surprising, for we have assumed that the relaxation time τ is the same for electrical and thermal conductivity; we now examine the nature of the relaxation processes in the two cases.

Fig. 4.7(a) is a diagram in **k**-space showing the occupied electron states at $T = 0$; on our free-electron model the states with $\varepsilon < \varepsilon_F$ fill a sphere of radius k_F in **k**-space called the **Fermi sphere**. The surface in **k**-space separating occupied from unoccupied states is called the **Fermi surface**; it still exists at finite temperature, but is blurred in **k** by an amount of order $k_B T/\hbar v_F$. If the Fermi sphere is displaced in **k**-space, as in Fig. 4.7(b), we obtain a distribution corresponding to a flow of current, for there are more electrons with momentum (and hence velocity) in the $+x$ direction than the $-x$ direction. For such a current to decay to the state in Fig. 4.7(a) electrons have to be removed from the right side of the Fermi sphere and added to the left side; typical transitions are shown in Fig. 4.7(b), and it can be seen that they involve a crystal momentum change for the electron of order $2\hbar k_F$.

Fig. 4.7(c) shows the electron distribution in the presence of a temperature gradient. Because of the finite temperature there will be some vacant states (open circles) below the Fermi surface and some occupied states (solid circles) above. If the specimen is hotter at the left hand end, electrons moving from the left (i.e. those with $k_x > 0$) will have a distribution corresponding to a higher temperature than those coming from the right (i.e. those with $k_x < 0$). This is the situation shown in Fig. 4.7(c), where the Fermi surface is more blurred for $k_x > 0$ than for $k_x < 0$. The relaxation process described by τ in this case is one in which the blurring of the Fermi surface is evened out; this can be done with quite small momentum changes, as indicated in the figure.

Thus the relaxation processes for electrical and thermal conductivity involve different momentum transfers and hence different types of collision; different relaxation times are therefore to be expected.

Note that the departures from equilibrium shown in Fig. 4.7 are enormously exaggerated. Thus, the total blurring of the Fermi surface at room temperature is of order 1 % of k_F, and the difference in blurring on the two sides due to a temperature gradient corresponds to the temperature difference in an electron mean free path, which is usually minute, since at room temperature

$$l = \tau v_F \sim 10^{-13} \times 10^6 \text{ m}$$

$$= 0.1 \ \mu\text{m}.$$

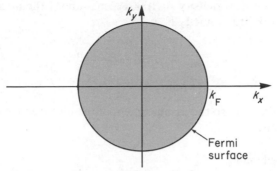

(a) Equilibrium Fermi sphere at $T = 0$.

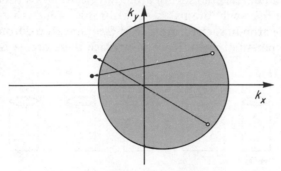

(b) Current-carrying state, with typical relaxation processes.

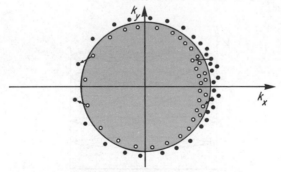

(c) State in a temperature gradient, with typical
relaxation processes.

Fig. 4.7.

Similarly, for a current density of 1000 A/cm² (about the largest normally used), the electron drift velocity is

$$v = \frac{j}{ne} \sim \frac{10^7}{10^{29} \times 10^{-19}} \, \text{m/s}$$

$$= 1 \, \text{mm/s,}$$

which is about $10^{-9}v_F$, so the displacement of the Fermi sphere in Fig. 4.7(b) is actually minute.

4.2.3   Metallic binding

Our free-electron model of a metal consists of an electron gas in a uniform potential box. We can construct this by assembling 'atoms' if we invent a very special (and unrealistic) sort of atom: an electron in a square potential well. A row of five such 'atoms' is shown in Fig. 4.8(a); the ground state energy of an electron in such an atom is $h^2/8ma^2$ (measured from the bottom of the infinite potential well). If we now place these atoms in contact we

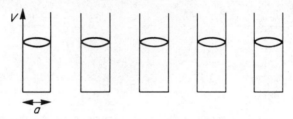

(a) A row of 5 square-well 'atoms' and their
       ground-state wavefunctions.

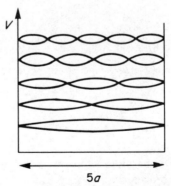

(b) The same row assembled into a 'crystal',
       showing the 5 lowest-energy wavefunctions.

Fig. 4.8.

obtain the one-dimensional 'crystal' shown in Fig. 4.8(*b*)—a square well of width 5*a*. The five lowest electron wavefunctions are shown, and we can see that the *highest* of these has the same wavelength (and hence the same energy) as the ground state of a single 'atom'. The *mean* energy of the five electrons is therefore lowered by forming the crystal, even without allowing for electron spin, which permits two electrons in each energy level. This quantum mechanical effect, the reduction of kinetic energy by delocalizing electrons, is an important contribution to metallic binding; though in a real metal the electrostatic contributions of electron–ion attraction and ion–ion repulsion are of comparable importance.

4.3   THE EFFECT OF A PERIODIC LATTICE POTENTIAL

We now attempt to improve the free–electron model by taking into account the fact that the positive ions produce not a uniform attractive potential, but one which has strong negative peaks at the lattice sites. A one-dimensional example of such a periodic potential is shown in Fig. 4.9(*a*); the period is equal to the lattice spacing *a*. We shall estimate the correction to the free

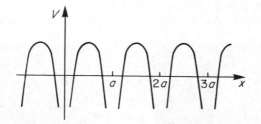

(*a*)   A periodic potential in one dimension.

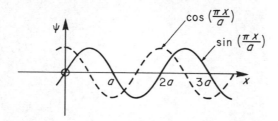

(*b*)   Standing waves $\sin\left(\frac{\pi x}{a}\right)$ and $\cos\left(\frac{\pi x}{a}\right)$, with nodes and antinodes, respectively, at the lattice sites.

Fig. 4.9.

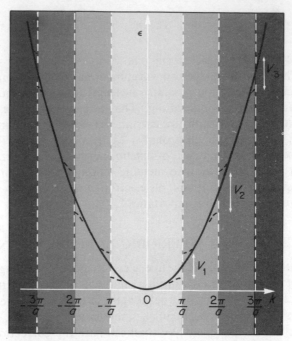

Fig. 4.10. Free electron parabola $\varepsilon = \hbar^2 k^2/2m$, with perturbed energies for standing waves at $|k| = n\pi/a$; continuation to other k values shown by broken lines.

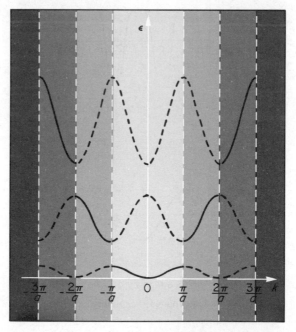

Fig. 4.11. Perturbed energies of Fig. 4.10 continued to physically equivalent k values, to show the correspondence to a series of energy bands like Fig. 3.1.

electron energy $E = \hbar^2 k^2/2m$ by the standard formula of first order perturbation theory

$$\Delta E = \frac{\int \psi^* V \psi \, dx}{\int \psi^* \psi \, dx}. \tag{4.8}$$

For convenience we take our zero of energy as the mean value of the potential; then in our one-dimensional example the potential can be written as a Fourier series in the form

$$V = -\sum_{n=1}^{\infty} V_n \cos \frac{2\pi n x}{a}, \tag{4.9}$$

where we expect all the V_n to be positive numbers for a potential with strong negative peaks at lattice sites, as shown in Fig. 4.9(a). For travelling waves $\psi = e^{\pm ikx}$, $\psi^* \psi = 1$ and Eq. (4.8) gives $\Delta E = 0$. But the degenerate linear combinations $\sin kx$ and $\cos kx$ do have their degeneracy removed by the perturbation V for certain special k values for which their periodic charge density is in synchronism with the periodicity of the lattice (Fig. 4.9(b)).

Thus, for $\psi = \sin kx$ we calculate the perturbation as

$$\Delta E = -\frac{\sum\limits_{n=1}^{\infty} \int dx \sin^2 kx V_n \cos (2\pi n x/a)}{\int dx \sin^2 kx}$$

$$= -\frac{\sum\limits_{n=1}^{\infty} \int dx (1 - \cos 2kx) V_n \cos (2\pi n x/a)}{\int dx (1 - \cos 2kx)}. \tag{4.10}$$

All periodic terms will vanish asymptotically on integrating over all space, so that the result is non-zero only if $k = (n\pi/a)$, in which case Eq. (4.10) reduces to

$$\Delta E = \frac{\int dx \, V_n \cos^2 (2\pi n x/a)}{\int dx}$$

$$= \tfrac{1}{2} V_n, \tag{4.11}$$

all other terms in the series having integrated to zero. Similarly, for $\psi = \cos kx$, by using $\cos^2 kx = \tfrac{1}{2}(1 + \cos 2kx)$, we find

$$\Delta E = -\tfrac{1}{2} V_n.$$

The physical reason for these results, for the case $n = 1$, becomes obvious on inspection of Fig. 4.9. $\psi = \sin (\pi x/a)$ has antinodes where the potential is repulsive, so its energy is raised; and $\psi = \cos (\pi x/a)$ has antinodes where the potential is attractive, so its energy is lowered.

In Fig. 4.10 we show these perturbed energies in relation to the free electron parabola $E = \hbar^2 k^2/2m$. Since the stationary states for $k = (n\pi/a)$ are *standing* waves it follows that the group velocity, and hence $d\varepsilon/dk$, must be zero at these points. To calculate other points on the $\varepsilon(k)$ curve requires second order perturbation theory, since the first order result is zero. But knowing ε and $d\varepsilon/dk$ at $k = (n\pi/a)$, we can easily sketch in the rest of the curve by continuity, as shown by the dotted curves in Fig. 4.10. We see that the continuous parabola is broken into a number of separate branches by energy gaps. The lowest branch is strongly reminiscent of the result we obtained in section 3.2 by a quite different approach (compare Fig. 3.1). This resemblance is enhanced if we redraw Fig. 4.10 as in Fig. 4.11, where we have continued each branch of the $\varepsilon(k)$ curve periodically in k. Then, from the viewpoint of section 3.2, we can regard the higher energy branches as derived from atomic excited states, just as the lowest branch was derived from an atomic ground state. On this viewpoint, addition of any multiple of $(2\pi/a)$ to k has no physical significance, because it gives the same values of the probability amplitudes c_n; but from the point of view of perturbed free electrons the k values indicated by the solid parts of the curve are to be preferred.

The important point that emerges is that we obtain qualitatively the *same* $\varepsilon(k)$ curves whichever of two opposite extremes we start from: strongly bound atoms perturbed by coupling, as in section 3.2; or free electrons perturbed by ionic attraction, as in this section. The main feature, bands of allowed energy separated by energy gaps, is in fact, as we shall see in Chapter 6, a geometrical property of wave propagation in a periodic lattice. This feature is of major importance in distinguishing conductors from insulators, as we shall see in the next section.

Although the qualitative form of $\varepsilon(k)$ is thus independent of our approximation scheme for calculating the effect of the lattice potential, it must be emphasized that $\varepsilon(k)$ is a *single particle* energy. We are therefore still making an independent electron approximation, and the only justification we have for this at the moment is the empirical one that metals appear to behave like that; we defer any attempt to give a theoretical justification to Chapter 9.

4.4   CLASSIFICATION INTO METALS, SEMICONDUCTORS, AND INSULATORS

We have already seen in section 4.1 that because there are N allowed states uniformly distributed in the range $-(\pi/a) < k \leqslant (\pi/a)$, each of which can accommodate two electrons of opposite spin, the lowest energy band of Fig. 4.11 will be half occupied if we have one electron per atom, as in Fig. 4.1(a). Because there are vacant states immediately adjacent in energy to the occupied states we can easily construct a current carrying state by moving electrons from states near $-k_F$ to states near $+k_F$* (Figs. 4.1(b) and 4.12(a)), essentially as for free electrons; we therefore have metallic behaviour.

The energy bands of Fig. 4.11 are often labelled by the region of **k**-space to which they correspond on a free electron model, and these regions are known as **Brillouin zones**. Thus, in our one dimensional example the region $|k| < (\pi/a)$ is the first Brillouin zone, the region $(\pi/a) < |k| < (2\pi/a)$ is the second Brillouin zone, and so on. We say that the first Brillouin zone is half-filled by one electron per atom.

If on the other hand our atoms have two electrons outside a closed shell, this is just enough to fill the first Brillouin zone exactly, and because of the energy gaps at $k = \pm(\pi/a)$ a current carrying state can be produced only by expending a finite amount of energy V_1 for each electron shifted from $k = -(\pi/a)$ to $k = +(\pi/a)$ (Fig. 4.12(b)). Since this energy is not available from an electric field such a material is an insulator at absolute zero. At finite temperature, if V_1 is sufficiently small, some electrons are thermally excited into the second Brillouin zone, leaving a few holes in the first zone, and we have semiconducting behaviour; for larger V_1 (several eV) the material continues to behave as an insulator.

For trivalent atoms we expect the *second* Brillouin zone to be half filled, giving metallic behaviour again. In general, we expect metallic behaviour from odd valency atoms, insulating or semiconducting behaviour from even valency atoms—provided the crystal is made of single atoms, and not, as with the halogens, of diatomic molecules.†

However, divalent metals exist: the alkaline earths Ca, Sr, Ba . . . . To see how this can occur we need to extend our energy band structure of Fig. 4.11 to a two dimensional example. Consider a square lattice of $N \times N$ atoms, spacing a. For simplicity we confine our attention to the energy gaps produced by V_1, i.e. the boundary between the first and second Brillouin zones. By the argument of section 4.3 energy gaps will be produced whenever standing waves are in synchronism with a periodicity of the lattice. Thus there will be synchronism with the periodicity in the x-direction if $k_x = \pm(\pi/a)$, and with the periodicity in the y-direction if $k_y = \pm(\pi/a)$. The boundaries of the first Brillouin zone are therefore the lines $k_x = \pm(\pi/a)$, $k_y = \pm(\pi/a)$, as shown in Fig. 4.13. To show electron energy on this diagram we draw contours of electron energy as a function of k_x and k_y. Thus, the concentric circles shown in Fig. 4.13 mark equal intervals of free electron energy $\varepsilon = \hbar^2 k^2/2m$.

---

* Or, equivalently, by moving all electrons a small amount in **k**, as happens when a current is accelerated by an electric field.

† From the arguments leading to Fig. 4.11 we can see that what really controls the Brillouin zone is the size of the *primitive unit cell*. If our structure has more than one atom in a primitive unit cell, it is the number of *valence electrons per primitive unit cell* that determine how full the Brillouin zones are.

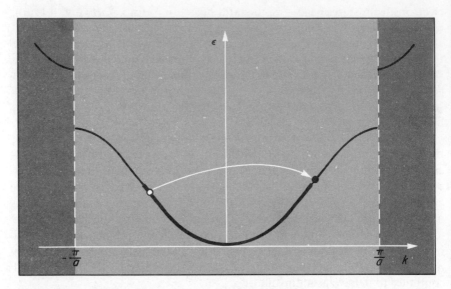

(a)   Producing a current-carrying state in a metal.

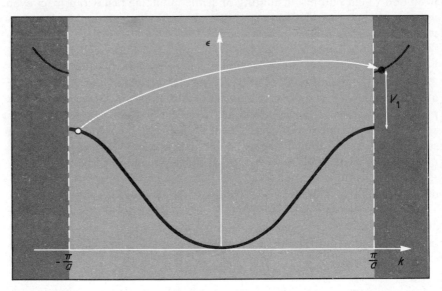

(b)   Producing a current-carrying state in an insulator.

Fig. 4.12.

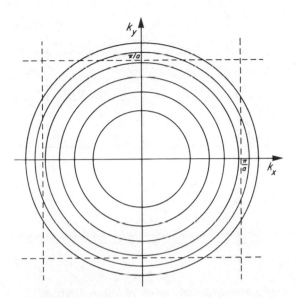

Fig. 4.13. Boundaries of the first Brillouin zone (dashed line) of a square lattice with circular free electron energy contours (at equal energy intervals) superposed.

To find how these energy contours are perturbed by the lattice potential we note from Fig. 4.10 that depression of the energy just inside a zone boundary moves the energy contours out towards the boundary, and similarly the increased energy just outside a zone boundary moves a constant energy contour in towards the zone boundary. Thus, zone boundaries 'attract' energy contours. In this way we obtain perturbed energy contours as shown in Fig. 4.14(a). Note that the energy contours meet the zone boundary at right angles; because $(d\varepsilon/dk)$ is zero normal to a zone boundary (Fig. 4.10), the *gradient* of ε in **k**-space is *parallel* to the boundary and the *contour* of ε is perpendicular to the boundary.

For our lattice of $N \times N$ atoms we have N^2 states in the first zone, each of which can be occupied by two electrons. The occupied states at $T = 0$ for a lattice of divalent atoms are therefore determined by two conditions:
(i) we must fill an area in **k**-space equal to that of the first zone;
(ii) we must fill all levels below some fixed energy ε_F.

The solution to this is shown in Fig. 4.14(b), where the occupied area of **k**-space is shaded. Although the Fermi surface is very different from a free electron sphere, the essential point is that it does have some free area, and can be slightly displaced as indicated by the dotted lines to give a current carrying state. We therefore have a metal.

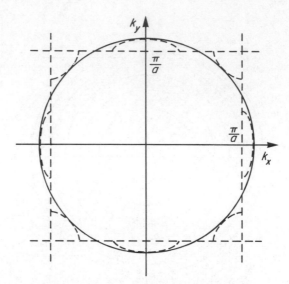

(*a*)   Perturbation (dotted curve) of a free electron
energy contour by a lattice potential.

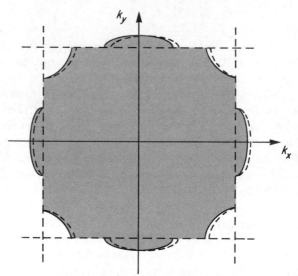

(*b*)   Occupied states at $T = 0$ for a divalent metal
(in two dimensions). The Fermi surface for a
current-carrying state is shown dotted.

Fig. 4.14.

The essential reason for this behaviour is that for a free electron (Fig. 4.13) the energy at a zone corner is twice that at the centre of a zone edge, because $|\mathbf{k}|$ is $\sqrt{2}$ times larger. Therefore, for small perturbations the lowest states in the second zone are *below* the highest states in the first zone, and we have **overlapping bands,** the situation portrayed in Fig. 4.14. As the energy gap increases we reach a situation where the bands no longer overlap, and for a divalent atom the first zone is entirely full and the second zone entirely empty. The occupied region of $\mathbf{k}$-space is then entirely bounded by Brillouin zone boundaries and there is no free area of Fermi surface. Electrons now have to be given a finite amount of energy to create a current-carrying state, and we have an insulator.

The Fermi surface of Fig. 4.14(b) consists of a number of disconnected pieces, broken up by zone boundaries. It is convenient to bring these together by translations of $(2\pi/a)$ in $\mathbf{k}$-space, as was done in constructing Fig. 4.11. The result is shown in Fig. 4.15; we have a roughly circular surface of holes

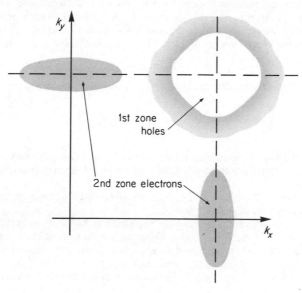

Fig. 4.15. Occupied states of Fig. 4.14(b) remapped by moving portions of the Fermi surface by $(2\pi/a)$ in k_x or k_y.

in the first zone and two almost elliptical surfaces containing electrons in the second zone. The numbers of electrons and holes are equal, as in an intrinsic semiconductor; the wide variety of Hall coefficients observed in group II metals (Table 4.1) is accounted for by differences in electron and hole mobilities.

The conclusions of this section may be summarized as follows:
  (i) odd numbers of electrons per primitive unit cell give a metal;
  (ii) even numbers of electrons per primitive unit cell give a metal if there
       is band overlap, a semiconductor if there is a small band gap ($\lesssim 1$ eV),
       and an insulator if there is a large band gap.

PROBLEMS 4

4.1 From the measurements shown in Fig. 4.6, calculate the Fermi temperature and
    Debye temperature for potassium.

4.2 A hypothetical monovalent metal has a simple cubic space-lattice with lattice
    constant a. Assume that the valence electrons are free and calculate the radius of
    the Fermi sphere. Calculate the distance of closest approach of this sphere to the
    Brillouin zone boundary; is the free electron sphere contained within the first
    Brillouin zone?
       How would you expect this Fermi surface to be modified by the periodic potential
    of the lattice? (Compare Fig. 4.14.)

4.3 Metallic lithium has a body-centred cubic structure with cell side 3.5 Å. Calculate
    on the free electron model the width of the K emission band of soft x-rays from
    lithium. How would you expect the width of emission to depend on temperature?

4.4 Show that the kinetic energy of a free electron gas at absolute zero is

$$E_0 = \tfrac{3}{5} N \varepsilon_F,$$

where ε_F is the Fermi energy. Derive expressions for the pressure p and the bulk
modulus

$$B = -V\left(\frac{\partial p}{\partial V}\right).$$

Estimate the contribution of conduction electrons to the bulk modulus of lithium,
using information from Problem 4.3. Compare your answer with typical experimental bulk moduli.

4.5 Calculate the Hall coefficient of sodium on a free electron model, given that sodium
    has a bcc structure of cell side 4.28 Å.
       Calculate also the Hall coefficient of pure InSb at 300 K, given that $E_G = 0.15$ eV,
    $m_e = 0.014m$, $m_h = 0.18m$, and that electrons are the only effective carriers (why?).
       Estimate the EMFs generated in each case when a current of 100 mA passes
    along a sample 5 mm wide and 1 mm thick in a perpendicular field of 0.1 T.

4.6 In semiconductors, holes tend to be heavier and less mobile than electrons. Why?

CHAPTER

Magnetism

5.1 MAGNETISM IS A QUANTUM EFFECT

Although classical derivations of formulae for the magnetic susceptibility of assemblies of atoms or ions are often given, they are none of them completely self-consistent. Thus, prior to quantum mechanics, Langevin obtained an expression for the magnetism due to the alignment of small permanent magnetic dipoles in a magnetic field (paramagnetism). But if we regard matter as made up of charged particles, the existence of *permanent* magnetic moments is an assumption outside classical physics. Classically, magnetic moments due to circulating charges would be altered by an applied field so as to give zero magnetic susceptibility.

Similarly, we shall see that diamagnetism, the tendency of all matter to exclude an applied magnetic field, depends on the existence of discrete quantum states that have a certain stability against external perturbations, such as magnetic fields.

The impossibility of magnetism in classical mechanics may be seen as follows. In thermal equilibrium the distribution of particles over velocities depends only on the Boltzmann factor $\exp(-E/k_B T)$ where E is the *total* energy, kinetic and potential, of the system. E is exactly the same function of the particle positions and velocities as in the absence of a field, since the Lorentz force $e\mathbf{v} \times \mathbf{B}$ is perpendicular to $\mathbf{v}$ and therefore does no work; consequently the distribution of velocities is unchanged. There can thus be no net circulating current and no magnetic moment, because there was none before the magnetic field **B** was applied. Each individual electron trajectory is curved by the magnetic field, but there is no net circulating current after summing over all particles. As the field **B** is switched on circulating currents with associated angular momentum and magnetic moment *are* induced. But classically these currents are completely destroyed by collisions as thermal equilibrium is reestablished.

5.2   PARAMAGNETISM

5.2.1   Free ion paramagnetism*

Transition metal atoms or ions have a permanent magnetic moment as a result of a partly-filled *d* shell. According to Hund's rule the various orbital states are filled first by electrons of one spin, then by the other, so that pairing of electron spins is the least possible and the spin magnetic moment is maximized. This is an example of exchange interaction tending to produce parallel electron spins (see section 1.4.1). Similar effects occur in the incomplete *f* shells of the rare earth elements.

In the ionic salts of transition elements the paramagnetic ions are commonly kept fairly well apart by large anions (such as nitrate or sulphate) and many molecules of water of crystallization. In these circumstances the interactions between the ions are quite weak and may be estimated as follows. Each ion will see a local magnetic field due to the magnetic dipole moments μ of its neighbours; the order of magnitude of this field will be†

$$B \sim \mu_0 \mu / 4\pi a^3, \tag{5.1}$$

where a is the distance between ions. A typical interaction energy between ion pairs is therefore

$$\Delta E \sim \mu B = \mu_0 \mu^2 / 4\pi a^3. \tag{5.2}$$

If we take μ as a Bohr magneton ($\approx 10^{-23}$ A m^2) and $a \approx 3$ Å Eq. (5.2) gives

$$\Delta E \sim \frac{10^{-7} \times 10^{-46}}{3 \times 10^{-29}}$$

$$\approx 3 \times 10^{-25} \text{ joule}$$

$$\approx 0.03 \text{ K};$$

---

* For a full account of the statistical mechanics of an assembly of permanent magnetic dipoles see, for example, Mandl,[2] Chapter 3.

† Beware of the confusing conventional notation here. μ is an atomic magnetic moment in A m^2 (*not* a permeability). The magnetic constant $\mu_0 = 4\pi \times 10^{-1} \Omega \, \text{s m}^{-1}$ has nothing to do with this; it is the constant of proportionality relating a magnetic field measured in V s m^{-2} (or tesla) to the current that causes it.

if the separation is increased to 10 Å (for example, by including some anions without permanent moments) the temperature at which $k_B T = \Delta E$ falls to about 10^{-3} K. We may therefore expect such salts to approximate very closely the behaviour of independent ions.

It is because the magnetic dipolar interaction energy we have just estimated is so small that it was quite impossible to understand the ferromagnetism of iron classically, even allowing the existence of permanent magnetic dipoles. Alignment of spins to give spontaneous magnetization at room temperature can only be achieved with the much stronger exchange interaction discussed in section 1.4.1 and 5.4; for this to take effect the wavefunctions of different ions must overlap. We defer consideration of ferromagnetism to section 5.4 and consider first the simpler situation of independent dipoles in an external field.

For simplicity we consider the case of spin $\frac{1}{2}$, so that there are only two energy levels: $\varepsilon = -\mu B$ for magnetic moment parallel to the applied field; and $\varepsilon = +\mu B$ for magnetic moment antiparallel to the applied field. The mean component of magnetic moment per ion parallel to $\mathbf{B}$ is then calculated as a weighted average with the Boltzmann factor $\exp(-\varepsilon/k_B T)$ as a weighting factor

$$\langle\mu\rangle = \frac{\sum \mu(\varepsilon)\,e^{-\varepsilon/k_B T}}{\sum e^{-\varepsilon/k_B T}} = \frac{\mu\,e^{\mu B/k_B T} - \mu\,e^{-\mu B/k_B T}}{e^{\mu B/k_B T} + e^{-\mu B/k_B T}}$$

$$= \mu \tanh(\mu B/k_B T); \qquad (5.3)$$

this relation is plotted in Fig. 5.1. The magnetic moment $M = N\langle\mu\rangle$ of unit volume is proportional to field at low fields and saturates at high fields. It is important to remember that B in Eq. (5.3) is the *local* magnetic field *at the atom*, which may not be either the macroscopic average field $\mathbf{B}$ or the macroscopic average $\mathbf{H} = \mathbf{B} - \mu_0\mathbf{M}$.* In weak fields $\mu_0 M \ll H$ and this distinction is unimportant; we may then take the susceptibility χ as

$$\chi = \frac{M}{H} = \frac{N\mu^2}{k_B T} \qquad (5.4)$$

with H the macroscopic field. By comparison with Eq. (5.2) we see that $\mu_0\chi$ is small at temperatures above about 1 K. At lower temperatures the distinction between the macroscopic fields B and H and the local field

---

* Note that the field that we have defined as $\mathbf{H}$ is usually defined as $\mu_0\mathbf{H}$ in SI units, and consequently the quantity we define as χ in Eq. (5.4) is usually defined as (χ/μ_0). We define $\mathbf{H}$ in this way primarily to discourage the misconception that there is some mystical distinction between $\mathbf{B}$ and $\mathbf{H}$ *in vacuo*. Our definitions have the added advantage that they are the same as in c.g.s. electromagnetic units, which are the units used in most published magnetic measurements; in unrationalized c.g.s. electromagnetic units $\mu_0 = 4\pi$. Since $\mu_0\chi$ is a dimensionless number on our definition, susceptibilities tabulated in e.m.u. may be converted to our units (A m^{-1}T^{-1}) by multiplying by 10^7. For further details see Appendix E.

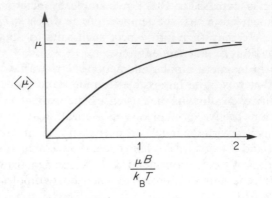

Fig. 5.1. Magnetization of an ideal paramagnetic.

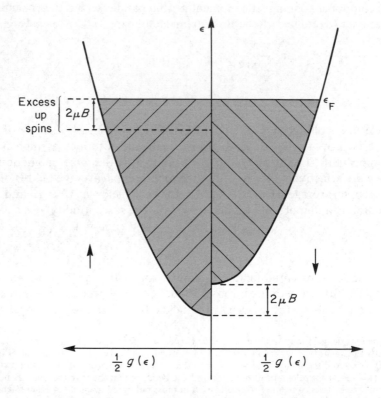

Fig. 5.2. Occupied states for spin-up and spin-down conduction electrons in a magnetic field.

matters; it turns out that for a spherical sample the contributions of other dipoles to the local field just cancel out, so that the local field is just the applied field (see Appendix C).

Eq. (5.4) is the Curie law for an ideal paramagnetic, susceptibility inversely proportional to absolute temperature. We have seen that $\mu_0\chi$ becomes of order unity at about 10^{-2} K, so that at room temperature it is quite small, 10^{-4} to 10^{-5}.

5.2.2 Conduction electron paramagnetism

The Curie law Eq. (5.4) applies to paramagnetic atoms in a low density gas, just as to well separated ions in a solid, for in this case also the orientation of any one magnetic moment is independent of the orientation of the others.

However, Eq. (5.4) does not apply to the alignment of conduction electron spins in a metal, even on a free electron model. The reason for this is that the Pauli exclusion principle behaves like an interaction between electrons: it restricts the states that an electron may occupy, and therefore hinders spin alignment so that the susceptibility is reduced below the Curie law value.

The effect may be calculated by reference to Fig. 5.2, which shows the conduction electron density of states curve split into two halves, one for electrons with spin parallel to the applied field (spin up) and one for electrons of opposite spin (spin down). In a field B the energy levels of spin up electrons are lowered by μB and the energy levels of spin down electrons are raised by μB, giving the density of states curves as drawn. The Fermi level ε_F must, of course, be the same for both spins in thermal equilibrium. It can be seen from Fig. 5.2 that there is an excess of up spins over down spins equal to

$$\tfrac{1}{2}g(\varepsilon_F)\cdot 2\mu B,$$

provided that $\mu B \ll \varepsilon_F$, so that the magnetic moment is

$$M = \mu^2 g(\varepsilon_F)B,$$

and the spin susceptibility is (since $B \approx H$)

$$\chi = M/H = \mu^2 g(\varepsilon_F). \tag{5.5}$$

For a free electron model Eq. (5.5) can be combined with our result

$$N = \tfrac{2}{3}\varepsilon_F g(\varepsilon_F) \tag{4.1}$$

to obtain

$$\chi = \frac{3N\mu^2}{2\varepsilon_F}. \tag{5.6}$$

It is interesting to note that Eq. (5.6) can be obtained from Eq. (5.4) by replacing the thermal energy of a classical gas, $\tfrac{3}{2}k_B T$, by ε_F. For a Fermi gas Eq. (5.6) applies for $k_B T \ll \varepsilon_F$ (as is the case for conduction electrons) and

Eq. (5.4) applies for $k_B T \gg \varepsilon_F$, when the exclusion principle becomes in-effective. The effect we have just considered is known as the Pauli spin paramagnetism of conduction electrons; we shall see in section 5.3 that there are other contributions to the susceptibility of conduction electrons, but nevertheless Eq. (5.6) does give the right order of magnitude for the total.

5.3   DIAMAGNETISM

When a magnetic field is applied to a system of charged particles, induced EMFs are set up which accelerate the particles so as to produce currents in such a sense as to tend to exclude the applied magnetic field—Lenz's law. In classical mechanics these currents are destroyed by collisions as thermal equilibrium is restored, and the magnetic field penetrates completely. This is a very good approximation to the observed situation in most cases, but we shall see that quantum mechanics gives a certain stability to screening currents on an atomic scale, resulting in a weak tendency of all matter to exclude applied fields, known as diamagnetism. To understand this effect it will be convenient to begin by considering in more detail what happens when a charged particle is accelerated by a changing magnetic field.

5.3.1   Momentum in a magnetic field

Consider a particle of charge e tethered by a string of length r to a point on the axis of a long solenoid carrying a current i, as in Fig. 5.3. For simplicity, suppose that the solenoid is superconducting, so that it will carry a current i indefinitely, without external power supply. Then heat the solenoid above its transition temperature so that the current, and the magnetic field, decays. There will be an induced EMF round the circle C which will accelerate the particle giving it a certain angular momentum mvr. Where has this angular momentum come from, for we certainly gave no angular momentum to the system in heating the coil so as to cause the field to decay? This paradox can be resolved only by attributing to the initial state, a charged particle at rest in the presence of a magnetic field, a certain electromagnetic momentum.

To evaluate this electromagnetic momentum we apply Faraday's law of induction. The induced EMF round a circle C of radius r is $-d\phi/dt$, where ϕ is the magnetic flux through C. From the cylindrical symmetry of the problem the electric field E at the charge e is

$$E = -\frac{1}{2\pi r}\frac{d\phi}{dt}$$

in a direction tangential to the circle C, so that the angular momentum mvr

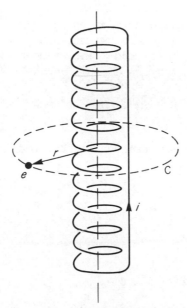

Fig. 5.3.   A charged particle
accelerated by a decaying magnetic field.

acquired by the particle during decay of the field is given by

$$mvr = er \int E \, dt = -\frac{e}{2\pi} \int \frac{d\phi}{dt} \, dt = \frac{e\phi}{2\pi},$$

since the flux *decreases* from ϕ to zero as the particle accelerates. But since
$\mathbf{B} = \text{curl } \mathbf{A}$ we can express ϕ in terms of the vector potential $\mathbf{A}$ as

$$\phi = \oint_C \mathbf{A} \cdot d\mathbf{l} = 2\pi r A,$$

so that the final velocity is related to the vector potential of the initial magnetic field by

$$mv = eA.$$

Thus, if we redefine momentum $\mathbf{p}$ as

$$\mathbf{p} = m\mathbf{v} + e\mathbf{A}, \tag{5.7}$$

the law of conservation of angular momentum is saved; in our initial state
the momentum is all in the electromagnetic form $e\mathbf{A}$,* and in the final

---

* This electromagnetic momentum, which we have associated with a charged particle, can
alternatively be regarded as distributed throughout space with a density $\varepsilon_0 \mathbf{E} \times \mathbf{B}$; see Feynman,
Lectures on Physics, Vol. II, Addison–Wesley (1964), sections 17-4 and 27-6.

state it is all in the kinematic form $m\mathbf{v}$. When we make the transition from classical mechanics to quantum mechanics it is the momentum $\mathbf{p}$ defined by Eq. (5.7) that has to be replaced by the operator $-i\hbar\nabla$.

This result can be derived more mathematically by manipulation of the acceleration equation

$$m\frac{d\mathbf{v}}{dt} = e(\mathbf{E} + \mathbf{v} \times \mathbf{B}).\tag{5.8}$$

By substituting $\mathbf{B} = \text{curl }\mathbf{A}$ in the Maxwell equation

$$\text{curl }\mathbf{E} = -\frac{\partial \mathbf{B}}{\partial t}$$

and integrating, we have

$$\mathbf{E} = -\left(\text{grad } V + \frac{\partial \mathbf{A}}{\partial t}\right)$$

so that Eq. (5.8) becomes

$$m\frac{d\mathbf{v}}{dt} = e\left[-\text{grad } V - \frac{\partial \mathbf{A}}{\partial t} + \mathbf{v} \times \text{curl }\mathbf{A}\right]$$

$$= e\left[-\text{grad } V - \frac{\partial \mathbf{A}}{\partial t} - (\mathbf{v} \cdot \nabla)\mathbf{A} + \text{grad }(\mathbf{v} \cdot \mathbf{A})\right],\tag{5.9}$$

where we have obtained the last line by use of the vector identity

$$\text{grad }(\mathbf{A} \cdot \mathbf{B}) = (\mathbf{A} \cdot \nabla)\mathbf{B} + \mathbf{A} \times \text{curl }\mathbf{B} + (\mathbf{B} \cdot \nabla)\mathbf{A} + \mathbf{B} \times \text{curl }\mathbf{A}$$

and the fact that $\mathbf{v}$ is a function of t only.

The combination $(\partial\mathbf{A}/\partial t) + (\mathbf{v} \cdot \nabla)\mathbf{A}$ is the time derivative of $\mathbf{A}$ following the motion of a particle with velocity $\mathbf{v}$, the second term representing the change in $\mathbf{A}$ at the particle due to the fact that it has moved a distance $\mathbf{v}\,\delta t$ in time δt. These two terms may therefore be written as $(d\mathbf{A}/dt)$, so that Eq. (5.9) simplifies to

$$\frac{d}{dt}(m\mathbf{v} + e\mathbf{A}) = e\,\text{grad }(\mathbf{v} \cdot \mathbf{A} - V).$$

It is thus the rate of change of the momentum defined by Eq. (5.7) that is given by the gradient of a scalar in classical mechanics, and consequently we obtain the correct classical limit only if this momentum is replaced by a gradient operator in quantum mechanics.

5.3.2   Screening by induced currents

Classically, the velocity $\mathbf{v}$ averages to zero in thermal equilibrium, so that the average momentum $\langle\mathbf{p}\rangle$ is $e\mathbf{A}$ and changes as the magnetic field changes. Quantum mechanically, $\mathbf{p}$ is determined by the pattern of nodes in the electron wavefunction. For a weak magnetic field this pattern of nodes is not perturbed for an atomic wavefunction, so that $\mathbf{p}$ is constant and $\mathbf{v}$ changes as a field is applied, giving permanent screening currents. The reason for this stabilization of momentum for an electron bound in an atom is that the node pattern of a wavefunction can only change discontinuously; it is thus reasonable to guess that an atomic wavefunction cannot change much until the applied field is large enough to make free electron orbits of atomic dimensions.

Classically, the current density for n electrons in unit volume is

$$\mathbf{j} = ne\mathbf{v} = \frac{ne}{m}(\mathbf{p} - e\mathbf{A}). \tag{5.10}$$

To evaluate screening currents in an atom we need the corresponding quantum expression; this is obtained by replacing the number density n by the probability density $\psi^*\psi$ and $\mathbf{p}$ by the operator $-i\hbar\nabla$, so that

$$\mathbf{j} = -\frac{e}{2m}[\psi^*(i\hbar\nabla + e\mathbf{A})\psi + \text{complex conjugate}]$$

$$= -\frac{i\hbar e}{2m}(\psi^*\nabla\psi - \psi\nabla\psi^*) - \frac{e^2}{m}\mathbf{A}\psi^*\psi, \tag{5.11}$$

where we have taken the mean of a quantity and its complex conjugate to obtain a real answer; it is shown in books on quantum mechanics that Eq. (5.11) is the current density required to satisfy conservation of probability. Eq. (5.11) can be put in a form closer to Eq. (5.10) by setting $\psi = n^{1/2}e^{i\theta}$; this gives

$$\mathbf{j} = \frac{ne}{m}(\hbar\nabla\theta - e\mathbf{A}). \tag{5.12}$$

Suppose as a first approximation that ψ is completely unperturbed by a magnetic field; then the first term in Eq. (5.12) is unchanged. For a purely real wavefunction it is zero; for a wavefunction of non-zero l_z (which contains a factor $\exp(il_z\phi)$) it gives a current which, by the method we use in the next section, integrates over space to give the magnetic moment $(e/2m)l_z$ associated with that state. When the total magnetic moment of the sample is calculated by weighting each l_z state with its thermal population, we thus find that the first term in Eq. (5.12) gives the orbital contribution to the paramagnetism we have already considered in section 5.2.1. Diamagnetic screening currents therefore come entirely from the second term

$$\mathbf{j} = -\frac{ne^2}{m}\mathbf{A}. \tag{5.13}$$

With Maxwell's equation (for static fields)

$$\text{curl } \mathbf{B} = \mu_0 \mathbf{j}$$

and $\mathbf{B} = \text{curl } \mathbf{A}$ we have

$$\text{curl curl } \mathbf{A} = -\frac{\mu_0 n e^2}{m} \mathbf{A},$$

or since we can always choose $\mathbf{A}$ so that $\text{div } \mathbf{A} = 0$,

$$\nabla^2 \mathbf{A} = \frac{\mu_0 n e^2}{m} \mathbf{A}. \tag{5.14}$$

A typical solution of this equation, for constant n, is

$$\mathbf{A} = \mathbf{A}_0 \, e^{-x/\lambda},$$

with $\lambda = (m/\mu_0 n e^2)^{1/2}$; $\mathbf{A}$ (and also $\mathbf{B}$) decays exponentially as we go into the interior of a region occupied by electrons. For a spatially varying n, as in an atom, the field does not decay exponentially, but the length scale of decay is still determined by λ calculated for a typical value of n, so that the field would be completely screened from the interior of a large region.

Let us compare λ with atomic dimensions. Taking $n \sim 10^{29}/\text{m}^3$ we have

$$\lambda^2 = \frac{m}{\mu_0 n e^2} \sim \frac{10^{-30}}{10^{-6} \times 10^{+29} \times 10^{-38}}$$

$$= 10^{-15} \, \text{m}^2;$$

$$\lambda \sim 300 \, \text{Å}.$$

Therefore, the magnetic field is hardly screened out from an atom at all, and we can evaluate the screening currents (and hence the magnetic moment) by using Eq. (5.13) with $\mathbf{A}$ the vector potential of the *applied* field. It is because λ is so large that atomic screening currents are rather ineffective, even if ψ is not perturbed at all, and diamagnetism is a weak effect. On an atomic scale, the inertia of electrons is too great for them to provide effective screening currents. When electronic wavefunctions extend over a larger region, as for conduction electrons in a metal, they are usually strongly perturbed by a magnetic field so that the diamagnetism remains weak; for conduction electrons it is rather smaller than the Pauli spin paramagnetism. Superconductors are an exception to this rule; we shall see in Chapter 11 that superconducting electron wavefunctions are only slightly perturbed by a magnetic field, so that behaviour similar to that predicted by Eq. (5.14) is seen on a macroscopic scale.

5.3.3 Calculation of magnetic moment

Consider a cylindrical system of coordinates with origin at the centre of the atom or molecule under consideration and z-axis parallel to $\mathbf{B}$ as in Fig. 5.4. The ring current element illustrated, $di = j\,dr\,dz$, contributes a magnetic moment $dM = \pi r^2\,di$, by the magnetic-shell theorem. The total moment of the electron wavefunction is therefore

$$M = \int \pi r^2 j \, dr \, dz ;$$

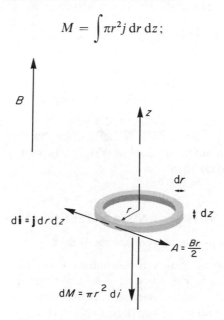

Fig. 5.4. Current element for calculating the magnetic moment due to screening currents.

with Eq. (5.13) and the fact that $A = Br/2$* in the direction illustrated for a uniform field this becomes

$$M = -\int \frac{ne^2}{m} \frac{Br}{2} \pi r^2 \, dr \, dz$$

$$= -\frac{e^2 B}{4m} \int nr^2 2\pi r \, dr \, dz$$

$$= -\frac{e^2 B}{4m} \int nr^2 \, dV$$

* Many choices of $\mathbf{A}$ such that curl $\mathbf{A}$ = const. are possible. It is convenient for the present problem to choose one with cylindrical symmetry. Compare the velocity field $\mathbf{v}$ for a rigid body rotating with angular velocity $\boldsymbol{\omega}$, for which curl $\mathbf{v} = 2\boldsymbol{\omega}$.

where dV is an element of volume. Since $\int n\, dV = 1$ for a single electron we can write

$$M = -\frac{e^2 B}{4m}\langle r^2 \rangle, \tag{5.15}$$

where $\langle r^2 \rangle$ is the mean square radius of the electron's probability distribution. With n electrons in unit volume the susceptibility is therefore*

$$\chi = -\frac{ne^2}{4m}\langle r^2 \rangle, \tag{5.16}$$

so that

$$\mu_0 \chi = -\langle r^2 \rangle/4\lambda^2 \sim 10^{-5}.$$

Since atomic sizes are fairly constant, diamagnetic susceptibilities are fairly constant; though organic molecules such as benzene and anthracene, which have extended wavefunctions only weakly perturbed by a magnetic field, show stronger diamagnetism.

5.4   FERROMAGNETISM AND ANTIFERROMAGNETISM

5.4.1   Weiss model of spontaneous magnetization

Before the advent of quantum mechanics Weiss had suggested that the remanent magnetism of iron was due to the alignment of molecular magnets, and had postulated a 'molecular field' proportional to the magnetization to explain this alignment. In order to explain the observed retention of a permanent magnetic moment up to temperatures of the order of 1000 K, it was necessary to invoke a molecular field of the order of 1000 T (10 megagauss). Such a field could not possibly arise from the dipolar fields of the individual magnets, and it remained unexplained until the advent of quantum mechanics. The mechanism of spin alignment is not magnetic in origin; it is essentially the effect of the Pauli exclusion principle tending to keep parallel spins apart, and is called **exchange interaction.**†

The exclusion principle is a consequence of the fact that any two electron wavefunction is antisymmetric under the exchange of all electron coordinates, space and spin:

$$\psi(\mathbf{r}_1, \sigma_1 ; \mathbf{r}_2, \sigma_2) = -\psi(\mathbf{r}_2, \sigma_2 ; \mathbf{r}_1, \sigma_1).$$

---

* You may see this formula with a 6 instead of a 4 in the denominator; in that case r has been taken as a radial distance from the origin, rather than a distance from the z-axis, and spherical symmetry has been assumed.

† We have already discussed exchange interaction in section 1.4.1. Readers who have read that section may omit the present explanation. A fuller discussion is given in Appendix B.

Consequently, the amplitude for two electrons of the same spin ($\sigma_1 = \sigma_2$) to be found at the same place ($\mathbf{r}_1 = \mathbf{r}_2$) is identically zero—the overall anti-symmetry keeps parallel spins apart. For this reason the expectation value of the Coulomb repulsion energy $e^2/4\pi\varepsilon_0|\mathbf{r}_1 - \mathbf{r}_2|$ of two electrons with such a wavefunction is smaller for parallel spins than antiparallel spins. This can be formally represented by including a term $-2J\sigma_1 \cdot \sigma_2$ in the energy, called the exchange energy. Although this has the form of a magnetic dipole–dipole coupling, its origin is not magnetic but electrostatic, and typical energies are in the range 0.1–1 eV. Our example of Coulomb repulsion between electrons gives rise to a value of J that is positive, favouring parallel-spin alignment; but in a solid an atomic electron is perturbed not only by the repulsion of other electrons but also by the attraction of the nuclei of neighbouring atoms, so that negative as well as positive contributions to J occur.

It is a difficult and hazardous step, fraught with more or less dubious assumptions, to proceed from the exchange interaction of a pair of electrons to writing the exchange energy of a substance such as iron in the form*

$$E_{ex} = - \sum_i \sum_{j \neq i} J_{ij} \mathbf{S}_i \cdot \mathbf{S}_j \qquad (5.17)$$

where $\mathbf{S}_i$ and $\mathbf{S}_j$ are the *total* spins on the ith and jth atoms. Whether or not Eq. (5.17) is valid in a particular case, the fundamental property of exchange interaction is certainly present: *in a many electron system there are energies of Coulomb origin dependent on how the spins are arranged.* We may take Eq. (5.17) as a convenient model, which may even be true in some cases, to represent these effects.

If we write the magnetic moment of an atom as

$$\boldsymbol{\mu}_i = \gamma\hbar\mathbf{S}_i, \qquad (5.18)$$

where γ is what is usually called the gyromagnetic ratio (more correctly, but less usually, the magnetomechanical ratio), we can express Eq. (5.17) in the form†

$$E_{ex} = -\tfrac{1}{2} \sum_i \boldsymbol{\mu}_i \cdot \mathbf{B}_i, \qquad (5.19)$$

where

$$\mathbf{B}_i = \frac{2}{\gamma\hbar} \sum_{j \neq i} J_{ij} \mathbf{S}_j. \qquad (5.20)$$

Let us now take a thermal average of Eq. (5.20). We can then say that *on the average* each atom sees the *same* field, proportional to the average spin per

---

* We omit the factor 2 in the definition of J to allow for the fact that each pair of spins is counted twice in this sum.

† As in Eq. (5.17), there is a factor $\frac{1}{2}$ to compensate for each pair of spins being counted twice in the sum.

atom, or the average magnetization of the sample. When we add the exchange field (5.20) to the applied field and the demagnetizing field (see Appendix C) we obtain a total effective field

$$\mathbf{B}_{\text{eff}} = \mu_0 \lambda \mathbf{M} + \mathbf{H} \qquad (5.21)$$

where λ is a constant, which we shall see shortly is much greater than 1.* Eq. (5.21) is the basic assumption of the Weiss model. From the way in which we have reached Eq. (5.21) we can see that by assuming an effective field at an atom depending on the average magnetization of the sample we have completely neglected fluctuations in magnetization. We shall discuss one aspect of this neglect in section 5.4.3, but for the moment we simply note that the model is approximate and proceed to analyse it. Fortunately, the model is simple enough for a complete solution to be obtained quite easily.

For simplicity we consider the case of spin $\frac{1}{2}$, rather than the unquantized magnetic moment considered by Weiss. Then, as in section 5.2.1, the magnetization is

$$M = N\mu \tanh\left(\frac{\mu B_{\text{eff}}}{k_B T}\right). \qquad (5.22)$$

To obtain the magnetization as a function of $\mathbf{H}$ and temperature we eliminate $\mathbf{B}_{\text{eff}}$ between Eqs. (5.21) and (5.22).

Consider first the case $M \ll N\mu$, which, from Eq. (5.22), applies at sufficiently high temperatures. Eq. (5.22) then takes the form

$$\mathbf{M} = \chi \mathbf{B}_{\text{eff}},$$

with

$$\chi = \frac{N\mu^2}{k_B T} = \frac{C}{T}.$$

Typically we have $N \approx 3 \times 10^{28}/\text{m}^3$, $\mu \approx 2 \times 10^{-23}\,\text{A m}^2$, so that

$$\mu_0 C = \frac{\mu_0 N\mu^2}{k_B} \approx \frac{10^{-6} \times 3 \times 10^{-28} \times 4 \times 10^{-46}}{10^{-23}} \approx 1\,\text{K}.$$

Substitution of Eq. (5.21) for $\mathbf{B}_{\text{eff}}$ then gives

$$\mathbf{M} = \frac{C}{T}(\mu_0 \lambda \mathbf{M} + \mathbf{H})$$

* Eq. (5.21) could equally be written as

$$\mathbf{B}_{\text{eff}} = \mu_0(\lambda - 1)\mathbf{M} + \mathbf{B};$$

we prefer to write it in terms of $\mathbf{H}$ because the externally applied field is approximately equal to $\mathbf{H}$ for the long rod samples normally used. Note that, excluding the exchange field, Appendix C gives, independently of sample shape,

$$\mathbf{B}_{\text{loc}} = \mathbf{H} + \tfrac{1}{3}\mu_0\mathbf{M} = \mathbf{B} - \tfrac{2}{3}\mu_0\mathbf{M}.$$

or

$$\mathbf{M} = \frac{C\mathbf{H}}{T - \mu_0\lambda C} \tag{5.23}$$

The term $\mu_0\lambda C$ in the denominator of Eq. (5.23) shows that the molecular field has the effect of a positive feedback, enhancing the magnetization. From Eq. (5.23), the measured susceptibility is

$$\chi = \frac{M}{H} = \frac{C}{T - T_c} \tag{5.24}$$

where the temperature $T_c = \mu_0\lambda C$ is known as the Curie temperature; Eq. (5.24) is the Curie–Weiss law, which describes fairly well the behaviour of ferromagnetic metals at high temperatures. At T_c the susceptibility diverges, and below T_c our assumption $M \ll N\mu$ is no longer true; we note that $T_c \approx 1000$ K so that $\lambda \approx 10^3$.*

Below T_c we have to solve Eqs. (5.21) and (5.22) without approximation, and this is best done graphically. In Fig. 5.5 the curve is Eq. (5.22) plotted in dimensionless variables, and the straight lines represent Eq. (5.21) under

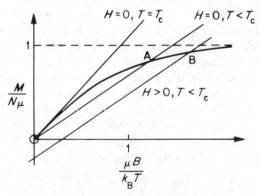

Fig. 5.5. Graphical solution of Eqs. (5.21) and (5.22).

various conditions. With zero applied field the straight line is just tangent to the tanh curve at the origin if $T = T_c$; for $T < T_c$ there are two solutions to our simultaneous equations given by the intersections at O and A. We can see that the intersection at O corresponds to an unstable equilibrium by the following argument. Suppose that, by a fluctuation, a small magnetization occurs; this will cause an effective field $\mathbf{B}_{\text{eff}}$ given by the straight line. Since

---

* Note that in unrationalized electromagnetic units, in which $\mu_0 = 4\pi$, our $\mu_0\lambda$ is usually called λ.

the slope of the straight line is less than that of the tanh curve, this value of $\mathbf{B}_{eff}$ will *increase* the original magnetization, which in turn will give a larger $\mathbf{B}_{eff}$. This process continues until the intersection at A is reached, which is stable by a similar argument. We therefore have a spontaneous magnetization below T_c, which increases from zero at T_c to $N\mu$ at absolute zero.

With $N \approx 3 \times 10^{28}/\text{m}^3$ and $\mu \approx 2 \times 10^{-23}$ Am^2 we have

$$N\mu \approx 10^6 \text{ A/m}$$

$$\approx 10^4 \text{ gauss,}$$

which is indeed of the order of the saturation magnetization of iron. But iron is not always magnetized to saturation below T_c; to explain the unmagnetized state we have to suppose that it is broken up into *domains* with spontaneous magnetization in different directions. Such a subdivision reduces the magnetic energy, but adds a surface energy where the direction of magnetization changes; this we shall consider in section 12.3.

Our simple model, even though it refers to a single domain, can also give a qualitative explanation of hysteresis. When a field H is applied parallel to the spontaneous magnetization the intersection A in Fig. 5.5 moves to B, increasing the magnetization slightly. Conversely, when a reverse field is applied the magnetization is reduced until the critical situation shown at C in Fig. 5.6(a) is shown. When the field is further increased in the reverse direction the only solution to Eqs. (5.21) and (5.22) is that represented by D in Fig. 5.6(a); consequently the magnetization suddenly reverses at a field H_c known as the coercive force. In this way the hysteresis loop shown by the full curves in Fig. 5.6(b) is generated. Hysteresis loops of this rectangular form are indeed found for particles so small as to consist of only one domain, but the magnitude of coercive force we predict is completely wrong. We can see from Fig. 5.6(a) that at low temperatures

$$H_c \approx \mu_0 \lambda N\mu$$

$$\approx 10^{-6} \times 10^3 \times 10^6 = 10^3 \text{ T}$$

$$= 10^7 \text{ gauss,}$$

which is the Weiss molecular field, and is four or five orders of magnitude too large—not very good even for astronomers! The reason for this discrepancy is that when there are two stable solutions to our equations the solution with magnetization parallel to the applied field always has lower energy; the hysteresis we have deduced above would be observed only if there were an enormous potential barrier between the two stable states, as would be the case if the magnetization had to be parallel or antiparallel to the applied field and could vary only in magnitude. However, in zero applied field there

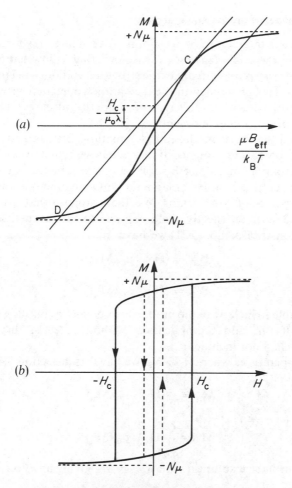

Fig. 5.6. The origin of hysteresis loops, according
to the Weiss model.

is no preferred direction in our model so that the magnetization is free to
rotate without any potential barrier, and we should in fact predict a sharp
reversal of magnetization at $\mathbf{H} = 0$ *without* hysteresis. In this respect our
model is wrong, for a crystal is not isotropic; thus, for iron, magnetization
along a (100) direction is a lower energy state than magnetization along other
directions. The resulting hindrance to rotation accounts for the observed
single-domain coercive forces of a 10–100 mT indicated schematically by
the broken lines in Fig. 5.6(b); in bulk samples the mechanism of domain
boundary motion further lowers the energy barrier to magnetization reversal
and coercive forces of a few μT can be obtained.

5.4.2 Néel model of antiferromagnetism

If the exchange interaction between ions is such as to favour antiparallel spin alignment, there is a tendency for an ordering at low temperatures in which up and down spins alternate in the structure, without a net macroscopic magnetization. This phenomenon is called antiferromagnetism; we shall see in Chapter 7 how evidence for this type of magnetic ordering can be obtained from neutron diffraction experiments.

Néel adapted the Weiss model to this situation. Let us divide the crystal lattice into an alternating sequence of sites which we assign to two sublattices, A and B; for example the A and B sublattices might be the Na and Cl sites in the NaCl structure; in antiferromagnetic MnO the sublattices are alternate close packed planes of Mn^{++} ions. We then suppose that atoms in each sublattice experience an effective field *opposite* to the magnetization of the other, so that instead of Eq. (5.21) we have

$$\mathbf{B}_{eff}^{A} = (\mathbf{H} - \mu_0 \lambda \mathbf{M_B})$$

$$\mathbf{B}_{eff}^{B} = (\mathbf{H} - \mu_0 \lambda \mathbf{M_A}). \tag{5.25}$$

It is also possible to include in the model ferromagnetic coupling to the same sublattice; but this adds complication without altering the qualitative results, so we shall not include it.

At high temperatures when $M \ll N\mu$ we have, as in section 5.4.1

$$\mathbf{M_A} = \frac{C}{T}(\mathbf{H} - \mu_0 \lambda \mathbf{M_B})$$

$$\mathbf{M_B} = \frac{C}{T}(\mathbf{H} - \mu_0 \lambda \mathbf{M_A});$$

these equations have a solution with $M_A = M_B$ giving an effective susceptibility

$$\chi = \frac{2C}{T + T_N} \tag{5.26}$$

where the Néel temperature $T_N = \mu_0 \lambda C$, and the factor 2 arises because $\chi = (M_A + M_B)/H$. Note that we now have a negative feedback from exchange forces reducing the susceptibility, in contrast to the positive feedback for the ferromagnetic case.

For $T < T_N$ and $\mathbf{H} = 0$ we have an energetically preferable solution in which $\mathbf{M_A} = -\mathbf{M_B}$; Eqs. (5.25) then reduce to Eq. (5.21), so that each sublattice now has a spontaneous magnetization equal in magnitude to the spontaneous magnetization in the ferromagnetic case.

The susceptibility of an antiferromagnetic is shown in Fig. 5.7. Above T_N it is given by Eq. (5.26); below T_N there are two values depending on whether

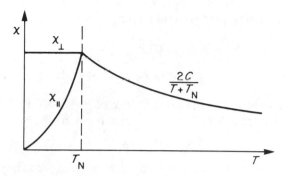

Fig. 5.7. Susceptibility of an antiferromagnetic (schematic).

the applied field is parallel or perpendicular to the sublattice magnetization. We shall not calculate these susceptibilities, but it is easy to see that the behaviour shown, which agrees fairly well with experiment, is qualitatively reasonable. When the field is perpendicular to the sublattice magnetization the spins are fairly easily tilted, even at absolute zero, giving the constant susceptibility shown. But magnetization parallel to the sublattice magnetization is opposed by the full molecular field, and cannot occur at absolute zero when the sublattice magnetizations are both $N\mu$; it is only near T_N that the molecular field weakens and the two susceptibilities become comparable.

5.4.3 Spin waves

We mentioned in section 5.4.1 that the energy of a spin should really depend on the magnetization in its immediate neighbourhood, not the average magnetization of the sample. Consider the ground state of a ferromagnet at absolute zero, with all spins aligned. On the Weiss model the lowest excited state is obtained by reversing a single spin; but because each spin is coupled to its neighbours reversal of a single spin is not a normal mode, and the motions of all spins are coupled together.

We shall first analyse this problem classically, by treating the one-dimensional problem of a chain of atoms; the method is very similar to our analysis of lattice vibrations in Chapter 2. We write the exchange interaction energy of a pair of spins as $-2J\mathbf{S}_1 \cdot \mathbf{S}_2$ in accordance with Eq. (5.17) and treat the spins $\mathbf{S}$ as classical vectors. For simplicity we consider only interactions between nearest neighbours in our chain of atoms so that the effective magnetic field $\mathbf{B}_n$ at the nth atom is given by Eq. (5.20) as

$$\mathbf{B}_n = \frac{2J}{\gamma\hbar}(\mathbf{S}_{n-1} + \mathbf{S}_{n+1}). \tag{5.27}$$

We now equate torque to rate of change of angular momentum to obtain the gyroscopic equation of motion of a spin:

$$\hbar \frac{d\mathbf{S}_n}{dt} = \boldsymbol{\mu}_n \times \mathbf{B}_n$$

$$= 2J\mathbf{S}_n \times (\mathbf{S}_{n-1} + \mathbf{S}_{n+1}), \tag{5.28}$$

where the last line has been obtained by use of Eq. (5.27). Eq. (5.28) is non-linear; for small amplitudes we may linearize it by writing

$$\mathbf{S}_n = \mathbf{S}_z + \boldsymbol{\sigma}_n \tag{5.29}$$

where $\mathbf{S}_z$ is a constant vector parallel to the mean magnetization and $\boldsymbol{\sigma}_n$ is a small vector in the x-y plane. This situation is illustrated in Fig. 5.8; note particularly that the magnetization direction z does not have any special relation to the direction of our chain of atoms. Substituting Eq. (5.29) in Eq. (5.28) and retaining only terms of the first order in $\boldsymbol{\sigma}_n$ we obtain

$$\hbar \frac{d\boldsymbol{\sigma}_n}{dt} = 2J\mathbf{S}_z \times (\boldsymbol{\sigma}_{n-1} + \boldsymbol{\sigma}_{n+1}) + 2J\boldsymbol{\sigma}_n \times 2\mathbf{S}_z$$

$$= 2J\mathbf{S}_z \times (\boldsymbol{\sigma}_{n-1} - 2\boldsymbol{\sigma}_n + \boldsymbol{\sigma}_{n+1}), \tag{5.30}$$

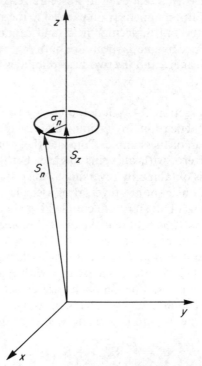

Fig. 5.8. Precession of a single spin in
a classical spin-wave.

which is rather reminiscent of our equation for lattice vibrations, Eq. (2.2); instead of a second derivative with respect to time we have a first derivative and a vector product. In component form Eq. (5.30) is

$$\hbar\left(\frac{d\sigma_n}{dt}\right)_y = 2JS_z(\sigma_{n-1} - 2\sigma_n + \sigma_{n+1})_x,$$

$$\hbar\left(\frac{d\sigma_n}{dt}\right)_x = -2JS_z(\sigma_{n-1} - 2\sigma + \sigma_{n+1})_y. \qquad (5.30)$$

If we multiply the second of these by i and subtract the first we obtain a single equation in the complex variable $\sigma^+ = \sigma_x + i\sigma_y$:

$$i\hbar\frac{d\sigma^+}{dt} = -2JS_z(\sigma^+_{n-1} - 2\sigma^+_n + \sigma^+_{n+1}). \qquad (5.31)$$

This is extremely similar to Eq. (3.1) for the propagation of an electron in a lattice, and it can in fact be obtained quantum mechanically by a similar method.

   If we apply Eq. (1.7) to the probability amplitude for a reversed spin and replace S in the interaction energy by a Pauli spin matrix (for the case of spin $\frac{1}{2}$) we obtain Eq. (5.31) for this probability amplitude; this is shown in Feynman,[1] Chapter 15 (remember in making the comparison that $|S| = \frac{1}{2}$). The correspondence between this probability amplitude and σ^+ is not surprising; for the operator equivalent of $\sigma_x + i\sigma_y$ reverses a spin from a low-energy state to a high-energy state.

   As with Eq. (3.1), Eq. (5.31) is satisfied by a wavelike solution (a spin wave) with

$$\sigma^+ \propto e^{i(kna - \omega t)} \qquad (5.32)$$

where a is the lattice spacing; substitution in Eq. (5.31) gives, on cancelling a factor $\exp^{i(kna-\omega t)}$,

$$\hbar\omega = -2JS_z(e^{-ika} - 2 + e^{ika})$$
$$= 4JS_z(1 - \cos ka); \qquad (5.33)$$

as with lattice waves and electrons we have a solution periodic in k with period $2\pi/a$. Note that Eq. (5.32) implies that

$$\sigma_x \propto \cos \omega t$$

$$\sigma_y \propto -\sin \omega t$$

so that each spin precesses as shown in Fig. 5.8 (on a classical picture), with phase changes from spin to spin given by Eq. (5.32). As we have mentioned before, the z-direction in Fig. 5.8 need not have any special relation to the direction of our chain of atoms; similarly, in a three-dimensional lattice, the wavevector **k** of a spin wave can have any direction with respect to the direction of spontaneous magnetization about which the spins precess. Spin waves with **k** parallel and perpendicular to z are shown schematically in Fig. 5.9.

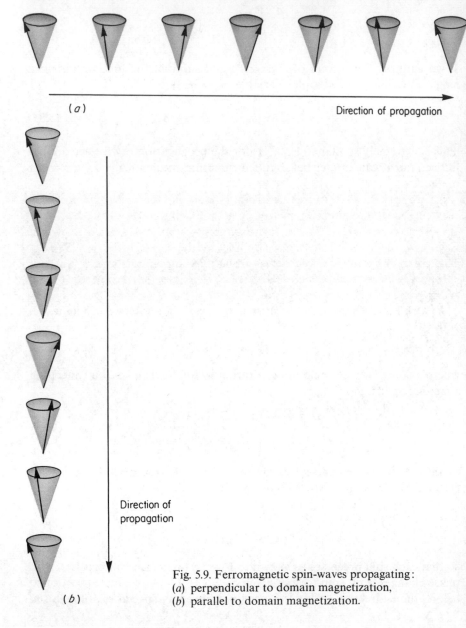

Fig. 5.9. Ferromagnetic spin-waves propagating:
(a) perpendicular to domain magnetization,
(b) parallel to domain magnetization.

Near $\mathbf{k} = 0$ Eq. (5.33) for the energy of a spin wave can be written as

$$\varepsilon = \hbar\omega = 2JS_z a^2 k^2$$
$$= \frac{\hbar^2 k^2}{2m^*} \tag{5.34}$$

with

$$m^* = \hbar^2 / 4JS_z a^2.$$

The energy is thus of just the same form as for electrons; but we have no exclusion principle. We can have as many spin-wave quanta (known as **magnons**) as we like with a given frequency; like phonons, magnons are bosons.

We may estimate roughly the effective mass m^* in Eq. (5.34) as follows. From Eq. (5.27) we can write the effective field at a spin, in the uniformly magnetized state, as

$$B = \frac{4J}{(\gamma\hbar)^2}\langle\mu\rangle$$
$$= \frac{4Ja^3}{(\gamma\hbar)^2}M \tag{5.35}$$

where we have written $a^3 = 1/N$ for the volume per spin. With Eq. (5.21) we thus have

$$\mu_0 \lambda = \frac{4Ja^3}{(\gamma\hbar)^2} = \frac{2a}{\gamma^2 m^*}$$

by use of Eq. (5.34) for spin $\frac{1}{2}$. By using $\gamma = (e/2m)$ for spin $\frac{1}{2}$, and $\varepsilon_0\mu_0 = 1/c^2$, this can be put in the form

$$\frac{m^*}{m} = \frac{mc^2}{\lambda(e^2/8a\varepsilon_0)}. \tag{5.36}$$

Since mc^2 for an electron $\approx 5 \times 10^5$ eV and with $a \approx 3$ Å, $e^2/4\pi\varepsilon_0 a \approx 3$ eV, and $\lambda \approx 10^3$, Eq. (5.32) gives very roughly

$$m^* \sim 100m.$$

We have already noted that magnons, like phonons, are bosons, so that we can have many quanta in a given mode. For phonons this leads to the macroscopic phenomenon of sound waves at small $\mathbf{k}$ (long wavelength). Similarly, we should expect long wavelength spin waves to be a macroscopically observable phenomenon. From Eq. (5.28) we expect the local intensity of magnetization $\mathbf{M}$ to obey the equation

$$\frac{d\mathbf{M}}{dt} = \gamma \mathbf{M} \times \mathbf{B}_{\text{eff}}. \tag{5.37}$$

We have already considered the exchange contribution to $\mathbf{B}_{\text{eff}}$, but in a macroscopic experiment other contributions can arise; however, any contributions to $\mathbf{B}_{\text{eff}}$ that are proportional to the local value of $\mathbf{M}$ have no effect, because $\mathbf{M} \times \mathbf{M} \equiv 0$.

The usual experimental geometry for the macroscopic observation of spin waves is shown in Fig. 5.10(a). A thin film of the material is given a static magnetization $\mathbf{M}_z$ by applying an external field $\mathbf{B}_e$ normal to the film. Spin waves are excited by applying a radiofrequency magnetic field in the plane of the film. For this geometry the non-exchange magnetic field is $[\mathbf{B}_e + \mu_0(\mathbf{M} - \mathbf{M}_z)]$, so that the contributions to this field that do not follow changes in $\mathbf{M}$ are the applied field $\mathbf{B}_e$ and the demagnetizing field $-\mu_0\mathbf{M}_z$. Eq. (5.34) for the spin wave frequency therefore becomes

$$\omega = \gamma(B_z - \mu_0 M_z) + \frac{\hbar k^2}{2m^*}. \tag{5.38}$$

The essential feature of this experimental geometry is that the spin direction is effectively pinned at the surface of the film by strong local anisotropy. Spin wave modes analogous to those of a stretched string can then be excited, with an integral number of half wavelengths in the thickness d of the film. If the film is thin compared with a radiofrequency penetration depth the radiofrequency magnetic field is uniform and only odd harmonics can be excited $(k = (2n + 1)\pi/d)$, but if the radiofrequency field is partly screened out even harmonics can also be excited. Fig. 5.10(b) shows experimental results for a thin nickel foil, showing several harmonics. From such measurements and a knowledge of the thickness of the film Eq. (5.38) can be used to find an experimental value of m^*.

5.4.4   Magnetic specific heat

When a ferromagnet is heated from absolute zero its saturation magnetization decreases; this requires energy and leads to a magnetic contribution to the specific heat. On a Weiss model the energy required can be calculated from the work done against the *internal* magnetic field $\mathbf{B}_{\text{eff}}$ given by Eq. (5.21); the change in energy due to spin reversal in the Weiss molecular field is

$$dU = -\mathbf{B}_{\text{eff}} \cdot d\mathbf{M}$$

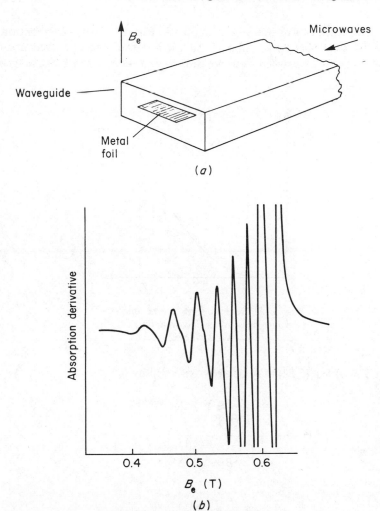

Fig. 5.10. (a) Geometry of a spin-wave experiment.
(b) Spin-wave resonances in a nickel film.
{after T. G. Phillips and H. M. Rosenberg, *Int. Magnetism Conf.*,
Nottingham (1964), p. 306}.

so that the specific heat

$$C_{H=0} = \left(\frac{\mathrm{d}U}{\mathrm{d}T}\right)_{H=0} = -\mu_0\lambda M\frac{\mathrm{d}M}{\mathrm{d}T}$$

$$= -\frac{\mu_0\lambda}{2}\frac{\mathrm{d}}{\mathrm{d}T}(M^2).$$ (5.39)

The general form of this function is shown in Fig. 5.11; this contribution to the specific heat disappears above the Curie temperature T_c because the disordering is then complete, and we thus have a second order phase transition.

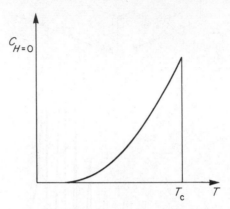

Fig. 5.11. General form of the magnetic specific heat of a ferromagnet on the Weiss model.

Near $T = 0$ Eq. (5.22) may be approximated by

$$\frac{M}{N\mu} \approx 1 - 2e^{-2\mu B/k_B T}$$

with

$$B \approx \mu_0 \lambda N\mu,$$

so that

$$\frac{\mu B}{k_B T} \approx \mu_0 \frac{\lambda N\mu^2}{k_B T} = \frac{T_c}{T};$$

thus

$$M^2 \approx (N\mu)^2(1 - 4e^{-2T_c/T}),$$

and by substitution in Eq. (5.39)

$$C_{H=0} \approx 4Nk\left(\frac{T_c}{T}\right)^2 e^{-2T_c/T}. \tag{5.40}$$

This exponential variation is characteristic of the Weiss assumption that each spin orients itself *independently* in a *uniform* molecular field. It is analogous to the exponential specific heat we obtain for lattice vibrations if we regard the atomic vibrations as *uncoupled* (see section 2.6.1). By contrast, our treatment of spin waves in section 5.4.3 corresponds to our treatment of lattice waves in section 2.2 and the analogue of a Debye theory of specific heats (section 2.6.3) can be constructed for the magnetic specific heat due to magnon excitation; the only difference is that we have $\omega \propto k^2$, whereas for phonons $\omega \propto k$.

We shall not work out the theory in detail, since we are interested only in the temperature dependence of the specific heat at low temperatures, which may be obtained without detailed calculation. At temperature T the magnon states will be occupied up to an energy

$$\varepsilon \sim k_{\rm B} T,$$

corresponding to a wavenumber

$$k \sim (m^* k_{\rm B} T)^{1/2} / \hbar;$$

this region of wavenumber space contains a number of states $\sim k^3$, so that the total number of magnons in unit volume is

$$n \sim (m^* k_{\rm B} T)^{3/2} / \hbar^3,$$

and the total energy is

$$U \sim n k_{\rm B} T = m^{*\ 3/2} (k_{\rm B} T)^{5/2} / \hbar^3.$$

By differentiation with respect to temperature we finally obtain the specific heat as

$$C \sim \left(\frac{m^{*\ 3/2} k_{\rm B}^{5/2}}{\hbar^3} \right) T^{3/2}. \tag{5.41}$$

Thus, as with lattice vibrations, the coupling produces a less rapid vanishing of the specific heat at low temperatures because the normal mode spectrum extends down to zero frequency. The $T^{3/2}$ law of Eq. (5.41) is found to be fairly well obeyed experimentally, and this is evidence in favour of the functional form $\omega \propto k^2$ of Eq. (5.34) for the magnon energy.

PROBLEMS 5

5.1 The most important contribution to the paramagnetism of copper sulphate comes from the copper ions which have spin $\frac{1}{2}$ and may be considered to be non-interacting. Show that the magnetization in a field B is given by

$$M = N \mu_{\rm B} \tanh (\mu_{\rm B} B / k_{\rm B} T),$$

where N is the number of ions per unit volume and $\mu_{\rm B}$ is the Bohr magneton. Derive the 'high' temperature form of the magnetic specific heat in a constant field B. What is a 'high' temperature if $B = 0.5\ {\rm T}$?

Compare qualitatively the behaviour of the specific heat of copper sulphate in a constant field with that of copper metal in zero field for temperatures up to a few kelvin.

5.2 On the independent particle model of electrons in a metal, obtain expressions relating specific heat and paramagnetic susceptibility to the density of states at the Fermi surface, for $T \ll T_F$.

What effects cause the *ratio* of magnetic susceptibility to electronic specific heat for a metal to differ from the value given by your expressions?

5.3 In benzene the carbon atoms form a regular hexagon of side 1.4 Å. One outer electron from each carbon atom has a wavefunction that extends round the whole ring of atoms (the other three are in sp^2 atomic orbitals). Estimate roughly the contribution of these electrons to the diamagnetic susceptibility of liquid benzene (density 0.88 g cm^{-3}, molecular weight $(C_6H_6) = 78$).

5.4 State the assumptions of the Weiss model of a ferromagnet. Use the expression $S = k \ln \Omega$ to calculate the entropy of a spin $\frac{1}{2}$ ferromagnet as a function of spontaneous magnetization. Form the free energy $F = U - TS$, and show that minimizing F gives the spontaneous magnetization deduced from Eqs. (5.21) and (5.22). Deduce also the specific heat. What is the thermodynamic nature of the transition at the Curie point, according to the Weiss model?

How are these results modified by the presence of an external magnetic field?

5.5 Calculate, according to the Néel model, the parallel and perpendicular magnetic susceptibilities of an antiferromagnet below the Néel temperature.

5.6 The experimental results illustrated in Fig. 5.10 were obtained at 9.3 GHz on a nickel film about 600 nm thick. Estimate the effective mass of magnons in nickel. (Remember the graph shows absorption *derivative*, and ignore the first two resonances, for which the detector saturates. Assume the saturation magnetization M_z is constant.)

5.7 The low temperature specific heat of $Y_3Fe_5O_{12}$ gives a straight line graph when $C/T^{3/2}$ is plotted as a function of $T^{3/2}$. What information may be obtained from the slope and intercept of this line?

6

Waves in periodic structures

6.1 WAVELIKE NORMAL MODES

We have so far considered, in one-dimensional examples, three instances of normal modes of excitation of a crystal lattice: atomic vibrations (Eq. 2.2); mobile electrons (Eq. 3.1); and spin waves (Eq. 5.31).* In each case we obtained an equation of motion for an amplitude—atomic displacement, probability amplitude, spin direction—in terms of the amplitude at neighbouring atoms. We thus had N coupled equations of motion for the N atoms of our lattice, but in each case we were able to satisfy all of these equations with a wavelike solution $\exp i(kx - \omega t)$, where the atomic sites are $x = na$.

Let us examine a little more closely why such a solution was possible. A normal mode, by definition, oscillates without change of form; its time dependence is therefore $e^{-i\omega t}$ but its spatial variation is required only to be independent of time. We therefore write a general normal mode in three dimensions as

$$e^{-i\omega t} f(\mathbf{r}).$$

However, the symmetry of a Bravais lattice imposes some restriction on $f(\mathbf{r})$. The lattice looks identical from each lattice point, and it is for this

---

* The reader who has omitted one of these topics may simply ignore references to it in what follows; but if all three have been studied it is an advantage to note the similarities between them in this section.

reason that our N equations of motion were all identical in form. Therefore, $f(\mathbf{r})$ must also look the same from each lattice point, so that if, for a lattice with one atom in a primitive unit cell, it changes by a factor A in going from the nth atom to the $(n + 1)$th, it must change by the *same* factor A in going from the $(n + 1)$th to the $(n + 2)$th; and so on. This is illustrated for a two dimensional example in Fig. 6.1, where we have supposed that $f(\mathbf{r})$ changes

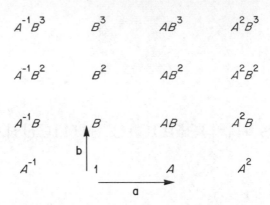

Fig. 6.1. Normal mode amplitudes in a rectangular
lattice.

by a factor A for displacement by a lattice vector $\mathbf{a}$ and a factor B for displacement by a lattice vector $\mathbf{b}$. It can be seen that for a general displacement $p\mathbf{a} + q\mathbf{b}$ we have

$$f(\mathbf{r}) \propto A^p B^q;$$

this unfortunately means that $f(\mathbf{r})$ increases indefinitely in some directions unless $|A| = |B| = 1$, so we therefore put

$$A = e^{i\theta} \qquad B = e^{i\phi},$$

as our only acceptable solution; more general solutions are acceptable near a boundary, but not in an infinite crystal. We thus have

$$f(\mathbf{r}) \propto \exp i(p\theta + q\phi).$$

It is now convenient to *define* a vector $\mathbf{k}$ such that $\mathbf{k} \cdot \mathbf{a} = \theta$, $\mathbf{k} \cdot \mathbf{b} = \phi$; remembering that the position vector $\mathbf{r}$ of an atom is

$$\mathbf{r} = p\mathbf{a} + q\mathbf{b}$$

we now have our complete solution as

$$e^{-i\omega t} f(\mathbf{r}) \propto \exp i(\mathbf{k} \cdot \mathbf{r} - \omega t). \tag{6.1}$$

The wavelike nature of the solution is thus a consequence of the translational symmetry of the lattice.

The result, Eq. (6.1), is not completely general, however, for we have so far confined our attention to amplitudes that are defined only at lattice sites, because we have only one atom in a primitive unit cell. A more general situation has been considered already in section 2.3, vibrations of a linear chain consisting of two types of atom. There we had two types of equation of motion and the solution was specified by an amplitude ratio $\alpha(\mathbf{k})$ for the motions of the two types of atom as well as by Eq. (6.1); correspondingly, there were two values of ω for a given $\mathbf{k}$ (Figs. 2.6 and 2.7). This result is easily generalized to n types of atom : there will be n distinct types of equation of motion; $(n - 1)$ amplitude ratios $\alpha_2 \ldots \alpha_n$ will be required for each $\mathbf{k}$; and there will be n branches to the frequency spectrum $\omega(\mathbf{k})$. The argument is not specific to lattice vibrations; it applies to any situation in which an amplitude has to be defined at n points in the primitive unit cell.

Now consider the case in which our amplitude is an electron wavefunction, as in section 4.3, which has to be known everywhere. Our amplitude is now

$$\Psi(\mathbf{r}, t) = e^{-i\omega t}\psi(\mathbf{r}).$$

The appropriate generalization of Eq. (6.1) for this case may be obtained by taking the limit $n \to \infty$; note that this gives an infinite number of branches of $\omega(\mathbf{k})$, as we found in section 4.3. In the limit $n \to \infty$ the set of numbers α_n tends to a continuous function $u(\mathbf{r})$ defined within a unit cell; the values of $u(\mathbf{r})$, like those of α_n, repeat within the next unit cell, so that $u(\mathbf{r})$ is a periodic function with the period of the lattice. The appropriate generalization of Eq. (6.1) is therefore

$$\psi(\mathbf{r}) = V^{-1/2} e^{i\mathbf{k}\cdot\mathbf{r}} u_{\mathbf{k}}(\mathbf{r}), \tag{6.2}$$

where we have added a suffix $\mathbf{k}$ to indicate that the form of the function $u(\mathbf{r})$ depends on $\mathbf{k}$; it is also different for each branch of the frequency spectrum. It is convenient to write explicitly in Eq. (6.2) a normalization factor $V^{-1/2}$, where V is the volume of the crystal. The wavefunction (6.2) is known as a **Bloch wavefunction** and the fact that wavefunctions in a crystal lattice can be expressed in this form is **Bloch's theorem.**

We have already seen that in one dimension $\omega(k)$ is periodic in k with period $(2\pi/a)$ when the amplitude is defined at only one point in a unit cell, because of a certain ambiguity in assignment of k value to a given motion (Fig. 2.4). We can now use Eq. (6.2) to show that this ambiguity occurs also in the more general case when the amplitude is required throughout the unit cell, as for electron wavefunctions. To do this for our one dimensional example of section 4.3 we write $u_k(x)$ as a Fourier series

$$u_k(x) = \sum_{n=-\infty}^{\infty} a_n(k) \exp i(2\pi nx/a),$$

so that from Eq. (6.2)

$$\psi(x) = V^{-1/2} e^{ikx} \sum_{n=-\infty}^{\infty} a_n(k) \exp i(2\pi nx/a). \tag{6.3}$$

But Eq. (6.3) can equally well be written, for any integer m,

$$\psi(x) = V^{-1/2} \exp i\left(k + \frac{2\pi m}{a}\right)x \sum_{n=-\infty}^{\infty} a_n(k) \exp i[2\pi(n-m)x/a]$$

$$= V^{-1/2} \exp i\left(k + \frac{2\pi m}{a}\right)x \sum_{l=-\infty}^{\infty} a'_l(k) \exp i(2\pi lx/a), \tag{6.4}$$

where we have simply defined $l = n - m$ and

$$a'_l(k) = a_n(k) = a_{l+m}(k).$$

With these definitions Eq. (6.4) is manifestly of the form of Eq. (6.2) just as much as Eq. (6.3). Since Eqs. (6.3) and (6.4) are just different ways of writing the *same* wavefunction, it is clear that any Bloch wavefunction (6.2) can equally well be assigned to $(k + 2\pi m/a)$, for any integer m. These different k labels all correspond to the *same* wavefunction, and hence $\varepsilon(k + 2\pi m/a) = \varepsilon(k)$. This is the justification for periodically continuing the $\varepsilon(k)$ curves of Fig. 4.10 to give Fig. 4.11. Of course, for nearly free electrons such a reassignment of k values may seem perverse. To write e^{ikx} as

$$e^{i(k-2\pi/a)x}u(x)$$

where

$$u(x) = e^{i2\pi x/a}$$

is undoubtedly possible, but not obviously useful. It is because of the strong perturbation of free electron wavefunctions near the energy gaps of Figs. 4.10 and 4.11 that such relabelling is sometimes useful; this will become apparent when we extend our arguments to three dimensions in the remainder of this chapter.

6.2 DIFFRACTION BY A CRYSTAL LATTICE

Since by Bloch's theorem, normal modes of a perfect crystal are wavelike in character, it is particularly important to study the propagation of waves in a crystal lattice. We have seen in the preceding chapters, for a number of special cases, that wave motion in a crystal lattice is strongly perturbed when a standing wave with an integral number of half wavelengths in a lattice spacing can be set up; energy gaps arise for these special wavenumbers, which thus separate different regions of **k**-space, called Brillouin zones. We now study the relevant geometry of **k**-space in a more general way.

We begin by considering the diffraction of a wave by a three-dimensional Bravais lattice. It will turn out that the diffraction condition is a generalized form of the standing-wave condition mentioned above. This is rather analogous to nuclear resonant scattering; a strong standing wave in the scatterer is set up on resonance. A single layer of a three-dimensional Bravais lattice is shown in Fig. 6.2(a). We shall now derive the Bragg diffraction condition Eq. (1.16) in a more general form by requiring in turn that waves scattered from lattice points along the x-, y- and z-axes* should be in phase. Consider first a line of lattice points along the x-direction as in Fig. 6.2(b). If a wave of wavelength λ is incident at an angle θ it reaches adjacent lattice points with path difference $a \sin \theta$ or phase difference $(2\pi a \sin \theta)/\lambda$. The scattered wave, of wavelength λ' (not necessarily equal to λ) leaves with path difference $a \sin \phi$ and phase difference $(2\pi a \sin \phi)/\lambda'$ in the opposite sense. The net phase difference is thus

$$\frac{2\pi a \sin \theta}{\lambda} - \frac{2\pi a \sin \phi}{\lambda'} = ka \sin \theta - k'a \sin \phi$$

$$= (\mathbf{k} - \mathbf{k}') \cdot \mathbf{a}, \tag{6.5}$$

where the vectors $\mathbf{k}$, $\mathbf{k}'$ (of magnitude $2\pi/\lambda$, $2\pi/\lambda'$) are normal to the respective wavefronts. The scattered waves from the row of lattice points have a large resultant only when this phase difference is an integral multiple of 2π, i.e.

$$(\mathbf{k} - \mathbf{k}') \cdot \mathbf{a} = 2\pi l, \tag{6.6}$$

and this equation is true for every row of lattice points in the x-direction. Eq. (6.6) is illustrated geometrically in Fig. 6.3: a set of planes are drawn normal to the x-direction with spacing $2\pi/a$; then the vector $\mathbf{K} = (\mathbf{k} - \mathbf{k}')$ must end on one of these planes. Similar considerations apply to a line of atoms in the y-direction,† and we have

$$(\mathbf{k} - \mathbf{k}') \cdot \mathbf{b} = 2\pi m, \tag{6.7}$$

where m is an integer; this can likewise be illustrated geometrically by drawing a set of planes $2\pi/b$ apart normal to the y-direction. The sets of planes corresponding to Eqs. (6.6) and (6.7) are superposed in Fig. 6.4. It can be seen from the figure that *both* diffraction conditions are satisfied along the lattice of lines where these planes intersect. The third diffraction condition is clearly

$$(\mathbf{k} - \mathbf{k}') \cdot \mathbf{c} = 2\pi n, \tag{6.8}$$

---

* Note that in general these axes are *not* mutually perpendicular. The scattering from a lattice point is obtained by summing the scattered amplitudes from all the atoms in the basis of the primitive unit cell; in general both the amplitude and phase of this scattering varies with the scattering angle.

† Remember that this is not in general perpendicular to the x-direction.

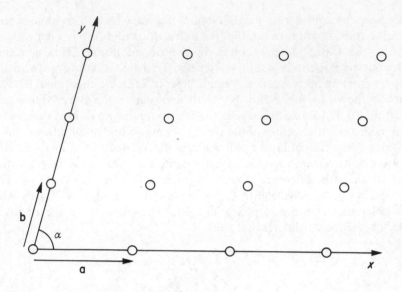

(*a*) A crystal lattice in two dimensions.

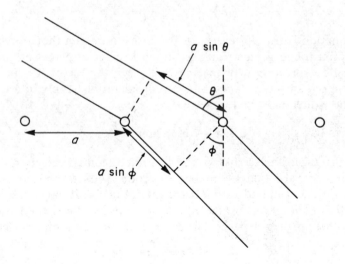

(*b*) Diffraction by a line of atoms in the *x*-direction.

Fig. 6.2.

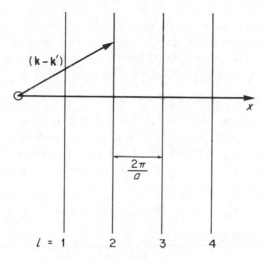

Fig. 6.3. Diffraction condition for a line of atoms
in the x-direction.

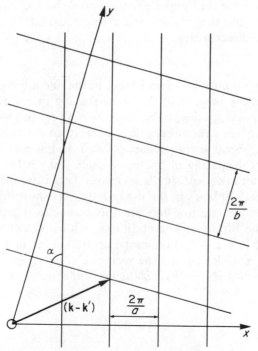

Fig. 6.4. Diffraction condition for a line of atoms
in the y-direction superposed on Fig. 6.3; at the
intersections the diffraction conditions for both
x- and y-directions are satisfied.

where n is an integer, and can be represented by a series of planes $2\pi/c$ apart normal to the z-direction. All three conditions, Eqs. (6.6), (6.7) and (6.8), are satisfied where this last set of planes intersects the lines formed by the intersection of the first two sets; this occurs on a lattice of points

$$\mathbf{G} = l\mathbf{a}^* + m\mathbf{b}^* + n\mathbf{c}^*, \tag{6.9}$$

called the **reciprocal lattice**. The layer $n = 0$ and the unit vectors $\mathbf{a}^*, \mathbf{b}^*$ of this lattice are shown in Fig. 6.5; $\mathbf{G}$ is a general **reciprocal lattice vector**. From the method of construction we can see that $\mathbf{a}^*$ lies in planes of the second and third set, and is therefore perpendicular to $\mathbf{b}$ and $\mathbf{c}$; it is thus defined by

$$\mathbf{a} \cdot \mathbf{a}^* = 2\pi, \qquad \mathbf{b} \cdot \mathbf{a}^* = 0, \qquad \mathbf{c} \cdot \mathbf{a}^* = 0, \tag{6.10}$$

with the formal solution

$$\mathbf{a}^* = \frac{2\pi \mathbf{b} \times \mathbf{c}}{\mathbf{a} \cdot (\mathbf{b} \times \mathbf{c})}.$$

The definitions of $\mathbf{b}^*$ and $\mathbf{c}^*$ may be written down by cyclic permutation of $\mathbf{a}, \mathbf{b},$ and $\mathbf{c}$. Note that the relation of the reciprocal lattice to the original space lattice—the **direct lattice**

$$\mathbf{r} = p\mathbf{a} + q\mathbf{b} + r\mathbf{c} \tag{6.11}$$

—is a completely symmetrical one. Planes in the one are perpendicular to rows of points in the other, and the plane spacing in one is 2π times the reciprocal of the point spacing in the other. We can see this for special cases from our method of constructing the reciprocal lattice. A less special example of this is that the general reciprocal-lattice vector $\mathbf{G}$, Eq. (6.9), has magnitude $2\pi/d$, where d is the spacing of planes of index (lmn) in the direct lattice; thus, the coordinates of points in the reciprocal lattice are the indices of the corresponding sets of planes in the direct lattice, as shown in Fig. 6.5. We now demonstrate the relation between this geometrical fact and Bragg's law, Eq. (1.17). The diffraction condition is $(\mathbf{k} - \mathbf{k}') = \mathbf{G}$, which is illustrated for $(lmn) = (210)$ in Fig. 6.5. For elastic scattering, as in x-ray crystallography, we have $|\mathbf{k}| = |\mathbf{k}'|$, so that the vectors $\mathbf{k}, \mathbf{k}'$ are as shown in Fig. 6.5; the vector $\mathbf{k}$ ends on a plane which is the perpendicular bisector of $\mathbf{G}$. From Fig. 6.5

$$2k \sin \theta = |\mathbf{G}| = \frac{2\pi}{d}$$

or

$$2d \sin \theta = \lambda,$$

which is equivalent to Eq. (1.17) if we remember that l, m and n in Eq. (6.9)

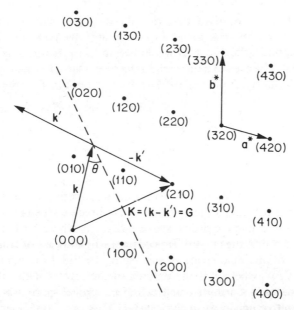

Fig. 6.5. Geometry of Bragg reflection in the reciprocal
lattice. The reflection is by (210) planes.

may have a common factor. For example, the reciprocal-lattice vector (2, 2, 2)
gives $(\mathbf{k} - \mathbf{k}')$ for the second order reflection from (111) planes; in x-ray
crystallography this is often called a (222) reflection.

We now summarize our conclusions algebraically. From Eq. (6.5) the
phasor (vector giving amplitude and phase) of a wave scattered from a point
r, when observed at a point distant from the scattering crystal so that the
radiation is effectively parallel, contains a factor

$$\exp i(\mathbf{k} - \mathbf{k}') \cdot \mathbf{r}. \tag{6.12}$$

With $(\mathbf{k} - \mathbf{k}') = \mathbf{G}$ and **r** a lattice point given by Eq. (6.11) this becomes

$$\exp i(l\mathbf{a}^* + m\mathbf{b}^* + n\mathbf{c}^*) \cdot (p\mathbf{a} + q\mathbf{b} + r\mathbf{c}) = \exp 2\pi i(lp + mq + nr),$$

where the last step follows from Eqs. (6.10). Thus all lattice points scatter
in phase, since $\exp(2\pi ni) = 1$. For a full discussion of the way the phasor
sum rapidly falls off when the diffraction condition is not satisfied, see Smith
and Thomson,[13] Chapter 11.

6.3   THE RECIPROCAL LATTICE AND BRILLOUIN ZONES

The reciprocal lattice is a construction in wavenumber space, of dimension
$(\text{length})^{-1}$, the same space in which we evaluated the density of states in

section 2.6.2. Let us compare that density of states with the density of reciprocal lattice points in k-space. For simplicity we consider first a two-dimensional example, the direct lattice shown in Fig. 6.2(a) and its reciprocal lattice in Fig. 6.4. The area of a unit cell in the direct lattice is, from Fig. 6.2(a), $A = ab \sin \alpha$. Similarly, from Fig. 6.4, the area of a unit cell in the reciprocal lattice is

$$a^*b^* \sin \alpha = a^* \cdot \frac{2\pi}{b}$$

$$= \frac{2\pi}{a \sin \alpha} \cdot \frac{2\pi}{b} = \frac{(2\pi)^2}{A}.$$

This result may also be obtained by vector algebra from Eqs. (6.10), and this is the easiest way to obtain its generalization to three dimensions: the volume of a primitive unit cell in reciprocal space is $(2\pi)^3/v$, where v is the volume of a primitive unit cell of the crystal. By contrast, the number of running wave states in unit volume of k-space is, according to the discussion following Eq. (2.28), $V/(2\pi)^3$, where V is the volume of the *crystal*. Thus, **there are N running wave states in a primitive unit cell of reciprocal space, where N is the number of primitive unit cells in the crystal.** This is in agreement with the result we had already obtained for a two dimensional special case in section 4.4, and is essential to the discussion of that section.

We may now also generalize our Fourier expansion of the Bloch wave-function Eq. (6.2) to three dimensions. The quantities $(2\pi n/a)$ in Eq. (6.3) are the reciprocal lattice vectors of a one-dimensional lattice. Similarly, because of the Fourier transform relationship (Smith and Thomson,[13] Chapter 12), between an object (the direct lattice) and its diffraction pattern (the reciprocal lattice), the periodic function $u_k(\mathbf{r})$ in Eq. (6.2) can be written in three dimensions as a Fourier series in the set of reciprocal lattice vectors $\mathbf{G}$:

$$u_k(\mathbf{r}) = \sum_{\mathbf{G}} a_{\mathbf{G}}(\mathbf{k}) \exp i\mathbf{G} \cdot \mathbf{r}. \tag{6.13}$$

Moreover, by an argument just like that leading to Eq. (6.4), a factor $\exp(i\mathbf{G} \cdot \mathbf{r})$ can be taken out of this sum so that the *same* wavefunction Eq. (6.2) is labelled as belonging to wavevector $(\mathbf{k} + \mathbf{G}_1)$ rather than $\mathbf{k}$, where $\mathbf{G}_1$ is any particular reciprocal-lattice vector. Therefore, adding a reciprocal-lattice vector to $\mathbf{k}$ is devoid of physical content. All information is contained in a single unit cell of the reciprocal lattice; because $u_k(\mathbf{r})$ has the periodicity of the lattice in $\mathbf{r}$-space, $\omega(\mathbf{k})$ has the periodicity of the reciprocal lattice in k-space.

It does not matter how we choose our fundamental unit cell in reciprocal space, but it is convenient to choose it in the same way as we have done in the one- and two-dimensional examples we considered in the preceding chapters: we let it be bounded by wavenumbers at which energy gaps occur, i.e. by wavenumbers for which Bragg reflection occurs. From Fig. 6.5 this means that **the bounding planes of our primitive unit cell in reciprocal space are the perpendicular bisectors of reciprocal lattice vectors.** In other words our unit cell is the **coordination polyhedron** (defined in section 1.5.1) of a reciprocal lattice point; it is called the **first Brillouin zone,** and is illustrated for our general two-dimensional lattice in Fig. 6.6. These polyhedra are clearly unit cells, since they stack together to fill space, as illustrated in the figure.

6.3.1   The extended, reduced, and repeated zone schemes

Because of the ambiguity of **k** value implied by the Bloch function, Eq. (6.2), there are various conventional ways, called **zone schemes,** of assigning **k** labels to states in different energy bands. It is convenient to explain these conventions first in one dimension, by reference to our electron energy diagram of Fig. 4.11, and then to proceed to the two dimensional example for which the reciprocal lattice and first Brillouin zone are shown in Fig. 6.6:

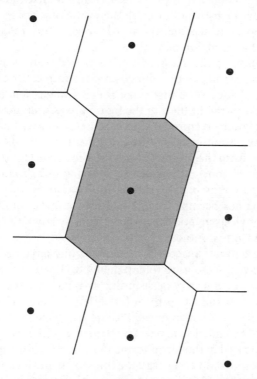

Fig. 6.6. The first Brillouin zone (shaded), with boundaries of adjacent reciprocal lattice unit cells.

the three-dimensional case involves no further new ideas, but is less readily illustrated.

The **extended zone scheme** is indicated by the full curves in Fig. 4.11; it is that assignment of **k** which corresponds most closely to free-electron behaviour, or, for other waves, to the behaviour of a wave in a continuous medium without atomic structure. In this scheme the first zone is that region of **k**-space inside the first energy discontinuity from the origin; the second zone is the region between the first and second energy gaps; and so on. This is the description we used in section 4.4.

In the **reduced zone scheme** advantage is taken of Bloch's theorem to remap the separate parts of higher zones into the first zone; thus, in Fig. 4.11, we make the convention that $|k| < \pi/a$ in all cases. The lowest energy band is said to be in the first zone, the next in the second zone, and so on.

It is sometimes desirable to emphasize that opposite sides of the reduced zone represent the same state. One then uses the **repeated zone scheme,** represented by the whole diagram of Fig. 4.11, in which the periodic variation of ε with **k** is allowed to extend through the whole of **k**-space. In this scheme, as in the reduced zone scheme, the word 'zone' really refers to a band of energies rather than a region of **k**-space.

The convenience of the reduced and repeated zone schemes becomes apparent when we consider a two-dimensional example, the reciprocal lattice of Fig. 6.6. Fig. 6.7 shows the first three zones in the extended zone scheme for this lattice. The straight lines in the figure are perpendicular bisectors of lines from the origin to reciprocal lattice points, and are thus, from Fig. 6.5, surfaces of energy discontinuity. As in one dimension, points in the first zone are those reached from the origin without crossing a surface of discontinuity; points in the second zone are reached by crossing one discontinuity; points in the third zone by crossing two discontinuities; and so on. It can be seen that the higher zones become fragmented into a large number of separate parts with no obvious relation between them, so that the extended zone scheme becomes quite impracticable.

In Fig. 6.8 the separate pieces of the second zone have been translated by reciprocal lattice vectors so as to remap them into the first zone. By comparison of Figs. 6.7 and 6.8 it may be seen that two pieces have been translated by $\pm\mathbf{a}^*$, two by $\pm\mathbf{b}^*$ and two by $\pm(\mathbf{a}^* + \mathbf{b}^*)$. Where the pieces join $\varepsilon(\mathbf{k})$ is continuous and $\partial\varepsilon/\partial\mathbf{k}$ is zero normal to the join (since these joins are zone boundaries in the extended scheme). The structure of the $\varepsilon(\mathbf{k})$ surface is therefore more complicated in the second zone that the first; for slightly perturbed free electrons there would be energy maxima at M in the zone and minima at m on the zone edge. If one were dealing with states near the minimum it would therefore be convenient to use either the repeated zone scheme or a reduced zone with a displaced origin.

The remapping of the third zone, shown in Fig. 6.9, is more complicated, in that the pieces cannot be remapped into the first zone without dividing them; this is a consequence of the lack of symmetry in our original lattice. The pieces of the third zone are shown remapped into a unit which can be repeated to fill **k**-space—the only requirement of Bloch's theorem—but this natural unit is no longer the Brillouin zone of Fig. 6.6, even if the origin is displaced. The Brillouin zone can still be used as a unit, but, because of the lack of symmetry, it is no longer true that $\partial\varepsilon/\partial\mathbf{k}$ is zero normal to a zone boundary. In fact, as for the second zone, $\partial\varepsilon/\partial\mathbf{k}$ is zero normal to the joins between different zone segments, and for nearly free electrons there would be energy maxima and minima at M and m respectively. In the higher zones symmetry contraints such as this determine quite a lot about the $\varepsilon(\mathbf{k})$ surface.

The possibility of remappings such as we have just discussed is a consequence of Bloch's theorem. In a continuous medium (or free space) a wave has modes for all **k** values; in the reduced zone scheme the periodicity of the lattice breaks this up into a number of branches each with a **k** within the first Brillouin zone; each branch is specified by a different $u_{\mathbf{k}}(\mathbf{r})$ in Eq. (6.2). Imagine the periodic potential of the lattice 'turned on' slowly. For reasons of continuity each mode of a given branch in the periodic lattice must arise from a definite mode in the continuous medium, with a definite **k** value.* Therefore, each branch of the $\varepsilon(\mathbf{k})$ curve arises from a definite region of **k** space for the wave in a continuous medium; this region is the extended zone, and is equal in volume to the reduced zone because it must contain the same number of states. In any zone scheme, a state is the nth zone if $(n-1)$ energy gaps are crossed to reach it from the lowest state.

6.3.2   Zones for structures in which the atoms form a Bravais lattice

The reciprocal lattice of a Bravais lattice is another Bravais lattice; if the direct lattice is one of the special Bravais lattices of high symmetry, the reciprocal lattice is in general another of these special Bravais lattices. An example of particular importance in the study of metals is that the face-centred and body-centred cubic lattices are reciprocal to each other. This can be seen by reference to Fig. 6.10. The vectors **a, b** to face centering points of the right hand cube lie in a (111) plane, so that the reciprocal lattice vector **c*** perpendicular to both of them is directed towards the body centering point of the left hand cube. Similarly the vector **a*** normal to **b** and **c** is directed at the centre of a cube below the right hand one, and the vector **b*** normal to **a** and **c** is directed at the centre of a cube behind the right hand one. The vectors **a***, **b***, **c*** make up a rhombic primitive unit cell of the bcc lattice.

Because of this reciprocity the Brillouin zone of the bcc structure is the coordination polyhedron of fcc, the rhombic dodecahedron of Fig. 1.17; and the Brillouin zone of fcc (close-packed cubic) is the truncated octahedron of

---

* Except actually *at* the zone boundaries, where things are not continuous.

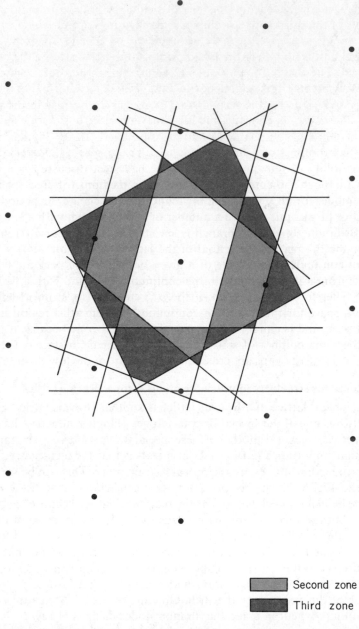

Second zone

Third zone

Fig. 6.7. Two-dimensional reciprocal lattice, showing the extended zone
scheme.

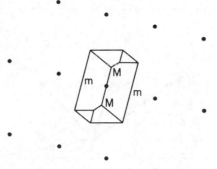

Fig. 6.8. Second zone of Fig. 6.7 remapped into the first zone.

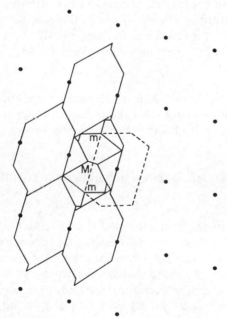

Fig. 6.9. Third zone of Fig. 6.7 remapped
to fill **k**-space periodically; first zone
shown by broken lines.

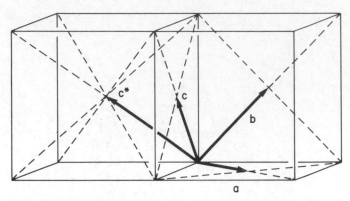

Fig. 6.10. Pairs of primitive vectors of a fcc lattice lie in a (111) plane, so that the reciprocal lattice vectors are in [111] directions— primitive vectors of a bcc lattice.

Fig. 1.22. We shall see in Chapter 10 that the nearness of the hexagonal faces of the fcc zone to the zone centre has an important effect on the electronic structure of the noble metals.

The reciprocity of the fcc and bcc lattices may alternatively be shown by considering diffraction of a wave by one of these structures, referred to the conventional cubic unit cell. Thus, bcc has a basis of atoms at $(0, 0, 0)$ and $(\frac{1}{2}, \frac{1}{2}, \frac{1}{2})$; expressed in full, there are atoms at $\mathbf{r} = 0$ and

$$\mathbf{r} = \tfrac{1}{2}\mathbf{a} + \tfrac{1}{2}\mathbf{b} + \tfrac{1}{2}\mathbf{c}$$

in the unit cell; the vectors $\mathbf{a}$, $\mathbf{b}$, $\mathbf{c}$ are equal in magnitude but mutually perpendicular. With Eq. (6.12) for the phasor of the wave scattered from a lattice point we thus have for the total scattering from a unit cell

$$1 + \exp[i(\mathbf{k} - \mathbf{k}') \cdot (\tfrac{1}{2}\mathbf{a} + \tfrac{1}{2}\mathbf{b} + \tfrac{1}{2}\mathbf{c})];$$

with $(\mathbf{k} - \mathbf{k}') = \mathbf{G} = l\mathbf{a}^* + m\mathbf{b}^* + n\mathbf{c}^*$ and Eqs. (6.10) this becomes

$$1 + \exp \pi i(l + m + n). \tag{6.14}$$

The second term in the expression (6.14) is -1, making the total zero, if $(l + m + n)$ is odd. In x-ray crystallography this phenomenon is referred to as 'characteristic absences' of x-ray reflections for special structures. For our present purpose, it means that we construct our reciprocal lattice by constructing a simple cubic reciprocal lattice for the conventional unit cell and omitting those points for which $l + m + n$ is odd; this amounts to omitting alternate (111) planes in the reciprocal lattice, since these are the planes $l + m + n = $ const., and the result is an fcc lattice (compare Fig. 1.24 of the NaCl structure, which can be regarded as a simple cubic lattice with alternate (111) planes Na$^+$ and Cl$^-$).

6.3.3   Zones for structures in which the atoms do not form a Bravais lattice

We now attempt to repeat the procedure we have just followed for the body-centred cubic structure for the hexagonal close-packed structure (Figs. 1.19 and 6.11(a)). This structure, as we have noted before, is not a Bravais lattice and its conventional unit cell is a primitive unit cell; but we can still use the method of the preceding paragraph to construct a reciprocal structure.* The basis of the hcp structure is $(0, 0, 0)$ and $(\frac{2}{3}, \frac{1}{3}, \frac{1}{2})$, so that the phasor sum analogous to (6.14) is

$$1 + \exp \pi i(\tfrac{4}{3}l + \tfrac{2}{3}m + n)$$
$$= 1 + \exp \pi i[(l + m + n) + \tfrac{1}{3}(l - m)]. \tag{6.15}$$

This is zero for odd l if $(l - m)$ is a multiple of 3, and gives the reciprocal structure shown in Fig. 6.11(b), alternating layers of close-packed points and hexagonal nets. This structure, like the direct structure, is not a Bravais lattice. Consequently, if zones are defined as previously, in the extended zone scheme, by the number of energy discontinuities crossed to reach a specific point, the results are altered. In particular the first zone is enlarged by the absence of the (001) and (00$\bar{1}$) discontinuities, because there are planes of atoms half way up the primitive unit cell of hcp; such a zone is called a Jones zone. This zone scheme has the serious defect that, because it is not based on a Bravais lattice, a reduced zone scheme cannot be constructed from it, and it is not now used. The only moderately satisfactory general solution is to base the reciprocal lattice and the Brillouin zone scheme on the Bravais lattice of the structure, in this case the simple hexagonal lattice. This reciprocal lattice has a unit cell whose base is defined by the vectors $\mathbf{a}^*$, $\mathbf{b}^*$ in Fig. 6.11(b), and the corresponding Brillouin zone is the hexagonal prism whose base is shown dotted in Fig. 6.11(b). The disadvantages of this scheme is that the first two zones are 'tied together' on the 001 faces; there is no energy discontinuity here, and the use of a reduced zone scheme becomes rather artificial. This difficulty can be avoided by using a 'double zone scheme' (Harrison,[14] section 3.4), in which the first and second zones are plotted in extended form along the $\mathbf{c}^*$ axis but in reduced form in the $\mathbf{a}^*$, $\mathbf{b}^*$ plane; the double zone is a hexagonal prism of twice the height. This resolution works for the first few zones (those of practical importance) because only the absence of (001) and (00$\bar{1}$) energy gaps matters, and consequently the zones are only tied together in pairs (first and second, third and fourth) along a single face.

Similar problems arise with the diamond structure (Fig. 1.25), which is so important in semiconductors. This structure is based on a face-centred cubic Bravais lattice, so that the Brillouin zone is the truncated octahedron

---

* We are careful not to call it a reciprocal lattice for reasons that will shortly become apparent.

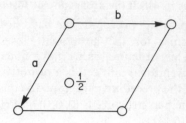

(*a*) Plan of hcp structure.

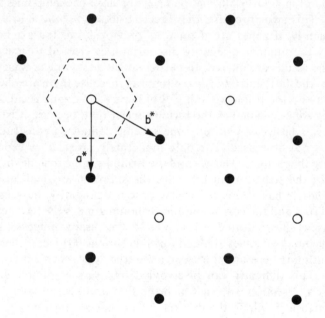

(*b*) Plan of reciprocal structure to hcp:

points ● occur in all layers;

points ○ occur only in even layers.

Fig. 6.11.

of Fig. 1.22. Here the second atom in the primitive unit cell prevents any energy discontinuity across the square faces of the zone. Because there are six such faces a double zone scheme such as is used for hexagonal close packed cannot be constructed; the first and second zone energy levels are inextricably entangled. Fortunately, this entanglement is not of practical importance since the top of the valence band is at the zone centre and the bottom of the conduction band, though sometimes in a (100) direction, does not reach the zone edge at (100). Since the problem is thus an academic one, and we lack the clear visualization of three dimensional surfaces in four-space necessary for its appreciation, we shall leave the matter here.

PROBLEMS 6

6.1 Draw the reciprocal lattice and the first two Brillouin zones in the extended and reduced zone schemes for a rectangular lattice with $a = 2\,\text{Å}, b = 4\,\text{Å}$.

Draw a free-electron Fermi surface for two electrons per unit cell, and indicate on your diagram how it is perturbed by the periodic lattice potential.

Replot the first and second zone Fermi surfaces in the repeated zone scheme, and shade the regions of k-space occupied by electrons.

6.2 Use the general expression for a reciprocal lattice vector

$$\mathbf{G} = \frac{2\pi}{V}(l\mathbf{b} \times \mathbf{c} + m\mathbf{c} \times \mathbf{a} + n\mathbf{a} \times \mathbf{b})$$

where $\mathbf{a}, \mathbf{b}, \mathbf{c}$ are *primitive* translation vectors of the direct lattice, with a primitive unit cell of volume V, to obtain an expression for $\mathbf{G}$ for a fcc lattice in cartesian coordinates appropriate to the *conventional* cubic unit cell of side a.

Calculate the lengths of the reciprocal lattice vectors for the following choices of l, m, and n.

l	m	n
1	0	0
1	−1	0
1	1	0
2	1	−1

6.3 Calculate the ratio k_F/k_M for metals of valence 1 for both the bcc and fcc lattices, where k_F is the free electron Fermi wavevector and k_M is the minimum distance in k-space from the origin to the boundary of the first Brillouin zone.

What is the relevance of your results to the Fermi surfaces of sodium (bcc) and copper (fcc)?

CHAPTER

7

Neutron crystallography

7.1 COMPARISON OF X-RAYS AND NEUTRONS

7.1.1 Structure determination

Both x-rays and neutrons have wave properties, and can therefore be diffracted by a crystal to give structural information. In section 6.2 we generalized our considerations of section 1.7 to show that the positions of diffraction maxima are given by the reciprocal lattice in **k**-space, and thus serve to determine the unit cell of the direct lattice. The distribution of atoms within a unit cell determines the total scattered amplitude from a unit cell, and hence the intensities of the diffraction maxima. In attempting to deduce the crystal structure from these intensities x-rays and neutrons are complementary; because x-ray and neutron scattering cross sections vary from one element to another in a different way, the two types of experiment emphasize different features of a structure.

Thus, x-rays are scattered primarily by extranuclear electrons and therefore give a scattering that increases steadily with atomic number. It also falls off with increasing scattering angle because of destructive interference between radiation scattered by different parts of the electron cloud of an atom. Neutrons, on the other hand, are scattered primarily by the nucleus. Because of the possibility of resonant scattering, the scattering amplitude shows no obvious systematic variation with atomic number; neutrons are therefore particularly useful for the detection of light elements (especially hydrogen) and for distinguishing between elements of similar atomic number. Also, the nucleus is so small that the scattered amplitude does not fall off with increasing scattering angle.

There is an additional type of scattering that occurs for neutrons but not for x-rays, magnetic scattering by interaction of the neutron magnetic moment with the electron magnetic moments in the scatterer. Since this interaction is sensitive to the relative orientation of neutron and electron spins it can be used to determine the distribution of **spin density** in a crystal, whereas x-rays are used to determine the distribution of charge density. We shall discuss the use of magnetic scattering to determine magnetic structure in section 7.4.1.

7.1.2   Inelastic scattering

The Bragg scattering discussed above is elastic ($|\mathbf{k}| = |\mathbf{k}'|$) because the energy of the crystal is not changed in the scattering process; no thermal excitations are created or destroyed.* It is also possible to have inelastic processes in which thermal excitations are created or destroyed; we shall discuss neutron scattering accompanied by the emission or absorption of phonons (section 2.5) in section 7.3, and the analogous processes for magnons (section 5.4.3) in section 7.4.2. Such processes also occur in photon scattering, but the energy changes are quite unmeasureable for x-rays; only for the much longer wavelength photons from a laser is it possible to measure energy (or frequency) changes with sufficient precision.

To see the reason for this difference between photons and neutrons consider a neutron of wavelength 2 Å ($k = 10^{10}\pi \, \mathrm{m}^{-1}$). Its energy is

$$E = \frac{\hbar^2 k^2}{2m} \approx \frac{10^{-68} \times 10^{21}}{3 \times 10^{-27}} \, \text{joule}$$

$$\approx 3 \times 10^{-21} \, \text{joule}$$

$$\approx 200 \, \text{K} \approx 0.02 \, \text{eV},$$

and we may therefore expect considerable fractional energy changes in inelastic scattering from solids, because the dominant excitations present in thermal equilibrium have energy $\sim k_{\mathrm{B}}T$; moreover, there will be plenty of neutrons of this sort of wavelength in room temperature moderated neutrons from a pile. For a photon of the same wavelength, on the other hand, the energy is

$$E = \hbar k c \approx 10^{-34} \times 10^{10}\pi \times 3 \times 10^8$$

$$\approx 10^{-15} \, \text{joule} \approx 10^4 \, \text{eV},$$

* The crystal does receive an impulse $\hbar(\mathbf{k} - \mathbf{k}')$, but because the crystal is very much more massive than the neutron the recoil energy associated with this is negligible.

so that higher resolution than is obtainable with x-rays would be necessary to resolve the energy changes characteristic of solids.

From what we have said it might be thought that neutrons are in all respects a superior tool to x-rays. However, the available neutron intensities are relatively low; the total thermal neutron flux obtainable from a pile is only 10^{14}–10^{15} cm^{-2} s^{-1}, and this is severely reduced by selecting a narrow band of energy. In practice, in a neutron scattering experiment there is always some sacrifice of collimation and energy resolution for the sake of intensity. For this reason precision structure determination is best done with x-rays.

7.1.3   Coherent and incoherent neutron scattering

The nuclear scattering of neutrons depends on the particular isotope of a given element, and is therefore not the same for all atoms of the same element in a crystal. Furthermore, for a nucleus with spin I the neutron scattering amplitude depends on whether the intermediate state in the process is a compound nucleus with spin $(I + \frac{1}{2})$ or $(I - \frac{1}{2})$; isotopes with spin consequently behave rather like a random mixture of two isotopes. Note that although the scattering depends on the relative orientation of neutron and nuclear spin, it is not magnetic in origin, but originates in nuclear forces. Magnetic scattering is important only for the much larger magnetic moments of electrons.

This variation of scattering amplitude means that even for a crystal of a single element we have to write the scattered intensity per unit solid angle as

$$\left| \sum_{i=1}^{N} b_i \exp[i(\mathbf{k} - \mathbf{k}') \cdot \mathbf{r}_i] \right|^2 = \left| \sum_{i=1}^{N} (\bar{b} + \beta_i) \exp[i(\mathbf{k} - \mathbf{k}') \cdot \mathbf{r}_i] \right|^2, \tag{7.1}$$

where b_i is the scattering amplitude of the ith atom at position $\mathbf{r}_i$ (compare Eq. (6.12)); $\bar{b}$ is the mean scattering amplitude and β_i is the deviation therefrom for the ith atom. In all neutron scattering experiments that have so far been done the distribution of isotopes and the nuclear spin orientation have been totally disordered. We may therefore consider the β_i as random phasors, so that by the properties of a random walk the expression (7.1) becomes

$$|\bar{b}|^2 \left| \sum_{i=1}^{N} \exp[i(\mathbf{k} - \mathbf{k}') \cdot \mathbf{r}_i] \right|^2 + \sum_{i=1}^{N} \beta_i^2. \tag{7.2}$$

The sum linear in β_i in (7.1) has vanished because of the randomness, and the phase factors are irrelevant in the sum quadratic in β_i because the β_i have random phases. The first term in the expression (7.2) is $N^2|\bar{b}|^2$ if the condition for Bragg reflection is satisfied and zero otherwise; it is called coherent scattering because the total amplitude is found by addition of amplitudes. The second term is $N\overline{\beta^2}$ at all scattering angles and is called incoherent scattering, because intensities rather than amplitudes are additive, and there are no interference effects. Note that the scattered wave from a single nucleus is always phase coherent with the incident wave; the incoherent addition of amplitudes in the second term in the expression (7.2) is a consequence not of truly incoherent scattering but of the random distribution of isotopes and nuclear spin orientations. The relative magnitudes of $|\bar{b}|^2$ and $\overline{\beta^2}$ vary considerably from element to element. For a single spinless isotope the scattering is wholly coherent, and for some elements (e.g. vanadium) it happens that the scattering is almost totally incoherent.

We have already mentioned that in addition to the nuclear scattering, a neutron may be magnetically scattered by the electrons in a solid. This scattering may be calculated by treating the potential energy of the neutron in the magnetic field of an electron as a perturbation. It is found that the magnetic scattering amplitude is proportional to the scalar product of the neutron spin and the component of electronic magnetic moment perpendicular to $(\mathbf{k} - \mathbf{k'})$. The magnetic scattering amplitude is therefore conventionally written as $p\boldsymbol{\lambda} \cdot \mathbf{S}$, where $\boldsymbol{\lambda}$ is a unit vector in the direction of the neutron spin and the **magnetic scattering vector S** is defined as

$$\mathbf{S} = \hat{\mathbf{K}} \times (\hat{\mathbf{K}} \times \hat{\boldsymbol{\mu}}) = \hat{\mathbf{K}}(\hat{\mathbf{K}} \cdot \hat{\boldsymbol{\mu}}) - \hat{\boldsymbol{\mu}}. \tag{7.3}$$

In Eq. (7.3) $\hat{\mathbf{K}}$ is a unit vector in the direction of $(\mathbf{k} - \mathbf{k'})$ and $\hat{\boldsymbol{\mu}}$ is a unit vector in the direction of the electronic magnetic moment. Because the neutron is a spin $\frac{1}{2}$ particle, $\boldsymbol{\lambda}$ can only be parallel or antiparallel to a chosen vector such as $\mathbf{S}$, so that

$$p\boldsymbol{\lambda} \cdot \mathbf{S} = \pm p|\mathbf{S}| = \pm p \sin \theta, \tag{7.4}$$

where θ is the angle between the electronic magnetic moment and $(\mathbf{k} - \mathbf{k'})$.* In a paramagnetic crystal the electronic magnetic moments are randomly oriented, so that the magnetic scattering is incoherent, and the scattered intensity per unit solid angle from N paramagnetic atoms is $\frac{2}{3}Np^2$; the factor $\frac{2}{3}$ is the average value of $\sin^2 \theta$.

In a ferromagnetically ordered crystal, however, the magnetic scattering amplitudes (7.4) add coherently to the nuclear scattering amplitude and contribute to the Bragg peaks. More importantly, we shall see in section 7.4.1 that in antiferromagnetic crystals the magnetic scattering amplitude can add coherently with a positive sign for some lattice sites and a negative sign for others. In this way coherent magnetic scattering of neutrons can give information about magnetic structure.

---

* For a proper discussion using the Pauli spin matrices, and for a derivation of the results quoted in this section, the reader is referred to O. Halpern and M. H. Johnson, *Phys. Rev.*, **55**, 898 (1939). This is a rare instance where the classic original paper is the best source of information; subsequent expositions in the literature seem to be either less clear, or less complete, or both.

7.2 NEUTRON ENERGY ANALYSIS TECHNIQUES

A scattering experiment in general requires some means of defining the incident energy; collimation to define the incident and scattered wavenumbers with respect to the crystal axes; and some means of analysing the energy of the scattered radiation. For structure determination the last step can be omitted, as in x-ray crystallography, but in this section we consider the most general type of experimental arrangement. For x-rays a monochromatic incident beam is easily obtained, since the output of an x-ray tube is a mixture of characteristic lines and continuous background; with a suitable choice of operating voltage a large proportion of line emission can be obtained. Neutrons in a pile, on the other hand, have an entirely continuous spectrum; the velocity distribution is quite closely Maxwellian at the temperature of the moderator. A monoenergetic source can therefore be obtained only by selecting a narrow band from this continuous spectrum.

Because of the limited source flux in a pile it pays to obtain a given solid angle of collimation with a rather large source area, and neutron experiments are therefore usually rather large: specimens are several centimetres in size and path lengths are of the order of metres.

7.2.1 Chopper and time of flight

The velocity of a neutron of 2 Å wavelength is given by

$$v = \frac{\hbar k}{m} \approx \frac{10^{-34} \times 10^{10}\pi}{1.6 \times 10^{-27}} \, \text{m s}^{-1}$$

$$\approx 2000 \, \text{m s}^{-1};$$

this velocity may be readily measured by timing the flight of short pulse of neutrons over a distance of a few metres. Fig. 7.1(a) shows two types of chopper which may be used to obtain pulses of neutrons about 10 μs long from a continuous beam. In the disc type, a disc opaque to neutrons, made of a material such as Cd, rotates about an axis parallel to the beam, allowing bursts of neutrons to pass through one or more slots near the periphery. The Fermi-type chopper consists of a multilayer sandwich of Cd (high neutron absorption) and Al (low neutron absorption) rotated about an axis perpendicular to the neutron beam; neutrons are transmitted only when the sandwich is parallel to the incident beam.

To define the incident velocity (and hence energy) two such choppers are used in series (Fig. 7.1(b)), separated by several metres and rotating in synchronism with a suitable phase difference between them. The first chopper thus passes a burst of neutrons of all velocities; but only those of the desired velocity reach the second chopper at the right time to be transmitted. The phase difference between the two rotors, and hence the incident velocity, may be adjusted either mechanically or electrically.

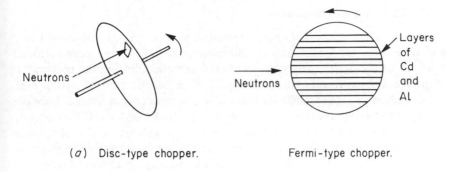

(*a*) Disc-type chopper.   Fermi-type chopper.

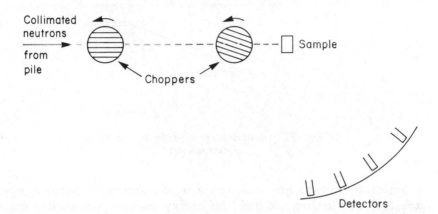

(*b*) Chopper and time-of-flight apparatus.

Fig. 7.1. Selection of monoenergetic neutrons.

The energy of the scattered neutrons is deduced from the time at which they reach a detector (usually a BF_3 proportional counter, see D. Halliday, *Introductory Nuclear Physics*, Wiley, New York, 2nd ed., 1955, section 9.8) several metres from the scatterer. To do this the time sweep of a multi-channel scaler is synchronized with the choppers, so that counts can be displayed as a function of arrival time. An advantage of this technique is therefore that data are collected at all outgoing energies at once, so making the maximum use of the incident neutrons. Furthermore, given sufficient counters and electronics (i.e. money) data can be collected at several scattering angles simultaneously, as indicated on Fig. 7.1(*b*).

7.2.2   Crystal monochromators

An alternative method of energy selection is to use Bragg reflection from a suitably oriented single crystal (the monochromator, Fig. 7.2) to scatter neutrons of the desired energy towards the sample, allowing the remainder of the thermal neutron beam to be absorbed in shielding, not shown in Fig. 7.2. Commonly, a (111) reflection from a germanium crystal is used, because the diamond structure (Fig. 1.25(b)) gives no second order (111) reflection, so that reflection of neutrons of half the desired wavelength is avoided. Such crystals are normally too perfect, and need to be strained a little to introduce some mosaic structure and avoid the effect known as extinction (Bacon[15]).

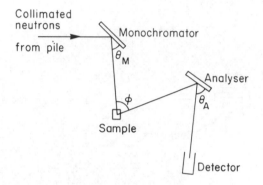

Fig. 7.2.   Geometry of a triple-axis neutron spectrometer.

Since this method gives a continuous monoenergetic beam it is not convenient to use time of flight for energy analysis, and another crystal adjusted for Bragg reflection, the analyser, is used to define the energy of those neutrons scattered from the sample which reach the detector. The whole arrangement is known as a triple axis spectrometer, since rotation about three axes is required to adjust the three scattering angles. If polarized neutrons are required, a magnetized ferromagnetic crystal can be used as a monochromator.

This arrangement collects data more slowly than the previous one, because only one outgoing energy and one scattering angle are examined at any particular time. On the other hand, the chopper and time of flight technique suffers from the difficulty that as the scattered energy changes so does the scattered wavevector $\mathbf{k}'$, and hence also the change in neutron wavevector $\mathbf{K} = \mathbf{k} - \mathbf{k}'$. It is often desirable to look at a particular value of $\mathbf{K}$ at various energies, and this can be done with a suitably automated triple axis spectrometer. Thus, if the incident energy is fixed by keeping θ_M constant and the outgoing energy adjusted by changing θ_A, the consequent change in

$|\mathbf{k}'|$ can be compensated by simultaneously altering the scattering angle ϕ to keep $|\mathbf{K}| = |\mathbf{k} - \mathbf{k}'|$ constant, and also rotating the crystal so that $\mathbf{K}$ maintains its orientation with respect to the crystal axes.

7.3   PHONON SPECTRA

7.3.1   Diffraction by a phase-modulated lattice

We considered in section 6.2 the diffraction of a wave by a perfect crystal lattice. At finite temperature a real crystal lattice is not perfect because lattice vibrations are thermally excited to some extent—a number of phonons are present. We now consider diffraction from such a modified lattice in a classical way; for simplicity we suppose that a single lattice wave of wave-vector $\mathbf{q}$ and frequency ω is present; we shall find that this periodic perturbation gives rise to additional diffraction maxima analogous to the 'ghosts' that occur with optical gratings with a periodic ruling error (R. W. Wood, *Physical Optics*, Macmillan, New York, 3rd ed., 1934, p. 254).

Let the equilibrium atomic positions be denoted by $\mathbf{R}_i$ and the displacements due to the lattice wave be given by $\alpha \cos(\mathbf{q} \cdot \mathbf{R}_i - \omega t)$;* the magnitude and direction of α give the amplitude and polarization direction of the lattice wave. The actual atomic positions are then

$$\mathbf{r}_i = \mathbf{R}_i + \alpha \cos(\mathbf{q} \cdot \mathbf{R}_i - \omega t). \tag{7.5}$$

From Eq. (6.12) the time dependent phasor sum giving the total scattered amplitude from the crystal is proportional to

$$A = \sum_i \exp i(\mathbf{K} \cdot \mathbf{r}_i - \Omega t), \tag{7.6}$$

where $\hbar\Omega$ is the energy of the incident neutron and $\mathbf{K} = (\mathbf{k} - \mathbf{k}')$. Substitution of Eq. (7.5) in Eq. (7.6) gives

$$A = \sum_i \exp i(\mathbf{K} \cdot \mathbf{R}_i - \Omega t)\{1 + i\mathbf{K} \cdot \alpha \cos(\mathbf{q} \cdot \mathbf{R}_i - \omega t) + \cdots\}, \tag{7.7}$$

where the term in curly brackets is obtained by expanding an exponential; this is justified if the atomic displacements are small. Eq. (7.7) can be written as

$$A = \sum_i \exp i(\mathbf{K} \cdot \mathbf{R}_i - \Omega t)$$

$$+ \tfrac{1}{2}i\mathbf{K} \cdot \alpha \sum_i \exp i[(\mathbf{K} + \mathbf{q}) \cdot \mathbf{R}_i - (\Omega + \omega)t]$$

$$+ \tfrac{1}{2}i\mathbf{K} \cdot \alpha \sum_i \exp i[(\mathbf{K} - \mathbf{q}) \cdot \mathbf{R}_i - (\Omega - \omega)t]$$

$$+ \text{terms of order } \alpha^2. \tag{7.8}$$

* Here and later we follow the usual practice of calling a phonon wavevector $\mathbf{q}$, to distinguish it from an electron or neutron wavevector $\mathbf{k}$.

From our arguments of section 6.2 the first term in Eq. (7.8) gives a sharp maximum for $\mathbf{K} = \mathbf{G}$, where $\mathbf{G}$ is any reciprocal lattice vector, with the resultant oscillating at frequency Ω; it is the usual elastic Bragg scattering from a perfect lattice. The second term gives a sharp maximum for $\mathbf{K} + \mathbf{q} = \mathbf{G}$, or

$$\mathbf{k}' = \mathbf{k} + \mathbf{q} - \mathbf{G}; \tag{7.9}$$

the maximum amplitude is proportional to $\mathbf{K} \cdot \boldsymbol{\alpha}$ and oscillates at a frequency Ω' given by

$$\Omega' = \Omega + \omega. \tag{7.10}$$

Eqs. (7.9) and (7.10), multiplied by $\hbar$, have the appearance of statements of the laws of conservation of momentum and energy for a process in which a neutron of wavevector $\mathbf{k}$ absorbs a **phonon** (quantum of lattice vibrations, section 2.5) of wavevector $\mathbf{q}$ to become a neutron of wavevector $\mathbf{k}'$. This is the quantum interpretation of our classical diffraction calculation. We should, however, be cautious in interpreting Eq. (7.9) as a momentum conservation equation. It is certainly true that in the scattering the crystal receives an impulse $\hbar(\mathbf{k} - \mathbf{k}') = \hbar\mathbf{K} = \hbar(\mathbf{G} - \mathbf{q})$, and that this impulse is transmitted to the crystal mounting. But it is pure convention to divide this into a part $\hbar\mathbf{G}$ associated with the whole lattice and a part $\hbar\mathbf{q}$ associated with the phonon absorbed. In fact, as we mentioned in section 2.5, in the harmonic approximation a phonon carries no momentum with it. The quantity $\hbar\mathbf{q}$ is called the **crystal momentum** or **quasi-momentum** of a phonon because it behaves like a momentum in equations such as Eq. (7.9).

Similarly, the third term in Eq. (7.8) represents neutron scattering in which a phonon of wavevector $\mathbf{q}$ is *emitted*. The terms of order α^2 and higher correspond to quantum processes in which two or more phonons are emitted or absorbed; we shall not consider them further here.*

We see from Eq. (7.8) that the amplitude of these single phonon emission and absorption processes is proportional to α, and hence the intensity is proportional to α^2, i.e. to phonon intensity or phonon number. Our classical calculation is not quite right here; if n is the number of phonons present initially, the emission probability is proportional to $(n + 1)$ and the absorption probability to n; this is shown for any particle obeying Bose–Einstein statistics in Feynman,[1] Chapter 4. In the particular case of photons (black body radiation, see Mandl[2]) the 1 in $(n + 1)$ corresponds to spontaneous emission. At low temperatures, when very few phonons are present ($n \ll 1$), it follows that only the phonon emission process can occur; there are no phonons to be absorbed.

---

* Note however that these higher order terms necessarily arise from expanding an exponential to obtain Eq. (7.7); they are analogous to the infinite series of sidebands for a single modulation frequency in FM radio. We shall see that they do not occur for the magnetic scattering problem in section 7.4.2, which is analogous to AM radio.

7.3.2   Experimental results

Eqs. (7.9) and (7.10) are the basis of the experimental determination of phonon spectra by inelastic neutron scattering. In such an experiment $\mathbf{k}, \mathbf{k}', \Omega, \Omega'$ are determined by collimation and energy measurement, so that $\mathbf{q}$ and ω for phonons emitted or absorbed to produce observed neutron scattering peaks can be deduced.

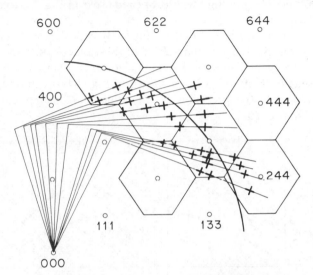

Fig. 7.3. Cross-section of the reciprocal lattice of Al, showing the location in reciprocal space of neutron groups scattered at 95.1° for various crystal orientations. The crosses indicate experimental accuracy.

Note that reciprocal lattice points are indexed according to the conventional cubic unit cell, not a primitive unit cell. Consequently, for example, the point (100) is absent (compare section 6.3.2).

{after B. N. Brockhouse and A. T. Stewart, *Rev. Mod. Phys.*, **30**, 236 (1959)}.

Fig. 7.3 is a cross-section of the reciprocal lattice of aluminium, drawn to illustrate the location in $\mathbf{k}$-space of various observed groups of inelastically scattered neutrons. Aluminium has the cubic close packed structure, so that its reciprocal lattice is body centred cubic; a cross-section normal to an $[0\bar{1}1]$ direction is shown (compare Figs. 6.5 and 10.5). The hexagons are cross-sections of the Brillouin zone, which is the truncated octahedron of Fig. 1.22. In the experiment, incident energy and scattering angle were kept fixed, but the orientation of the crystal with respect to the incident beam was varied. Lines parallel to $\mathbf{k}$ and $\mathbf{k}'$ are shown on the figure for several orientations, and the end of the vector $(\mathbf{k} - \mathbf{k}')$ for observed neutron groups is shown by a cross, the size of which represents the estimated experimental error. The circular arc is the locus of $(\mathbf{k} - \mathbf{k}')$ for elastic scattering, which is actually observed at one crystal setting, for a (333) Bragg reflection; the distance of a cross from this circle is a measure of the energy gained or lost by the neutron. This diagram illustrates the interdependence of $\mathbf{k}'$ and Ω' discussed in section 7.2. From Eq. (7.9) the $\mathbf{q}$ of the phonon emitted or absorbed is obtained by joining the appropriate cross in Fig. 7.3 to the nearest reciprocal lattice point.

It can be seen from Fig. 7.3 that in this early experiment the accuracy in $\mathbf{q}$ and ω was not much better than 20 % of the value at the zone boundary. Results of a more recent experiment on aluminium with an automated triple axis spectrometer (section 7.2.2) are shown in Fig. 7.4. It can be seen that in less than a decade the precision of this type of measurement has been enormously improved.

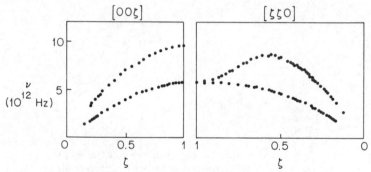

Fig. 7.4. Phonon spectra of Al in the [001] and [110] directions in reciprocal space by inelastic neutron scattering (after Egelstaff[16]). ζ is the coordinate of a point in reciprocal space, as indicated above the figures. Thus, in the [001] direction the phonon wavevector $\mathbf{q}$ is $(0, 0, \zeta a^*)$.

7.4 MAGNETIC SCATTERING

7.4.1 Magnetic structure

We have already seen in section 7.1.3 that the magnetic scattering of neutrons by electrons gives a scattering amplitude dependent on the electronic magnetic moment via the scattering vector $\mathbf{S}$ (Eqs. (7.3) and 7.4)), so that the disordered electronic moments in a paramagnetic contribute to a diffuse background of incoherent scattering. In a ferromagnetic substance the electron spins are no longer disordered, so that the magnetic scattering now behaves coherently and contributes to the Bragg peaks; the essential features of the Bragg peaks are however unmodified. In an antiferromagnetic more drastic effects are seen, because the differing electron spin orientation of magnetic atoms on the A and B sublattices (section 5.4.2) now means that they behave towards neutrons as *two different types of atom*, with coherent scattering amplitudes $(\bar{b} + p \sin \theta)$ and $(\bar{b} - p \sin \theta)$, from Eq. (7.4). This, in general, results in a more complicated structure that effectively has a larger unit cell.

Consider, for example, MnO, which has the NaCl structure, Fig. 1.24, and an antiferromagnetic Néel temperature of 120 K. The results of neutron scattering experiments on polycrystalline samples at liquid nitrogen temperature and room temperature are shown in Fig. 7.5. Below the Néel temperature

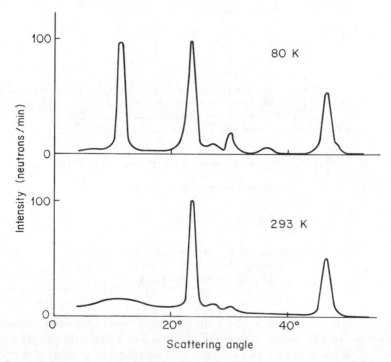

Fig. 7.5. Neutron powder patterns of MnO above and below the anti-ferromagnetic Néel temperature of 120 K. Note particularly the extra reflection at 12°, which is the (111) reflection of the doubled magnetic unit cell.
{after Bacon[15]}.

extra peaks appear, in particular a strong peak at about 12°, approximately half the scattering angle of the usual (111) peak. These results can be explained by the structure shown in Fig. 7.6, in which alternate close packed (111) planes of Mn are aligned oppositely in the (111) plane. Thus, the spins within a (111) plane are aligned ferromagnetically, but the (111) planes are arranged antiferromagnetically. Because of this opposite orientation a diffraction

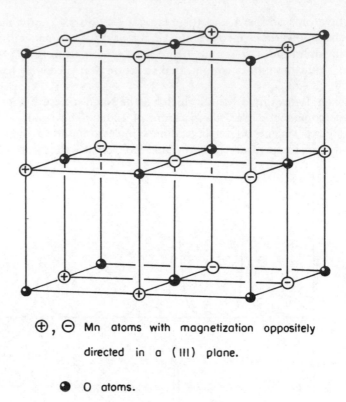

⊕ , ⊖ Mn atoms with magnetization oppositely

directed in a (III) plane.

● O atoms.

Fig. 7.6. The magnetic structure of MnO. A chemical unit
cell of the structure is shown; the magnetic unit cell has
twice the linear dimensions.

maximum can occur with a phase difference of only π between successive
(111) planes of Mn atoms; hence the reflection at a scattering angle of 12°
in Fig. 7.5. What is shown in Fig. 7.6 is one chemical or nuclear unit cell of
the structure; to obtain a repeat of spin orientation a unit of twice the linear
dimensions would have to be shown, giving the magnetic unit cell.

Precision x-ray measurements, of greater resolution than the neutron
measurements, show that the unit cell is no longer strictly cubic, having been
distorted by extension along a 111 direction; this is consistent with the
symmetry of the magnetic structure deduced from neutron measurements.

7.4.2   Magnon spectra

Because the magnetic scattering of neutrons depends on the orientation
of the atomic magnetic moment through Eqs. (7.3) and (7.4) the presence of
a spin wave (section 5.4.3) leads to a modulation of the scattering properties

of the lattice. We shall now show that this modified scattering makes it possible to determine experimentally the relation between magnon frequency ω and wave vector $\mathbf{q}$. In a ferromagnetic crystal containing a spin wave the magnetic scattering vector $\mathbf{S}$ defined by Eq. (7.3), for the ith atom may be written as

$$\mathbf{S}_i = \mathbf{S}_0 + \boldsymbol{\sigma}_0 \cos(\mathbf{q} \cdot \mathbf{R}_i - \omega t), \tag{7.11}$$

where the first term arises from the steady magnetization and the second from the fluctuating magnetization in the spin wave. From Eqs. (7.4) and (6.12) the total scattered amplitude from the crystal is given by

$$A = \sum p\lambda \cdot \mathbf{S}_i \exp i(\mathbf{K} \cdot \mathbf{R}_i - \Omega t), \tag{7.12}$$

where $\mathbf{K} = (\mathbf{k} - \mathbf{k}')$ and the incident neutron energy is $\hbar\Omega$. By use of Eq. (7.11) this can be put in the form

$$A = \sum_i p\lambda \cdot [\mathbf{S}_0 + \boldsymbol{\sigma}_0 \cos(\mathbf{q} \cdot \mathbf{R}_i - \omega t)] \exp i(\mathbf{K} \cdot \mathbf{R}_i - \Omega t)$$

$$= p\lambda \cdot \mathbf{S}_0 \sum_i \exp i(\mathbf{K} \cdot \mathbf{R}_i - \Omega t)$$

$$+ \tfrac{1}{2}p\lambda \cdot \boldsymbol{\sigma}_0 \sum_i \exp i[(\mathbf{K} + \mathbf{q}) \cdot \mathbf{R}_i - (\Omega + \omega)t]$$

$$+ \tfrac{1}{2}p\lambda \cdot \boldsymbol{\sigma}_0 \sum_i \exp i[(\mathbf{K} - \mathbf{q}) \cdot \mathbf{R}_i - (\Omega - \omega)t]. \tag{7.13}$$

Our arguments of section 6.2 show that the first term in Eq. (7.13) gives a sharp maximum for $\mathbf{K} = \mathbf{G}$; it is the ordinary elastic Bragg scattering. Similarly, the second term gives a sharp maximum for $\mathbf{K} + \mathbf{q} = \mathbf{G}$ or

$$\mathbf{k}' = \mathbf{k} + \mathbf{q} - \mathbf{G}, \tag{7.14}$$

and the scattered wave oscillates at a frequency given by

$$\Omega' = \Omega + \omega. \tag{7.15}$$

Eqs. (7.14) and (7.15), when multiplied by $\hbar$ have rather the appearance of laws of conservation of momentum and energy for a scattering process in which the neutron absorbs a magnon. This is indeed the quantum interpretation of the process we have calculated classically, but one peculiarity should be noted, the appearance of a momentum $\hbar\mathbf{G}$ in Eq. (7.14). This is a consequence of the ambiguity of wavevector we discussed in section 6.1. The quantity $\hbar\mathbf{q}$ is called the **crystal momentum** of a magnon; it is not a true momentum, though it behaves in a rather analogous way.

In an exactly similar way the third term in Eq. (7.13) represents a scattering in which a neutron creates a magnon. Eqs. (7.14) and (7.15) are the basis for the experimental determination of $\omega(\mathbf{q})$. For by the techniques described in section 7.2, $\mathbf{k}$, $\mathbf{k}'$, Ω, and Ω' can be measured and hence ω and $\mathbf{q}$ can be deduced.

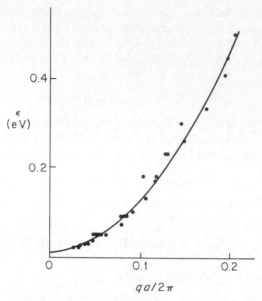

Fig. 7.7. Experimental spin-wave spectrum of $Co_{0.92}Fe_{0.08}$. Note the finite energy gap at $q = 0$ due to crystalline anisotropy. {after Egelstaff[16]}.

An example of a spin-wave spectrum determined in this way, for a cobalt-iron alloy, is shown in Fig. 7.7. Note that, in contrast to Eq. (5.34), the spin-wave energy is not quite zero at $\mathbf{q} = 0$; this is a consequence of crystalline anisotropy, which we neglected in section 5.4.3.

It is of interest to compare the emission and absorption of magnons in neutron scattering with the emission and absorption of phonons (section 7.3). Eqs. (7.9) and (7.10) are identical with Eqs. (7.14) and (7.15) respectively; but Eq. (7.8) contains terms of order α^2 whereas Eq. (7.13) contains no terms of order σ_0^2. This means that, in our approximation, processes involving several phonons occur, but only single magnon processes. This difference arises because a lattice vibration gives a phase modulation of the scattered wave whereas a spin wave gives an amplitude modulation.

PROBLEMS 7

7.1 Neutrons of energy 0.02 eV are scattered at an angle of 10° from solid helium (speed of sound = $300 \, \mathrm{m \, s^{-1}}$) with emission of a phonon. Estimate the energy loss of the neutrons. What is the time of flight over a 10 m path of unscattered and scattered neutrons?

7.2 Why will a small angle scattering experiment as in Problem 7.1 not work for a crystal such as sapphire, for which the speed of sound (10^4 m s^{-1}) is greater than the speed of the neutrons?

How would you use neutrons to investigate the phonon spectrum of sapphire?

7.3 Metallic dysprosium has a hexagonal structure. It is alleged that the atomic moments are aligned ferromagnetically in the basal plane, but that the direction of alignment rotates about the c-axis through an angle of order 40° from one layer to the next. What neutron scattering measurements would you make to confirm this, and what results would you expect?

CHAPTER

Thermal conductivity of insulators

8.1 PHONON COLLISIONS

In electrically insulating crystals there are no mobile electrons, and lattice waves are the only mechanism available for carrying thermal energy from one end of a crystal to the other; in metals this heat transport mechanism also exists, but is usually masked by electronic heat transport. Energy transport by lattice waves is most easily treated by considering fairly localized wavepackets or **phonons** (section 2.5); this is a convenient approach because the phonons behave as nearly independent particles, and the kinetic theory of gases can therefore be adapted to the problem. In the approximation of Chapter 2 the phonons are true normal modes and hence do not interact at all; the phonon gas is truly ideal, and there are no phonon collisions. Consequently the phonon mean free path is infinite and a coefficient of thermal conductivity cannot be defined; in this approximation the mechanism of thermal conduction is analogous to that in a Knudsen gas.

8.1.1 The scattering mechanism

Any real crystal resists compression to a volume smaller than the equilibrium value more strongly than expansion to a larger volume; this is a consequence of the general shape of interatomic potential energy curves, and is the origin of thermal expansion (see Flowers and Mendoza[3]). Our assumption of Hooke's law forces in Chapter 2 amounts to replacing the interatomic potential by a parabolic curve fitted to the minimum. This is known as the harmonic approximation, and is essential to our separation into normal modes. Deviations from this parabolic potential, expressed as terms cubic or quartic in the atomic displacements, give rise to coupling between the normal modes and consequently to a finite phonon lifetime. In a real crystal the phonon lifetime may also be limited by collisions with the crystal boundaries, lattice defects, or impurity atoms. But these processes are important only at very low temperatures. Normally, the phonon–phonon scattering, which we shall now discuss, is dominant.

The mechanism by which phonons of different frequency become mixed is best understood by considering a macroscopic consequence of anharmonicity, that the velocity of sound depends upon density, or, more generally, the state of strain of the crystal. Thus, if a sound wave passes through a crystal in which another sound wave is already present its wavefront will become phase modulated by this effect. To see this, consider a wave of frequency ω_1 and wavevector $\mathbf{q}_1$ which is phase modulated at frequency ω_2 and wavevector $\mathbf{q}_2$:

$$A = \exp i[\mathbf{q}_1 \cdot \mathbf{R} + C \cos (\mathbf{q}_2 \cdot \mathbf{R} - \omega_2 t) - \omega_1 t], \qquad (8.1)$$

where C is a constant specifying the amount of phase modulation. The phase velocity of this wave can be obtained by writing the equation for a wavefront, or surface of constant phase, as

$$\mathbf{q}_1 \cdot \mathbf{R} + C \cos (\mathbf{q}_2 \cdot \mathbf{R} - \omega_2 t) - \omega_1 t = \text{const.}, \qquad (8.2)$$

and differentiating with respect to time to obtain

$$\mathbf{q}_1 \cdot \frac{d\mathbf{R}}{dt} - \mathbf{q}_2 \cdot \frac{d\mathbf{R}}{dt} C \sin (\mathbf{q}_2 \cdot \mathbf{R} - \omega_2 t) + \omega_2 C \sin (\mathbf{q}_2 \cdot \mathbf{R} - \omega_2 t) - \omega_1 = 0,$$

from which the phase velocity $\mathbf{v}_s$ is given by

$$\mathbf{v}_s \cdot \frac{\mathbf{q}_1 - \mathbf{q}_2 C \sin (\mathbf{q}_2 \cdot \mathbf{R} - \omega_2 t)}{\omega_1 - \omega_2 C \sin (\mathbf{q}_2 \cdot \mathbf{R} - \omega_2 t)} = 1. \qquad (8.3)$$

Eq. (8.3) represents a velocity of sound modulated* at frequency ω_2 and wavevector $\mathbf{q}_2$, appropriate to a crystal containing a sound wave of this frequency and wavevector. Therefore, if a sound wave of frequency ω_1 and wavevector $\mathbf{q}_1$ is propagated through crystal containing another sound wave of frequency ω_2 and wavevector $\mathbf{q}_2$, we may expect its wavefront to become distorted according to Eq. (8.1) for a suitable value of C. For $C \ll 1$

---

* Unless $\mathbf{q}_1/\omega_1 = \mathbf{q}_2/\omega_2$, i.e. both sound waves are propagating in the same direction with the same velocity. In this special case the wave becomes non-sinusoidal, but the wavefronts are not distorted.

Eq. (8.1) can be expanded in powers of C as

$$A = \exp \mathrm{i}(\mathbf{q}_1 \cdot \mathbf{R} - \omega_1 t)[1 + \mathrm{i}C \cos(\mathbf{q}_2 \cdot \mathbf{R} - \omega_2 t) + \cdots]$$

$$= \exp \mathrm{i}(\mathbf{q}_1 \cdot \mathbf{R} - \omega_1 t)$$

$$+ \tfrac{1}{2}\mathrm{i}C \exp \mathrm{i}[(\mathbf{q}_1 + \mathbf{q}_2) \cdot \mathbf{R} - (\omega_1 + \omega_2)t]$$

$$+ \tfrac{1}{2}\mathrm{i}C \exp \mathrm{i}[(\mathbf{q}_1 - \mathbf{q}_2) \cdot \mathbf{R} - (\omega_1 - \omega_2)t]$$

$$+ \text{ terms of order } C^2. \tag{8.4}$$

The first term in Eq. (8.4) is our original sound wave; the second is a new wave of frequency

$$\omega_3 = \omega_1 + \omega_2 \tag{8.5}$$

and wavevector

$$\mathbf{q}_3 = \mathbf{q}_1 + \mathbf{q}_2. \tag{8.6}$$

If $\omega_3(\mathbf{q}_3)$ is a point on the phonon dispersion curve this new wave can propagate, and in terms of particles we have a process in which a phonon of wavevector $\mathbf{q}_1$ absorbs a phonon of wavevector $\mathbf{q}_2$ to become a phonon of wavevector $\mathbf{q}_3$. The third term in Eq. (8.4) corresponds to a process in which a phonon of wavevector $\mathbf{q}_1$ *emits* a phonon of wavevector $\mathbf{q}_2$ to become a phonon of wavevector $\mathbf{q}_3$. These two processes are therefore the inverse of each other: either two phonons coalesce to form one, or one phonon splits into two; they are both **three-phonon processes.** The terms of order C^2 and higher in Eq. (8.4) correspond to processes in which four or more phonons are involved; we shall not be concerned with these processes. Eqs. (8.5) and (8.6), multiplied by $\hbar$, give the laws of conservation of energy and crystal momentum for phonons. We should note, however, that Eq. (8.6) is not generally valid if, as is usual for phonons, we use a reduced zone scheme, for it may happen that the vector $\mathbf{q}_1 + \mathbf{q}_2$ lies outside the first zone. In this case, in order to obtain a $\mathbf{q}_3$ within the zone, Eq. (8.6) may be generalized to

$$\mathbf{q}_3 + \mathbf{G} = \mathbf{q}_1 + \mathbf{q}_2, \tag{8.7}$$

where $\mathbf{G}$ is a reciprocal lattice vector. This is equivalent to Eq. (8.6) because, by Bloch's theorem, the wavevectors $\mathbf{q}_3$ and $\mathbf{q}_3 + \mathbf{G}$ are alternative labels for the *same* phonon.

Note that the whole of the foregoing discussion is very similar to our treatment of neutron scattering by phonons in section 7.3.1; in particular, compare Eqs. (7.8) and (8.4). If we had considered the scattering of neutrons by atoms as giving the crystal a sort of refractive index (dependent on density) for neutrons, the similarity would have been even closer. In the present case the 'refractive index' approach is forced on us, because the idea of an atomic scattering cross-section for lattice waves is meaningless, but in the case of neutrons it was simpler to treat the atomic scattering directly.

8.1.2   Normal and Umklapp processes

Eq. (8.7) means that the crystal momentum $\hbar\mathbf{q}$ of phonons is not always conserved in phonon collisions; conservation is only within an amount $\hbar\mathbf{G}$. This is a general result for crystal momentum (sometimes called quasi-momentum), whether we are dealing with phonons, electrons, or magnons. Just as conservation of true momentum is a consequence of translational invariance, the limited conservation of crystal momentum expressed in Eq. (8.7) is a consequence of the limited translational invariance of a Bravais lattice, discussed in Chapter 6. It is usual to make a distinction between processes for which $\mathbf{G} = 0$, called **normal processes**, and those for which $\mathbf{G} \neq 0$, called **Umklapp processes**. An Umklapp process can be thought of as one in which a phonon is Bragg reflected simultaneously with absorbing or emitting another phonon. Both types of process are illustrated in a one dimensional example in Fig. 8.1. Transverse as well as longitudinal phonons are shown, because usually a three-phonon process cannot satisfy simultaneously Eqs. (8.5) and (8.7) unless both branches are used. If we have a single $\omega(\mathbf{q})$ curve which is everywhere concave downwards, two phonons cannot combine to give enough momentum for a phonon of their combined energy.

There is however, an essential distinction between an Umklapp process and Bragg reflection of neutrons or x-rays with phonon emission or absorption. In neutron or x-ray scattering the particles are incident on the crystal from outside and a momentum change $\hbar(\mathbf{k} - \mathbf{k}')$ is transmitted to the crystal as a whole, including its phonons. But in any phonon process, normal or Umklapp, nothing enters or leaves the crystal and the total momentum is constant. It is in some respects convenient to think of momentum $\hbar\mathbf{q}$ as associated with a phonon, but this is really an artificial division of the total momentum of the crystal, because the Bragg reflected phonon of momentum $\hbar(\mathbf{q} + \mathbf{G})$ is by Bloch's theorem, *identical* with the original phonon. Consequently the crystal momentum of a phonon is not uniquely defined; the only physically definable quantity is the momentum carried along by a phonon wavepacket, and if the vibrations are purely harmonic this is zero. Crystal momentum is really a quite distinct concept from true momentum. It is more correct to regard Eq. (8.7) as a geometrical interference condition on wavevectors, rather than a momentum conservation law.

8.2   THERMAL CONDUCTION

8.2.1   Kinetic theory

In the elementary kinetic theory of gases it is shown, by assuming a constant average velocity $\bar{v}$ for the molecules, that the steady state flux of a property P in the z-direction is given by

$$\text{flux} = \tfrac{1}{3}l\bar{v}\frac{\mathrm{d}P}{\mathrm{d}z}, \tag{8.8}$$

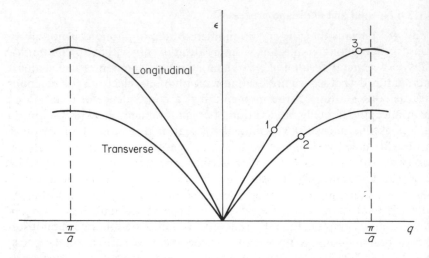

(*a*) A normal process; phonons 1 and 2 coalesce to give a phonon 3.

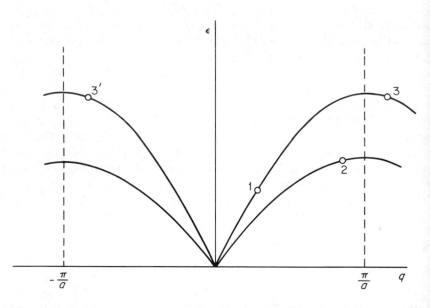

(*b*) An Umklapp process; phonons 1 and 2 give a phonon 3 outside the reduced Brillouin zone, which is equivalent to a phonon 3′.

Fig. 8.1.

where l is the mean free path, and the factor $\frac{1}{3}$ arises from an angular average (see, for example, Flowers and Mendoza[3]). In the simplest case where P is the number density of particles we obtain the diffusion coefficient $D = \frac{1}{3}l\bar{v}$. If P is the energy density E then the flux W is the heat flow per unit area and we have

$$W = \frac{1}{3}l\bar{v}\frac{\mathrm{d}E}{\mathrm{d}z}$$

$$= \frac{1}{3}l\bar{v}\frac{\mathrm{d}E}{\mathrm{d}T}\frac{\mathrm{d}T}{\mathrm{d}z}. \tag{8.9}$$

Now $\mathrm{d}E/\mathrm{d}T = C_v$, the specific heat *per unit volume*, so that the thermal conductivity K is given by

$$K = \frac{1}{3}l\bar{v}C_v, \tag{8.10}$$

a result we have already used to discuss electronic thermal conductivity in section 4.2.2.

We have not used particle conservation anywhere in the derivation of Eq. (8.10), so it may be applied to a phonon gas just as to a real gas; in fact the application to a phonon gas is particularly straightforward, because $\bar{v}$ really is almost a constant (the velocity of sound v_s) for phonons not too near the zone boundary. For a real gas of atoms the application of Eq. (8.10) is not so straightforward, for several reasons. First, $\bar{v}$ depends on temperature and so should really be included in the derivative in Eq. (8.8); second, the conservation of atoms imposes the constraint that there is no net particle flux; and finally, hydrostatic equilibrium requires that the pressure is uniform. A satisfactory kinetic theory of heat conduction in a real gas is therefore quite hard, and the correct numerical factor in Eq. (8.10) turns out to be rather different from $\frac{1}{3}$ in this case. But for phonons the simple theory is quite good.

The essential differences between the processes of heat conduction in a phonon gas and a real gas are illustrated in Fig. 8.2. For phonons the speed is approximately constant, but both the number density at the energy density are greater at the hot end; heat flow is primarily by phonon flow with phonons being created at the hot end and destroyed at the cold end. For a real gas (Fig. 8.2(b)) there is, in contrast, *no* flow of particles. The average velocity and the kinetic energy per particle are greater at the hot end, but the number density is greater at the cold end and the energy density is in fact *uniform* (because the pressure is uniform.) Heat flow is solely by transfer of kinetic energy from one particle to another in collisions—a rather minor effect in the phonon case.

(*a*) In a phonon gas there is a net phonon flow,
and there are more phonons at the hot end.

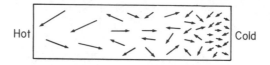

(*b*) In a real gas there is no flow of atoms;
the atoms are fewer but faster at the hot end.

Fig. 8.2. Heat conduction in a phonon gas and a real gas.

8.2.2   Phonon equilibrium and phonon flow

A heat current due to phonon flow is rather analogous to an electric current due to electron flow. Correspondingly, thermal resistance in an insulator is produced by similar types of collision to electrical resistance in a metal, namely those tending to reverse the momentum (or group velocity) of the flowing particles (Fig. 4.7(*b*)). For phonons, it is the Umklapp processes that conspicuously have this property (Fig. 8.1), and these are therefore important in producing thermal resistance. However, the rigid distinction between normal and Umklapp processes is a somewhat artificial one, since phonons just inside and just outside the Brillouin zone boundary are really very similar; both have a small group velocity and contribute little to the energy flow. The energy flow is proportional to a sum over all phonons of $\omega \mathrm{d}\omega/\mathrm{d}q$, and this is in fact reduced by *both* the three-phonon processes shown in Fig. 8.1. It is however true that if there were no Umklapp processes the energy flow would be *statistically* steady even in the absence of a temperature gradient; consequently the thermal conductivity would be infinite!

To see this we note that in the absence of Umklapp the total phonon crystal momentum

$$\mathbf{P} = \Sigma n(\mathbf{q})\hbar\mathbf{q} \tag{8.11}$$

is conserved in collisions. Thus, if we start with a heat flow, corresponding to non-zero $\mathbf{P}$, the final state of equilibrium produced by collisions will be one with this value of $\mathbf{P}$. An equilibrium distribution function for constant total crystal momentum may be evaluated analogously to one subject to the constraint of constant total number of particles. In Appendix D we show that this constraint leads to an equilibrium distribution function

$$n(\mathbf{q}) = \frac{1}{e^{(\beta\varepsilon(\mathbf{q}) - \hbar\mathbf{q}\cdot\boldsymbol{\gamma})} - 1},\qquad(8.12)$$

where $\beta = 1/k_B T$; the magnitude of the constant vector $\boldsymbol{\gamma}$ is to be chosen to make the total phonon crystal momentum right. This distribution function means that energy is effectively measured from a baseline $\hbar k_B T \boldsymbol{\gamma} \cdot \mathbf{q}$, as shown in Fig. 8.3; consequently there are excess phonons of positive group velocity and a net energy flow to the right. Since this is an *equilibrium* distribution in the absence of Umklapp, we have a steady heat flow with zero temperature gradient.

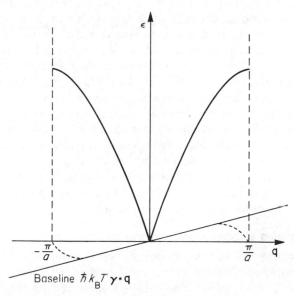

Fig. 8.3. $\varepsilon(\mathbf{q})$ for phonons, showing the energy baseline for a distribution of constant total phonon momentum.

The importance of Umklapp is that Eq. (8.12) cannot be a true equilibrium distribution function in a solid, by Bloch's theorem.* For opposite sides of the Brillouin zone ($\mathbf{q} = \pm\pi/a$) are equivalent, and hence must have the *same* $n(\mathbf{q})$; the only true equilibrium is therefore $\boldsymbol{\gamma} = 0$. Any actual heat flow must correspond not to the distribution function of Eq. (8.12), but to energy

* In liquid helium II Bloch's theorem does not apply, Umklapp processes do not exist, and Eq. (8.12) is a true equilibrium distribution function; this leads to the superfluid properties of helium II (Lane[20]).

measured from a baseline as indicated by the dotted curves in Fig. 8.3. *This differs from Eq.* (8.12) *only for the high energy phonons near the zone boundary.* If, at low temperatures, very few such phonons are present, there is almost no mechanism for relaxing from the distribution function of Eq. (8.12) to true equilibrium, and we should expect very large thermal conductivity. Thus, it is not so much Umklapp processes as such that are important in producing thermal resistance as *any* collisions involving high energy phonons. *The thermal resistance is determined by the relaxation time for high energy phonons near the zone boundary*, because it is only via these phonons that the distribution function of Eq. (8.12) can relax to true equilibrium.

8.2.3 Conduction at high termperatures

At high temperatures, greater than the Debye temperature Θ_D, the specific heat is essentially classical so that the only temperature dependent term in Eq. (8.10) for the thermal conductivity is the mean free path l. According to Eq. (8.4) the amplitude of a scattered sound wave is proportional to the amplitude of the scattering wave, and consequently the scattered intensity is proportional to the intensity of the wave causing the scattering, which in turn is proportional to the number of phonons present in the mode under consideration. Thus, our wave-scattering approach gives results of the same form as we would obtain by applying kinetic theory to the phonon gas: the scattering probability is proportional to phonon number and the mean free path l is inversely proportional to phonon number. At high temperatures the average phonon energy is constant and of order $k_B\Theta_D$, so that the phonon number is proportional to T. Consequently, we expect a thermal conductivity inversely proportional to T. Fig. 8.4(*a*) shows that the experimental results do tend towards this behaviour at high temperatures.*

8.2.4 Conduction at intermediate temperatures

Fig. 8.4 also shows that at temperatures below about Θ_D the conductivity rises more steeply with falling temperature, despite the fact that the specific heat is falling in this region. This can be understood on the basis of our result of section 8.2.2 that thermal resistance is controlled by the relaxation of the high energy phonons near the zone boundary. The relevant phonons are not very sharply defined, but we expect their population to vary roughly as $\exp(-\Theta_D/bT)$, where b is a constant somewhat greater than 1. This exponential factor will dominate any low power of T in the thermal conductivity, such as a factor T^3 from the specific heat. We therefore expect the thermal conductivity to be dominated by the factor l in Eq. (8.10), and to vary like $\exp(\Theta_D/bT)$. Experimentally, the variation is of this form over a fair range of temperature; the empirical values of b are of the order of 2 to 3.

---

* Curiously, there are almost no published results at or above room temperature since the beginning of the century.

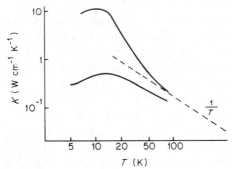

(a)   Thermal conductivity of a quartz crystal; the lower curve shows the effect of lattice defects in a neutron-irradiated sample.
[after R. Berman, *Proc. Roy. Soc. A*, **208**, 90 (1951)].

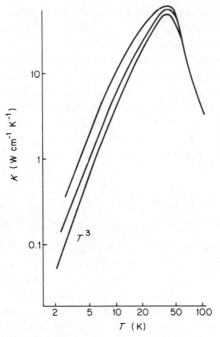

(b)   Thermal conductivity of artificial sapphire rods of different diameters.
[after R. Berman, E.L. Foster, and J.M. Zinan, *Proc. Roy. Soc. A*, **231**, 130 (1955)].

Fig. 8.4.

8.2.5   Conduction at low temperatures

Because of the exponentially decaying population of high energy phonons at low temperatures the mean free path rapidly becomes larger than the diameter of usual thermal conductivity specimens. The phonon free paths in a perfect crystal are then limited by collision with the specimen surface, and the flow of phonons becomes analogous to the flow of a gas in the Knudsen regime. The effective phonon free path to be used in Eq. (8.10) is then of the order of the specimen diameter; it may even be somewhat larger if the specimen surface is smooth enough for appreciable specular reflection of phonons to occur, since specular reflection does not contribute to thermal resistance. In these circumstances there is no true thermal conductivity, since the thermal conductance becomes proportional to the cube of the specimen diameter instead of the square; the variation in apparent conductivity with diameter is illustrated in Fig. 8.4(b). The only temperature dependence of the conductivity now comes from the specific heat, which obeys the Debye T^3 law (Eq. (2.43)) in this region.

We have so far assumed that we are dealing with a perfect single crystal, and have ignored phonon scattering by crystal imperfections, such as dislocations, grain boundaries, or impurities. At the very lowest temperatures the dominant phonon wavelength becomes so long that these imperfections are not effective scatterers, so the measured thermal conductance always has a T^3 variation at these temperatures. But the maximum conductivity between the T^3 region and the $\exp(\Theta_D/bT)$ region is largely controlled by imperfections. For an impure or polycrystalline specimen the maximum can be quite broad and low (Fig. 8.4(a)), whereas for a carefully prepared single crystal, as illustrated in Fig. 8.4(b), the maximum is quite sharp and the conductivity reaches a very high value, of the order of that of copper.

★   8.3   COLLECTIVE EFFECTS

8.3.1   Poiseuille flow

We have already emphasized that in a pure material thermal conduction is limited by the relaxation time or mean free path of the high energy phonons. It may however happen at low temperatures that most thermal phonons (energy $\sim k_B T$) have a short mean free path while the very few high energy phonons (energy $\sim k_B \Theta_D$) have a long mean free path. In these circumstances the thermal phonons rapidly come to equilibrium among themselves, with a distribution function of the form of Eq. (8.12), but the collective phonon flow to which this corresponds is only slowly destroyed by processes involving high energy phonons. As in section 8.2.5, the only effective restraint on thermal conduction is by collision of phonons with the specimen surface; but now phonons in the centre of the specimen are restrained only indirectly, by momentum conserving collisions with other phonons. In effect, conduction is restrained by viscosity of the phonon gas; we have a Poiseuille flow rather than a Knudsen flow, because viscosity is controlled by the mean free path of the thermal phonons, and this is small compared with the specimen diameter.

This effect is shown for a crystal of solid helium in Fig. 8.5. At the lowest temperatures all phonons have a mean free path larger than the sample

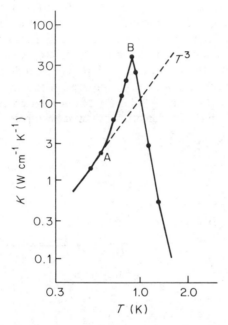

Fig. 8.5. Thermal conductivity of a single crystal of solid He, showing the Poiseuille flow effect.
{after L. P. Mezhov-Deglin, *Sov. Phys. JETP*, **22**, 47 (1966)}.

diamcter, and the conductance is proportional to T^3. Then, at A, the mean free path of thermal phonons becomes smaller than the tube diameter while that of high energy phonons remains long; the conductance rises *above* the T^3 law because of collective phonon flow in the centre of the sample, analagous to Poiseuille flow of gas in a tube. Finally, at B, the population of high energy phonons becomes large enough for their mean free path to become short, and the conductivity falls sharply.

The main reason that solid helium shows this effect well is that it is a highly anharmonic crystal, due to large amplitude zero point vibrations resulting from the small atomic mass and weak van der Waals binding. This anharmonicity produces a short mean free path for thermal phonons even at temperatures where no high energy phonons are excited.

8.3.2 Second sound

We have just seen that when the mean free path is short for thermal phonons but long for high energy phonons the phonon gas loses drift momentum only by viscosity, and exhibits flow like that of a gas in a tube. Under these conditions we would expect to be able to propagate density waves, analogous to sound, in the phonon gas. Just as sound can propagate in an ordinary gas if the wavelength is large compared with the mean free path, so that there is local thermal equilibrium, we expect phonon density waves to propagate if the wavelength is long compared with the mean free path of thermal phonons, and short compared with the mean free path of high energy phonons. Since a high density of phonons corresponds to a high temperature, such a phonon density wave would appear, macroscopically,

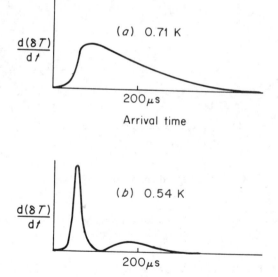

Fig. 8.6. Arrival of a temperature step in solid helium:
(a) in the Umklapp region,
(b) in the second sound region; the second pulse is a reflection that has travelled three times along the sample.
{after C. C. Ackermann, B. Bertman, H. A. Fairbank and R. Guyer, *Phys. Rev. Lett.*, **16**, 789 (1966)}.

as a temperature wave. In other words, in the temperature region between A and B in Fig. 8.5 we should expect heat to propagate according to a wave equation rather than a diffusion equation; such a temperature wave is known as second sound.* One consequence of a wave equation is that a pulse should propagate without change of form and with a definite velocity. Some results on the propagation of heat pulses in single crystals of solid helium are shown in Fig. 8.6; it can be seen that over a limited temperature range pulses do travel with relatively little distortion. A fuller discussion of these effects is given by Bertman and Sandiford.[21]

PROBLEMS 8

8.1 The thermal conductivity of an artificial sapphire crystal 3 mm in diameter reaches a sharp maximum at 30 K.
   Estimate roughly:
   (a) the maximum value of the thermal conductivity;
   (b) the thermal conductivity at liquid nitrogen temperature (80 K). Explain the reasoning behind your estimates.
   (For sapphire $\Theta_D = 1000$ K, speed of sound $= 10^4$ m s^{-1}, and for $T \ll \Theta_D$, $C_v = 10^{-1} T^3$ J m^{-3} K^{-1}.)

---

* The name was coined for a very similar phenomenon that occurs in superfluid liquid helium; see, for example, Lane.[20]

Real metals

9.1 WHY DO ELECTRONS BEHAVE INDEPENDENTLY?

Our discussion of metals in Chapter 4 depended on the assumption that electrons behave essentially independently. Bloch's theorem, as formulated in Chapter 6, is essentially a theorem about *single particle* wavefunctions;* consequently the whole idea of Brillouin zones and energy bands, which is essential to our distinguishing between metals and insulators in section 4.4, rests on our assumption of independent electrons. In this section we shall examine that assumption.

At first sight the idea of nearly free independent electrons is a very bad approximation. Consider the energies involved. For free electrons we have the Fermi energy of the free electron gas; interaction with the lattice can be represented by the potential of a positive ion at a radius ~ 1 Å; and electron–electron interaction can be represented by the mutual potential energy of two electrons ~ 1 Å apart. *All* these energies are of the order of a few eV, and therefore the neglect of any one of them seems ridiculous. Of these, the neglect of electron–electron interactions is the most serious, since our independent electron model is valid only if these interactions are small enough (in a sense yet to be defined) for the electrons to behave like an almost perfect gas.

* The consequences of lattice periodicity for a many-particle wavefunction are less straight-forward.

9.1.1   Plasma oscillations

A serious difficulty in trying to calculate the effect of electron–electron interactions is the long range of the Coulomb potential, which means that electrons do not interact only with their immediate neighbours. However, the long range nature of the force manifests itself on a macroscopic scale by the electric fields that arise if a crystal is not locally electrically neutral. This fortunately means that the otherwise intractable long range part of the force can be taken into account in a purely macroscopic way. To understand this let us consider a free-electron model in which the positive ions are replaced by a uniform distribution of positive charge of density numerically equal to the constant mean electron charge density ρ_0; if the actual electron charge density is $\rho(\mathbf{r}, t)$ we then have

$$\operatorname{div} \mathbf{E} = (\rho - \rho_0)/\varepsilon_0. \tag{9.1}$$

We have effectively assumed that the positive charge is fixed; this is a very good approximation because ions are so much heavier than electrons. As in section 3.4 (Eq. (3.21)), we shall assume that the electric field given by Eq. (9.1) accelerates the electrons according to

$$m^*\frac{d\mathbf{v}}{dt} = e\mathbf{E}, \tag{9.2}$$

so as to try and even out the charge distribution. In Eq. (9.2) we have ignored the effect of electron collisions, which lead in a steady state to Ohm's law, because it will turn out that we are concerned with motion on a time scale very much shorter than a typical collision interval. To solve these equations we need to relate velocity to charge density through the law of conservation of charge

$$\frac{\partial \rho}{\partial t} + \operatorname{div}(\rho\mathbf{v}) = 0.$$

If $(\rho - \rho_0) \ll \rho_0$ this may be approximated by

$$\frac{\partial \rho}{\partial t} + \rho_0 \operatorname{div} \mathbf{v} = 0, \tag{9.3}$$

which is linear in the variable quantities $(\rho - \rho_0)$ and $\mathbf{v}$. Differentiating Eq. (9.3) with respect to time and substituting in succession from Eqs. (9.2) and (9.1) we find

$$\frac{\partial^2 \rho}{\partial t^2} = -\rho_0\frac{\partial}{\partial t}\operatorname{div}\mathbf{v} = -\rho_0\operatorname{div}\frac{\partial \mathbf{v}}{\partial t}$$

$$= -\frac{\rho_0 e}{m^*}\operatorname{div}\mathbf{E} = -\frac{\rho_0 e}{m^*\varepsilon_0}(\rho - \rho_0),$$

or

$$\frac{\partial^2 (\rho - \rho_0)}{\partial t^2} + \omega_p^2 (\rho - \rho_0) = 0, \tag{9.4}$$

where the plasma frequency ω_p is given by

$$\omega_p^2 = \frac{ne^2}{m^* \varepsilon_0}, \tag{9.5}$$

in which n is the average number density of electrons, so that $\rho_0 = ne$.

Eq. (9.4) is a harmonic oscillator equation with angular frequency ω_p: it shows that any disturbance from uniform charge density, whatever its spatial form, oscillates at this frequency. Because of its long range nature, the electrostatic interaction of electrons manifests itself in this collective motion. From Eq. (9.5)

$$\omega_p^2 = \frac{ne^2}{m^* \varepsilon_0} \approx \frac{10^{29} \times 10^{-38}}{10^{-30} \times 10^{-11}},$$

$$\omega_p \approx 10^{16} \, \text{s}^{-1},$$

so that $\hbar \omega_p \approx 10^{-18}$ joule $\approx 10 \, \text{eV}$. Because of this high quantum energy the plasma oscillations of electrons in a metal are not normally excited, and this is the basic reason that the Coulomb interaction between metallic electrons is ineffective. However, quanta of plasma oscillations, called **plasmons,** can be excited if a beam of fast electrons is passed through a thin metal foil. The electrons are found to emerge with discrete energy losses corresponding to the excitation of one or more plasmons (Fig. 9.1). Note that since a typical collision interval τ is of order 10^{-12} s, $\omega_p \tau \gg 1$; this justifies our neglect of collisions in Eq. (9.2). The condition $\omega_p \tau \gg 1$ means that plasma oscillations are quite unlike sound waves, because they are a collective motion occurring in the absence of collisions.

Our replacement of electron interactions by plasma oscillations depends on two things. First, a continuum treatment of the charge distribution; this is not serious—more realistic models just give a plasma frequency that varies somewhat with the wavenumber of the disturbance. Second, we have assumed that the only velocity of the electrons is their drift velocity **v**. In fact they also have random velocities of order v_F, so that in one cycle of plasma oscillation an electron moves a distance of order (v_F/ω_p). This means that the electrons move too fast for the plasma modes to be effective in screening out the Coulomb interaction over this sort of distance. Consequently only the long range part of the Coulomb interaction is accounted for by the plasma modes: there remains a short range interaction effective over a distance of order

$$v_F/\omega_p \approx \frac{10^6}{10^{16}} \, \text{m} = 1 \, \text{Å}.$$

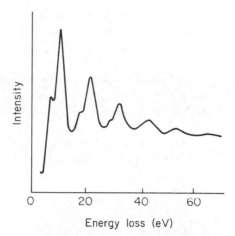

Fig. 9.1. Energy loss of 2020 eV electrons
after scattering through 90° by a Mg film.
After C. J. Powell and J. B. Swann, *Phys.
Rev.*, **116**, 81 (1961)}.

The peaks indicate a tendency for the
energy loss to be a multiple of $\hbar\omega_p$. The
minor peaks are a surface effect (see
Kittel[5]).

This distance, beyond which a plasma of mobile electrons can screen out
Coulomb interactions, is important to an understanding of how such mobile
electron states arise in the first place, as a collection of atoms is brought
together to form a metal. We saw in section 4.1.2, in the 'model system'
of impurity atoms in a semiconductor, that mobile states seem to appear
quite suddenly at a critical electron concentration. This sudden transition
may be understood as follows. If such mobile states *are* present, the electrons
will be screened from the attractive potential of the positive ions outside a
distance of order v_F/ω_p; and if this distance is less than about a Bohr radius
a_0 the effective potential is so weakened that a bound state cannot be formed,
justifying our original assumption that mobile states are present.

Now we have, from section 4.2,

$$v_F \propto k_F \propto n^{1/3},$$

and from Eq. (9.5)

$$\omega_p \propto n^{1/2},$$

so that v_F/ω_p decreases with increasing mobile electron concentration as
$n^{-1/6}$. Therefore, for large enough n, the plasma can screen the attractive
potential of the ions so as to prevent the formation of localized bound states.
Note particularly that this is a cooperative process; the more electrons are

free, the more effectively the ionic attraction is screened, and the easier it is for the remaining electrons to become free. It is because of this cooperative process that the transition to the metallic state is thought to be a sudden one, called the **Mott transition.**

The plasma oscillation we have just considered is very closely analogous to the longitudinal optical vibrations that we considered in section 2.4; the major difference is that in the present case the *only* restoring forces are electrostatic.

We may also, as in section 2.4, consider a transverse disturbance, for which $\rho = \rho_0$, so that Eqs. (9.1) and (9.3) are trivial. The acceleration equation, Eq. (9.2), now has to be solved in conjunction with Maxwell's equations in free space in the form used in section 2.4, namely

$$-\text{curl curl } \mathbf{E} = \frac{1}{c^2}\frac{\partial^2 \mathbf{E}}{\partial t^2} + \mu_0 \frac{\partial \mathbf{j}}{\partial t}.$$

With $\mathbf{j} = ne\mathbf{v}$ and Eq. (9.2) this gives

$$-\text{curl curl } \mathbf{E} = \frac{1}{c^2}\frac{\partial^2 \mathbf{E}}{\partial t^2} + \mu_0 \frac{ne^2}{m^*}\mathbf{E}$$

$$= \frac{1}{c^2}\left(\frac{\partial^2 \mathbf{E}}{\partial t^2} + \omega_p^2 \mathbf{E}\right). \tag{9.6}$$

For a longitudinal vibration curl $\mathbf{E} = 0$, so that we recover essentially Eq. (9.4); for a transverse vibration div $\mathbf{E} = 0$ so that Eq. (9.6) becomes*

$$\nabla^2 \mathbf{E} = \frac{1}{c^2}\left(\frac{\partial^2 \mathbf{E}}{\partial t^2} + \omega_p^2 \mathbf{E}\right). \tag{9.7}$$

This equation has wavelike solutions $e^{i(\mathbf{k}\cdot\mathbf{r} - \omega t)}$ with

$$k^2 = \frac{\omega^2 - \omega_p^2}{c^2}, \tag{9.8}$$

so that, for $\omega < \omega_p$, k is imaginary and the wave is an evanescent one; light is totally externally reflected, as in Reststrahlen. The difference from the situation discussed in section 2.4 is that there are now no elastic restoring forces so that the acoustic branch has disappeared; *all* frequencies below ω_p are totally reflected—hence the lustre of metals. Above ω_p, k is real and the crystal should be transparent;† thin films of metals are indeed transparent in the ultraviolet.

---

* By use of the vector operator identity curl curl = grad div $- \nabla^2$.

† Remember, however, that we have neglected electrical resistance, which will damp the wave.

9.1.2   The exclusion principle and scattering

We saw above that after the plasma oscillations have been taken into account we are left with a residual screened Coulomb interaction between electrons of range ~ 1 Å in a metal; we might therefore expect an electron–electron collision diameter of this order. For a collision cross section A elementary kinetic theory gives a mean free path

$$l = \frac{1}{nA} = v_F \tau,$$

so that the relaxation time τ is

$$\tau = \frac{1}{nAv_F}. \tag{9.9}$$

From this we may calculate the broadening of single particle energy levels due to the uncertainty principle as

$$\Delta \varepsilon = \frac{\hbar}{\tau} = nA\hbar v_F \approx \frac{nA}{k_F}\varepsilon_F. \tag{9.10}$$

With $A \approx 1$ Å^2, (nA/k_F) approaches unity, so that the level broadening is of order ε_F, even with the screened interaction, and the whole independent particle picture is apparently nonsense!!

What saves the situation is the exclusion principle. For a collision to occur the initial states of both particles must be occupied and the final states of both particles must be vacant. This can occur with conservation of energy and momentum only if *both* colliding particles have an energy within about $k_B T$ of the Fermi energy, since this is the only region where both occupied *and* vacant levels may be found. Without the exclusion principle the total collision rate per unit volume is

$$w = \frac{n}{\tau} = n^2 Av_F.$$

When the exclusion principle is allowed for, each n in this expression is multiplied by a factor of order (T/T_F) by the above energy restriction, so that the actual collision rate is of order

$$w = n^2 \cdot Av_F(T/T_F)^2,$$

and the relaxation time of a single particle is

$$\tau = n/w = \frac{(T_F/T)^2}{nA_F}. \tag{9.11}$$

From Eq. (9.10) the correct level broadening now becomes

$$\Delta\varepsilon = \frac{\hbar}{\tau} \approx \frac{nA}{k_F}\varepsilon_F \left(\frac{T}{T_F}\right)^2$$

$$\approx k_B T \left(\frac{T}{T_F}\right),$$

with $nA \sim k_F$ as before. The level broadening is thus small compared with $k_B T$, the thermal broadening of the Fermi function, and all is well. At room temperature the level broadening therefore corresponds typically to a temperature ~ 3 K ($k_B T \approx 5 \times 10^{-22}$ joules), so

$$\tau = \frac{\hbar}{\Delta\varepsilon} \sim 2 \times \frac{10^{-34}}{10^{-23}} = 2 \times 10^{-12} \text{ s}.$$

This is an order of magnitude longer than relaxation times deduced from electrical resistivities in section 4.1.1 (even more so at lower temperatures); consequently, electron–electron collisions are indeed negligible.

9.1.3   The exclusion principle and core wavefunctions

For many metals, especially the alkalis, the $\varepsilon(\mathbf{k})$ curve is only slightly altered from that for free electrons, with relatively small energy gaps at the zone boundary, despite the strong attractive potential of the positive ions. To understand this situation we again invoke the exclusion principle; the conduction electron wavefunctions must be 'different' from the occupied electron states of the ion core. In mathematical terms the conduction electron states must be *orthogonal* to the occupied states of the ion core. Thus for sodium the conduction band states must be orthogonal to 1s, 2s and 2p states; this means in practice that they must resemble 3s states quite closely near the ions. Roughly, the requirement of orthogonality puts 'wiggles' in a wavefunction near the ion-cores, as is shown for a one dimensional example in Fig. 9.2. The wavefunction of Fig. 9.2(a) is like a free-electron wave except near the ions, where it oscillates rapidly to make it orthogonal to the three lowest bound states of the ionic potential, shown in Fig. 9.2(b), which we suppose already occupied. These oscillations near the ions give a contribution to the kinetic energy which largely cancels the attractive potential energy, so that the total *energy* is relatively little affected by the potential, although, as we have seen in Fig. 9.2(a), the wavefunction is drastically different from that of a free electron.

The essentials of the effect we have just discussed can be seen already in the free atom. The potential of a Na^+ ion is that of a single positive charge at large distances, but that of a greater charge at short distances because the nuclear charge is not fully screened by extranuclear electrons. We should

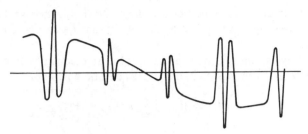

(*a*)   Real part of an electron wavefunction in a periodic potential.

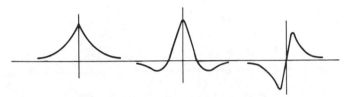

(*b*)   Atomic wavefunctions to which (*a*) must be orthogonal at each lattice site.

Fig. 9.2.

therefore expect, at first sight, that an electron would be more strongly bound by Na^+ than H^+. Yet the ionization potential of hydrogen is 13.6 V and that of sodium 5.1 V. The reason for this is that orthogonality requires the electron bound to Na^+ to have $3s$ character, and the consequent kinetic energy largely cancels the attractive potential. Thus, the ionization potential of the free atom is a better guide to the effective periodic potential in the solid than the Coulomb potential of the free ion.

9.2   CRYSTAL MOMENTUM AND EFFECTIVE MASS

At the beginning of this chapter and in section 3.4 we wrote the acceleration equation for the mean drift velocity $\mathbf{v}$ of electrons in a conduction band as

$$m^* \frac{d\mathbf{v}}{dt} = e\mathbf{E}. \tag{9.2}$$

With $\varepsilon = \hbar^2 k^2 / 2m^*$ for the mobile electrons, this is equivalent to assuming for each individual electron an acceleration equation

$$\hbar \frac{d\mathbf{k}}{dt} = e\mathbf{E}, \tag{9.12}$$

since the electron group velocities are $\hbar k/m^*$. In this section we examine more closely the assumption implicit in Eq. (9.12) that $\hbar k$ can be considered as a momentum.

We start by evaluating the momentum for an electron with a Bloch wavefunction which, from Eqs. (6.2) and (6.13) can be written as

$$\psi(r) = V^{-1/2} \sum_G a_G(\mathbf{k}) e^{i(\mathbf{k}+\mathbf{G})\cdot\mathbf{r}}. \tag{9.13}$$

This wavefunction is not a momentum eigenfunction, but we can evaluate the expectation value of the electron momentum as

$$
\begin{aligned}
\langle \mathbf{p}_{el} \rangle &= \int dV\, \psi^*(\mathbf{r})(-i\hbar\nabla)\psi(\mathbf{r}) \\
&= \frac{1}{V} \int dV \sum_G \sum_{G'} a_G^* e^{-i(\mathbf{k}+\mathbf{G})\cdot\mathbf{r}} \hbar(\mathbf{k}+\mathbf{G}') a_{G'} e^{i(\mathbf{k}+\mathbf{G}')\cdot\mathbf{r}} \\
&= \frac{1}{V} \sum_G \sum_{G'} a_G^* a_{G'} \hbar(\mathbf{k}+\mathbf{G}') \int dV\, e^{i(\mathbf{G}'-\mathbf{G})\cdot\mathbf{r}} \\
&= \sum_G \sum_{G'} a_G^* a_{G'} \hbar(\mathbf{k}+\mathbf{G}') \delta_{GG'} \\
&= \sum_G |a_G|^2 \hbar(\mathbf{k}+\mathbf{G}) \\
&= \hbar\left(\mathbf{k} + \sum_G \mathbf{G}|a_G|^2\right),
\end{aligned} \tag{9.14}
$$

where $\delta_{GG'} = 1$ if $\mathbf{G} = \mathbf{G}'$ and is zero otherwise, and the last step follows from the normalization condition $\sum_G |a_G|^2 = 1$. The result (9.14) is just what we should expect from the fact that the plane wave $e^{i(\mathbf{k}+\mathbf{G})\cdot\mathbf{r}}$ is a momentum eigenfunction of eigenvalue $\hbar(\mathbf{k}+\mathbf{G})$, and $|a_G|^2$ is the probability of the electron being found in this momentum eigenstate. If the electron is now accelerated so that its wavevector changes to $\mathbf{k}+\delta\mathbf{k}$ the expection value of its momentum changes by $\delta\mathbf{p}_{el}$, given by

$$\delta\mathbf{p}_{el} = \hbar\left(\delta\mathbf{k} + \sum_G \mathbf{G}\delta\mathbf{k} \cdot \frac{\partial}{\partial \mathbf{k}} |a_G(\mathbf{k})|^2\right). \tag{9.15}$$

A physical interpretation of the second term in Eq. (9.15) may be made by considering what happens when a free electron, incident on the crystal from outside, is Bragg reflected. The electron state changes from $e^{i\mathbf{k}\cdot\mathbf{r}}$ to $e^{i(\mathbf{k}+\mathbf{G})\cdot\mathbf{r}}$ and the lattice receives a recoil momentum

$$\delta\mathbf{p}_{latt} = -\hbar\mathbf{G}.$$

A more general process in which the coefficients $a_G(\mathbf{k})$ in Eq. (9.13) change by amounts other than 1 can be regarded as a series of partial Bragg reflections. The corresponding recoil momentum is then

$$\delta\mathbf{p}_{latt} = -\hbar \sum_G \mathbf{G}\delta(|a_G(\mathbf{k})|^2) \qquad (9.16)$$

which is just the negative of the second term in Eq. (9.15), so that the total change of momentum of the crystal is given by

$$\delta\mathbf{p}_{tot} = \delta\mathbf{p}_{el} + \delta\mathbf{p}_{latt} = \hbar\,\delta\mathbf{k}. \qquad (9.17)$$

This is the essential justification of Eq. (9.12); because $\hbar\,\delta\mathbf{k}$ is the change in the total momentum of the system, it is correct to equate the applied force to $\hbar\,d\mathbf{k}/dt$. The quantity $\hbar\mathbf{k}$ is called the **crystal momentum** of the electron; from Eq. (9.14) it is not the true momentum of the electron.

Having justified the equation of the motion Eq. (9.12) we now wish to define effective mass m^* so that Eq. (9.2) is valid for an electron with a general $\varepsilon(\mathbf{k})$ relation, not necessarily parabolic. We therefore define

$$m^*\delta\mathbf{v}_G = \hbar\,\delta\mathbf{k}, \qquad (9.18)$$

where $\mathbf{v}_G$ is the group velocity of an individual electron wavepacket, the mean value of which over all electrons gives the drift velocity previously denoted by $\mathbf{v}$. Since the group velocity is given by

$$\mathbf{v}_G = \frac{\partial\omega}{\partial\mathbf{k}} = \frac{1}{\hbar}\frac{\partial\varepsilon}{\partial\mathbf{k}}, \qquad (9.19)$$

Eq. (9.18) now gives the effective mass as

$$\frac{1}{m^*} = \frac{1}{\hbar^2}\frac{\partial^2\varepsilon}{\partial\mathbf{k}^2}. \qquad (9.20)$$

m^* is thus determined by the local curvature of $\varepsilon(\mathbf{k})$, and at the top of a band of allowed energy levels m^* is negative (Fig. 9.3(a)). Since, according to Eq. (9.2), the response of the electron to an applied field is determined by the ratio e/m^*, an electron on this part of the $\varepsilon(\mathbf{k})$ curve behaves in the same way as a particle with positive charge and positive mass when an electric field is applied. However, the statistical response of a single such electron to collisions differs from that of a particle of positive charge and positive mass. The statistical effect of collisions is that of a force opposed to the velocity, and this will accelerate a particle of negative mass. We can understand this effect by reference to Fig. 9.3(a). As a single electron near the top of a band collides it accelerates because of its negative effective mass and loses energy; this energy loss continues until it reaches a part of the $\varepsilon(k)$ curve occupied by other electrons and attains thermal equilibrium.

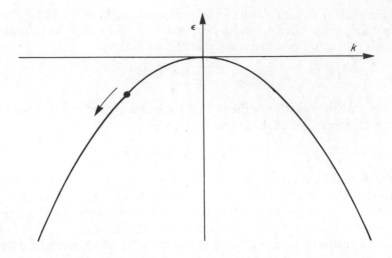

(a)   As an electron near the top of a band loses energy it accelerates.

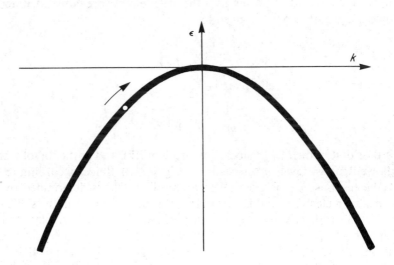

(b)   As many electrons near the top of a band lose energy the vacancy or hole rises to the top of the band.

Fig. 9.3.

Consider now a single *unoccupied* state near the top of a band, as in Fig. 9.3(*b*). As before, all the electrons in occupied states respond to an applied field according to Eqs. (9.2) and (9.12), so the vacant state also moves steadily in **k**-space according to Eq. (9.12). A single vacant state thus responds to an applied field just like a single occupied state—as a particle of positive charge and positive mass. But the statistical response to collisions is now altered. The electrons in occupied states tend to lose energy as before, and this causes the unoccupied state to 'float' to the top of the band; it is therefore retarded by collisions, like a particle of positive mass, not accelerated. We therefore conclude that a single unoccupied state near the top of a band behaves in all respects like a particle of positive charge and positive mass. Such an unoccupied state is called a hole; we introduced the idea of holes from a different point of view in Chapter 3.

In general δv may not be parallel to $\delta \mathbf{k}$, so we must allow for this by writing $(1/m^*)$ in tensor form and replacing Eq. (9.18) by

$$\delta v_{Gi} = \hbar \sum_j \left(\frac{1}{m^*} \right)_{ij} \delta k_j, \tag{9.21}$$

so that Eq. (9.20) becomes

$$\left(\frac{1}{m^*} \right)_{ij} = \hbar^{-2} \frac{\partial^2 \varepsilon}{\partial k_i \, \partial k_j}. \tag{9.22}$$

The effective mass tensor may also be expressed in terms of the electron wavefunction by use of Eq. (9.15), which we rewrite in component form as

$$\delta \langle p_{\mathrm{el}} \rangle_i = \hbar \sum_j \left(\delta_{ij} \delta k_j + \sum_{\mathbf{G}} G_i \, \delta k_j \frac{\partial}{\partial k_j} |a_{\mathbf{G}}(\mathbf{k})|^2 \right), \tag{9.23}$$

where $\delta_{ij} = 1$ if $i = j$, 0 if $i \neq j$.

We now make use of the fact that expectation values obey classical equations of motion, so that

$$\mathbf{v}_G = \frac{d \langle \mathbf{r} \rangle}{dt} = \frac{1}{m} \langle \mathbf{p}_{\mathrm{el}} \rangle \tag{9.24}$$

which with Eq. (9.21) gives

$$\frac{1}{m} \delta \langle p_{\mathrm{el}} \rangle_i = \delta v_{Gi} = \hbar \sum_j \left(\frac{1}{m^*} \right)_{ij} \delta k_j. \tag{9.25}$$

Elimination of $\delta \langle \mathbf{p}_{\mathrm{el}} \rangle$ between Eqs. (9.23) and (9.25), and comparison of coefficients of δk_j, now gives

$$\left(\frac{1}{m^*} \right)_{ij} = \frac{1}{m} \left(\delta_{ij} + \sum_{\mathbf{G}} G_i \frac{\partial}{\partial k_j} |a_{\mathbf{G}}(\mathbf{k})|^2 \right). \tag{9.26}$$

The meaning of a tensorial effective mass is just that the electron has more inertia for acceleration in some directions than others. If an electric field is applied in the directions of greatest or least inertia, acceleration is parallel to the field; but for a general field direction acceleration is preferentially along the direction of least inertia, and this is expressed by replacing Eq. (9.2) by

$$\frac{dv_i}{dt} = e \sum_j \left(\frac{1}{m^*}\right)_{ij} E_j. \tag{9.27}$$

9.3 TRANSPORT PROPERTIES

9.3.1 Electron–phonon scattering

We saw in section 4.1.1 that the electrical resistivity of a metal could empirically be divided into a temperature-independent part due to scattering of electrons by impurities and defects in the crystal structure, and a temperature dependent part which we attributed to scattering by lattice vibrations. This latter contribution we now consider in more detail.

Consider first the scattering of an electron incident on the crystal from outside. If thermal vibrations are present the atoms will be displaced from their equilibrium lattice positions R_i by an amount which is the sum of contributions from all the lattice waves present. To simplify the calculation, consider only one particular lattice wave of wavenumber q and angular frequency ω; the actual atomic positions can then be written as

$$r_i = R_i + \alpha \cos(q \cdot R_i - \omega t), \tag{9.28}$$

where α gives the amplitude and polarization direction of the lattice wave. From Eq. (6.12) the time dependent phasor sum giving the total scattered amplitude from the crystal is proportional to

$$A = \sum_i \exp i(K \cdot r_i - \Omega t), \tag{9.29}$$

where $K = (k_1 - k_2)$ is the change of electron wavevector on scattering and $\hbar\Omega$ is the incident electron energy. Substitution of Eq. (9.28) into Eq. (9.29) gives

$$A = \sum_i [1 + iK \cdot \alpha \cos(q \cdot R_i - \omega t) + \cdots] \times \exp i(K \cdot R_i - \Omega t) \tag{9.30}$$

where the term in square brackets is obtained by expanding an exponential; this is justified if the atomic displacements are small. Eq. (9.30) can be written in the form

$$A = \sum_i \exp i(\mathbf{K} \cdot \mathbf{R}_i - \Omega t)$$

$$+ \tfrac{1}{2} i \mathbf{K} \cdot \alpha \sum_i \exp i[(\mathbf{K} + \mathbf{q}) \cdot \mathbf{R}_i - (\Omega + \omega)t]$$

$$+ \tfrac{1}{2} i \mathbf{K} \cdot \alpha \sum_i \exp i[(\mathbf{K} - \mathbf{q}) \cdot \mathbf{R}_i - (\Omega - \omega)t]$$

$$+ \text{ terms of order } \alpha^2.$$

(9.31)

From our arguments in section 6.2 the first term gives a sharp maximum for $\mathbf{K} = \mathbf{G}$ (i.e. $\mathbf{k}_1 = \mathbf{k}_2 + \mathbf{G}$), where $\mathbf{G}$ is any reciprocal lattice vector; the resultant phasor oscillates at angular frequency Ω. This is the elastic Bragg scattering observed in electron diffraction experiments. It couples wavevectors $\mathbf{k}$ and $\mathbf{k} + \mathbf{G}$, and thus alters the stationary states for electrons in a crystal from plane waves to Bloch wavefunctions (Eq. (9.13)). The second term, however, gives a sharp maximum for $\mathbf{K} + \mathbf{q} = \mathbf{G}$, or

$$\mathbf{k}_2 = \mathbf{k}_1 + \mathbf{q} - \mathbf{G}, \tag{9.32}$$

and this therefore represents a scattering that couples together different Bloch functions. While the first term in Eq. (9.31) just describes the formation of Bloch wavefunctions as eigenstates of the non-vibrating crystal, the second term gives scattering from one Bloch state to another by lattice vibrations. Note that the scattered wave has a frequency Ω' given by

$$\Omega' = \Omega + \omega. \tag{9.33}$$

When Eq. (9.33) is multiplied by $\hbar$ it becomes a statement of the law of conservation of energy,

$$\varepsilon(\mathbf{k}_2) = \varepsilon(\mathbf{k}_1) + \hbar\omega(\mathbf{q}), \tag{9.34}$$

for a process in which an electron of wavevector $\mathbf{k}_1$ absorbs a phonon of wavevector $\mathbf{q}$ to become an electron of wavevector $\mathbf{k}_2$. This is the particle interpretation of the second term in Eq. (9.13); the third term can be similarly interpreted as scattering in which a phonon is emitted, and the higher order terms represent processes involving more than one phonon. Eq. (9.32), multiplied by $\hbar$, gives the conservation of crystal momentum, within $\hbar\mathbf{G}$, in the scattering process. From our discussion of electron momentum in section 9.2 and the argument leading to Eq. (9.32) it is clear that, despite its form, Eq. (9.32) is really a geometrical interference condition on wavevectors rather than a momentum conservation law. The results we have just obtained, and the calculations leading to them, are very similar to our previous results for the scattering of neutrons (section 7.3.1) and phonons (section 8.1.1) with emission or absorption of phonons. In fact, the laws of conservation of energy and of crystal momentum (within $\hbar G$) are true for all collision processes involving elementary excitations in solids.

When we considered phonon–phonon scattering in Chapter 8 we made a distinction between normal processes with conservation of crystal momentum ($G = 0$) and Umklapp processes ($G \neq 0$); this was important because only the latter processes can destroy phonon drift. However, any type of electron–phonon collision can remove drift momentum from the electrons, so we shall not need to make a corresponding distinction in this chapter.

9.3.2   Temperature dependence of electrical and thermal resistance

The types of electron scattering process that produce electrical and thermal resistance were illustrated in Fig. 4.7. We see that electrical resistance requires large momentum transfers of order $\hbar k_F$. Since k_F is of the order of, but less than, a zone boundary wavevector, this means that high energy phonons (energy $\sim k_B \Theta_D$) near the zone boundary are most effective. A typical value of the velocity of sound in a metal is of order $10^{-3} v_F$; consequently the fractional change in electron energy when a phonon is created or destroyed is very small. We may therefore think of electron scattering by phonons as almost elastic; the direction of $\mathbf{k}$ is changed, but its magnitude remains at k_F, as in Fig. 9.4. The effectiveness of a collision in producing electrical resistance may be measured by the loss of momentum by the electron along its original direction of motion. From Fig. 9.4 this is $\hbar k_F (1 - \cos \theta) = 2 \hbar k_F \sin^2 \tfrac{1}{2}\theta.$

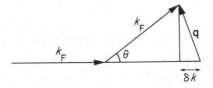

Fig. 9.4. Change in momentum when an electron is scattered through an angle θ by absorbing a phonon of wavevector $\mathbf{q}$.

At high temperatures ($T \gg \Theta_D$) a typical phonon energy is of order $k_B \Theta_D$ so that the large momentum transfers required to produce electrical resistance are easily achieved. Consequently the relaxation times for electrical and thermal resistance are quite similar and the Wiedmann–Franz law (Eq. (4.7)) is quite well obeyed. The actual electron mean free path l is inversely proportional to phonon number; since the internal energy is $3RT$ for $T \gg \Theta_D$ and the phonon energy is constant, the number of phonons is proportional to T. Therefore

$$1/l \propto T,$$

the electrical resistivity is approximately proportional to T and the thermal conductivity independent of T.

At low temperatures ($T \ll \Theta_D$) the average phonon energy is of order $k_B T$, and since the internal energy is proportional to $T^4(C_v \propto T^3$, Eq. (2.36)) the phonon number is proportional to T^3. For thermal resistance the momentum transfer is unimportant (Fig. 4.7(c)) and the effective mean free path l_{th} is therefore just inversely proportional to phonon number, so that

$$1/l_{th} \propto T^3. \tag{9.35}$$

But in the case of electrical resistance at low temperature we must also consider the magnitude of the momentum transfer. Since a phonon of energy $k_B\Theta_D$ has a wavevector of order k_F, a low temperature phonon of energy kT has $q \ll k_F$, and a typical scattering angle θ (Fig. 9.4) is small. The fractional momentum transfer k/k_F is then of order θ^2, which is proportional to q^2; with a constant velocity of sound for low energy phonons, this is in turn proportional to T^2. In calculating electrical resistance, collisions are therefore weighted by a factor of order $(T/\Theta_D)^2$ for the fraction of momentum lost, so that

$$\frac{1}{l_{el}} \sim \left(\frac{T}{\Theta_D}\right)^2 \frac{1}{l_{th}} \propto T^5 \tag{9.36}$$

From Eqs. (9.36) and (4.7) the Wiedemann–Franz ratio now becomes

$$\frac{K}{\sigma T} \sim \left(\frac{k_B}{e}\right)^2 \times \left(\frac{T}{\Theta_D}\right)^2,$$

and the simple law is no longer obeyed. This, however, is only true as long as electron–phonon collisions are dominant. At the lowest temperatures electron–impurity collisions are dominant, and these are capable of producing large momentum changes so that the Wiedemann–Franz law is again obeyed. It is thus only in the intermediate temperature region that low Wiedemann–Franz ratios are observed. Fig. 9.5(a) shows the electrical resistance of sodium at low temperatures; the residual value at the lowest temperatures is due to impurity scattering, and the initial rise with temperature is $\propto T^5$ in accordance with Eq. (9.27). Fig. 9.5(b) shows corresponding values of the Wiedemann–Franz ratio, with a minimum occurring when low energy phonons are the most important scatterers of electrons.

The preceding discussion is actually rather oversimplified. At temperatures below those shown in Fig. 9.5(a) the electrical resistance is found to

(*a*) Difference between observed
and residual resistance of
sodium at low temperatures.
[after D.K.C. Macdonald and
Mendelssohn, *Proc. Roy. Soc. A,*
202, 103 (1950)].

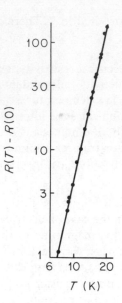

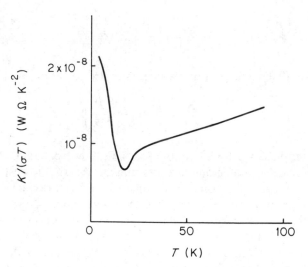

(*b*) Wiedemann – Franz ratio of sodium at low temperatures.
[after R. Berman and D.K.C. Macdonald, *Proc. Roy. Soc. A,*
209, 368 (1951)].

Fig. 9.5.

fall towards its residual value even more rapidly than T^5. This has recently been explained* as a consequence of an effect called **phonon drag.**

We assumed in the argument leading to Eq. (9.36) that when an electric current flows, the phonon distribution is that appropriate to thermal equilibrium, so that electron drift momentum is removed by electron–phonon collisions. But if the temperature is so low that phonon–phonon Umklapp processes are ineffective the phonons can take up a distribution of the form of Eq. (8.12). Exchange of crystal momentum between electrons and phonons will then tend to establish a distribution in which the phonons drift along with the same mean velocity as the electrons. If there were no Umklapp processes this phonon drag effect would be complete, and in a statistically steady state there would be no average exchange of crystal momentum between electrons and phonons, and consequently no electrical resistance due to electron–phonon collisions. In practice, as the temperature is reduced and Umklapp processes become less probable, the phonon drag effect causes the electrical resistance to fall progressively further below the value expected from Eq. (9.36).

9.3.3 Relation of electrical conductivity to the Fermi surface

Eq. (9.12) may be modified to allow for a relaxation time τ due to collisions so as to read

$$\hbar\left(\frac{d\mathbf{k}}{dt} + \frac{\delta\mathbf{k}}{\tau}\right) = e\mathbf{E}, \tag{9.37}$$

analogously to Eq. (3.22). Eq. (9.37) may be applied to electrons at any point on the Fermi surface, and $\delta\mathbf{k}$ represents the local displacement of the Fermi surface from its thermal equilibrium position. It is often convenient to calculate transport properties directly from the perturbation of the Fermi surface without introducing the idea of effective mass; we illustrate this with a calculation of the electrical conductivity.

Consider an element of area $d\mathbf{S}$ of the Fermi surface, which in a steady state is displaced an amount $\delta\mathbf{k}$ by the electric field. Eq. (9.28) gives $\delta\mathbf{k} = e\mathbf{E}\tau/\hbar$; this may vary over the Fermi surface if τ is not constant. The electrons in the volume $d\mathbf{S}\cdot\delta\mathbf{k}$ of k-space, shown in Fig. 9.6, move with the local Fermi velocity $\mathbf{v}_F$ and thus carry a current density

$$d\mathbf{j} = \frac{2}{(2\pi)^3}\, e\mathbf{v}_F(d\mathbf{S}\cdot\delta\mathbf{k}), \tag{9.38}$$

since there are $2/(2\pi)^3$ electrons per unit volume of metal in unit volume of k-space, allowing for spin degeneracy. Since the equilibrium electron distribution without a field carries no current, the current is obtained by integrating

* M. Kaveh and N. Wiser, *Phys. Rev. Letters,* **26,** 635–6 (1971).

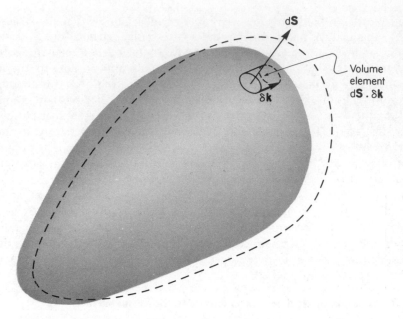

Fig. 9.6. Displacement of Fermi surface by $\delta\mathbf{k}$ when a current flows.

d$\mathbf{j}$ over the Fermi surface. Since $\mathbf{v}_F$ is $\hbar^{-1}\,\partial\varepsilon/\partial\mathbf{k}$, it is perpendicular to the Fermi surface, and thus parallel to d$\mathbf{S}$. The vectors $\mathbf{v}_F$ and d$\mathbf{S}$ in Eq. (9.38) can therefore be exchanged to give

$$\mathbf{j} = \frac{1}{4\pi^3} \int e\,d\mathbf{S}(\mathbf{v}_F \cdot \delta\mathbf{k})$$

$$= \frac{e^2}{4\pi^3\hbar} \int \tau\,d\mathbf{S}(\mathbf{v}_F \cdot \mathbf{E})$$

$$= \sigma\mathbf{E}.$$

The conductivity is therefore*

$$\sigma = \left| \frac{e^2}{4\pi^3\hbar} \int \tau\,d\mathbf{S}(\mathbf{v}_F \cdot \hat{\mathbf{E}}) \right|, \qquad (9.39)$$

where $\hat{\mathbf{E}}$ is a unit vector in the direction of $\mathbf{E}$. The integral in Eq. (9.39) can be regarded as defining an average mean free path for electrical conductivity so that

$$\sigma = \frac{e^2 l_{el} S}{4\pi^3\hbar}, \qquad (9.40)$$

* Strictly, of course, σ, like m^*, is a tensor.

where S is the total free area of the Fermi surface. This formula is more informative than Eq. (3.24); when the Fermi surface is restricted by contact with zone boundaries (e.g. Fig. 4.14(b)) it is the total free area that matters rather than the total number of electrons.

PROBLEMS 9

9.1 Obtain an equation for damped plasma oscillations by considering a medium in which the electron equation of motion is Eq. (3.22) rather than Eq. (9.2). Show that for a sufficiently small electron concentration n the plasma oscillations become critically damped, and estimate the critical electron concentration for $\tau = 10^{-12}$ s.

Show that for $\omega_p \tau \ll 1$ charge fluctuations decay exponentially with a dielectric relaxation time ε_0/σ, where σ is the electrical conductivity of the material.

9.2 From the discussion of the Mott transition to the metallic state in section 9.1.1 estimate the minimum value of the parameter (nA/k_F) in section 9.1.2 for a metal.

CHAPTER

10

Fermi surfaces

10.1 CONSTRUCTION OF FERMI SURFACES FOR NEARLY FREE ELECTRONS

We saw in section 9.1.3 that because a conduction electron wavefunction has to be orthogonal to the ion core wavefunctions its energy may be quite close to the free electron value. This together with the fact that $\varepsilon(\mathbf{k})$ is periodic with the period of the reciprocal lattice (section 6.3), gives a simple method, due to Harrison,[14] for constructing approximate Fermi surfaces from free electron Fermi spheres. Even if nearly free electrons are not a very good approximation, this construction often gives the correct topology of the Fermi surface. In this section we describe Harrison's construction, and show that it leads us to expect Fermi surfaces of complicated topology. The experimental determination of such Fermi surfaces we describe in section 10.2.

10.1.1 Square Bravais lattice

We illustrate this construction first for the model of a two dimensional square lattice with two electrons per unit cell, which we have already considered in section 4.4. By Bloch's theorem the free electron Fermi surface can be represented by a circle centred on any reciprocal lattice point, not just the origin. We therefore draw, in Fig. 10.1, the repeated zone scheme for our lattice with a circle of area equal to the reduced zone (so as to contain two electrons per unit cell) centred on each reciprocal lattice point. This

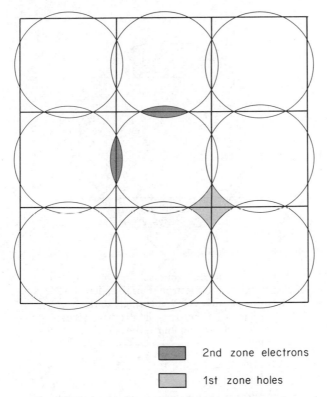

2nd zone electrons

1st zone holes

Fig. 10.1. Repeated zones of a square lattice, with free electron
Fermi surface for 2 electrons/atom.

divides the plane up into regions covered by $0, 1, 2 \ldots$ circles, and the
Fermi surface is found by applying two rules:

(1) The Fermi surface in the nth zone is the boundary dividing regions
covered by n circles from regions covered by $(n - 1)$ circles (spheres in 3
dimensions).

(2) If the region covered by a larger number of circles is *inside* this boundary we have an electron Fermi surface, and if it is outside we have holes.

Fig. 10.1 shows that these rules give the first zone holes and second zone
electrons that we already found in section 4.4. Comparison with Fig. 4.15
shows that the major effect of the energy perturbation due to the periodic
potential is to round off the corners of the Fermi surface determined by our
construction. The general effect of a stronger perturbation is to 'shrink'
the Fermi surface, whether it be electrons or holes, so that a strong enough
perturbation may remove one or more groups of carriers. The limit of this
process is an insulator, with no free Fermi surface; the occupied region of
k-space is entirely defined by zone boundaries.

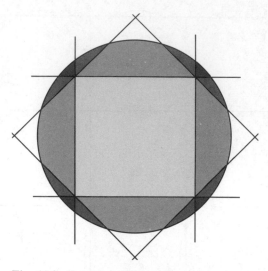

Fig. 10.2. Extended zone scheme of a square lattice with free electron Fermi sphere for 4 electrons per unit cell. Increasingly dark shades of grey indicate electrons in the first, second, third and fourth zones.

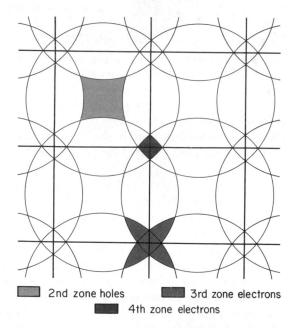

2nd zone holes   3rd zone electrons
4th zone electrons

Fig. 10.3. Repeated zones of a square lattice with free electron Fermi surface for 4 electrons/atom.

We now consider the more complicated case of four electrons per unit cell. Fig. 10.2 shows the free electron Fermi surface in the extended zone scheme of a square lattice. According to our rule of section 6.3.1, the nth zone is the region reached from the origin by crossing $(n - 1)$ zone boundaries; the electrons may therefore be assigned to the first four zones as indicated in Fig. 10.2. The Harrison construction in the repeated zone scheme provides a simple prescription for fitting together the separate pieces of Fermi surface in the higher zones. This is illustrated in Fig. 10.3, which differs from Fig. 10.1 only in that the circles are now equal in area to two reduced zones. This gives holes in the second zone and an equal number of electrons shared between the third and fourth zones; a general feature illustrated by this example is that in the higher zones very non-circular shapes of Fermi surface result. Paradoxically, this effect is greatest for nearly free electrons; stronger perturbation produces greater rounding off.

10.1.2   Face-centred cubic Bravais lattice

We have seen in section 6.3.2 that the reciprocal lattice of face-centred cubic is body-centred cubic, so that the reduced Brillouin zone is the truncated octahedron, Fig. 1.21. Fig. 10.4 shows a few of these zones stacked together to fill space in the repeated zone scheme. In three dimensions we do the Harrison construction of the Fermi surface by drawing a sphere centred on each reciprocal lattice point and applying the same rules as before. We shall illustrate the procedure by considering the particular case of 4 electrons per unit cell (as in lead, for example); this is the three-dimensional analogue of Fig. 10.3. The central geometrical fact, in three dimensions as in two, is that the radius of the free electron Fermi sphere is slightly greater than the radius of the corners of the reduced zone, so that the zone is wholly contained in the Fermi sphere. This means that two spheres overlap near zone faces, three near zone edges, and four near zone corners. Most of these features can be seen on a [110] cross-section of the repeated zone scheme; the plane of such a cross section is indicated in Fig. 10.4 and the cross-section is drawn in Fig. 10.5. Cross-sections of a second zone surface of holes and a third zone surface of electrons can be seen on Fig. 10.5; the cross-section misses the zone corners, where small pockets of electrons in the fourth zone occur.

Fig. 10.6 shows schematic views of these Fermi surfaces after the sharp corners have been rounded off by the periodic lattice potential. The most interesting feature is the third zone surface, known as the 'monster'. The three-dimensional view makes it clear that this cannot unambiguously be called 'electrons' or 'holes' because it is multiply connected; cross-sections can be drawn that look electron-like (as in Fig. 10.5) or hole like. We shall consider the experimental consequences of this type of Fermi surface in section 10.2. Note that monsters are essentially three-dimensional creatures; they cannot exist in a smaller number of dimensions. The third zone monster

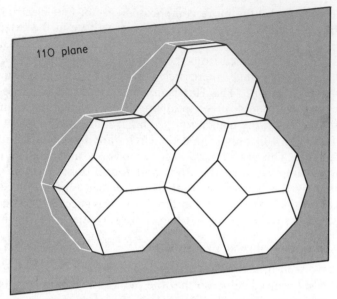

Fig. 10.4. Repeated zones of a face-centred cubic lattice, sectioned by a (110) plane.

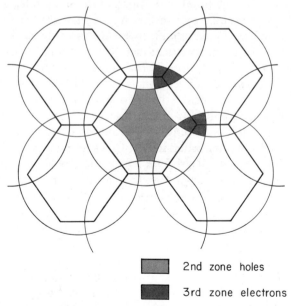

2nd zone holes

3rd zone electrons

Fig. 10.5. (110) section of repeated zone scheme for fcc, with section of free electron Fermi surface for 4 electrons/atom.

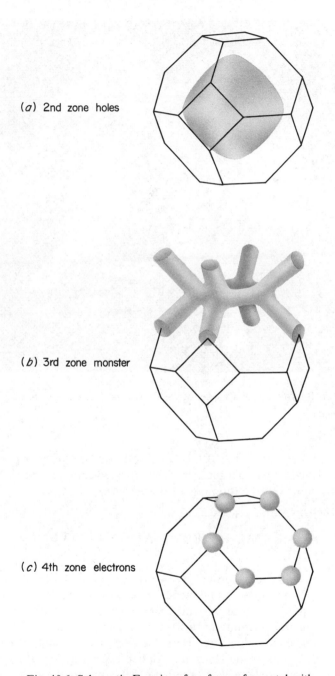

(*a*) 2nd zone holes

(*b*) 3rd zone monster

(*c*) 4th zone electrons

Fig. 10.6. Schematic Fermi surface for an fcc metal with
4 electrons/atom (e.g. Pb).

Fig. 10.7. Photograph of a model of the third-zone monster shown in Fig. 10.6.

is further illustrated in Fig. 10.7, by means of a model of intersecting rings. This model enables a larger section of the repeated zone scheme to be shown, thereby bringing out the topology of this surface.

10.2   EXPERIMENTAL DETERMINATION OF FERMI SURFACES

A magnetic field causes individual electrons to occupy a sequence of states around the Fermi surface, and hence gives rise to effects that can yield geometrical information about the Fermi surface. We shall consider two such effects in this section: cyclotron resonance, which we have already considered for semiconductor carriers describable by a single effective mass in section 3.4.3; and the de Haas-van Alphen effect, an oscillatory variation of electron diamagnetism with magnetic field due to the quantization of electron orbits.

10.2.1   Dynamics in a magnetic field

In the absence of collisions the equation of motion of an electron in a magnetic field **B** is

$$\hbar\frac{d\mathbf{k}}{dt} = e\mathbf{v} \times \mathbf{B} = e\frac{d\mathbf{r}}{dt} \times \mathbf{B}. \tag{10.1}$$

Since $\mathbf{v} = \hbar^{-1}\,\partial\varepsilon/\partial\mathbf{k}$ is normal to the Fermi surface, $d\mathbf{k}/dt$ is parallel to the Fermi surface and perpendicular to **B**. The component of **k** parallel to **B** is therefore constant, so that the electron orbit in **k**-space is obtained by taking a section through the Fermi surface with a plane normal to **B**. From Eq. (10.1) the orbit in **r**-space is similar in shape; it differs by a scale factor of $\hbar/e\mathbf{B}$ and a rotation of $\pi/2$; this relationship is depicted in Fig. 10.8.

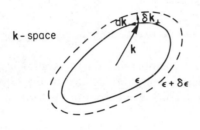

Fig. 10.8. Orbit of an electron in a
magnetic field.

The period T of an electron orbit in a magnetic field can be calculated as

$$T = \oint dt = \oint \frac{dr}{v}$$

in which

$$\mathbf{v} = \frac{d\mathbf{r}}{dt} = \hbar^{-1}\frac{\partial\varepsilon}{\partial\mathbf{k}} = \hbar^{-1}\frac{\delta\varepsilon}{\delta\mathbf{k}_{\perp}},$$

where, from Fig. 10.8, $\delta\mathbf{k}_{\perp}$ is the normal distance in **k**-space, projected on a plane perpendicular to **B**, between constant energy surfaces of energy ε and

$\varepsilon + \delta\varepsilon$. This gives

$$T = \hbar \oint \frac{dr \cdot \delta k_\perp}{\delta\varepsilon} = \frac{\hbar^2}{eB} \oint \frac{|dk \times \delta k_\perp|}{\delta\varepsilon}$$

$$= \frac{\hbar^2}{eB} \frac{\delta A_k}{\delta\varepsilon},$$

where A_k is the area of the electron orbit in **k**-space. The cyclotron frequency is therefore

$$\omega_c = \frac{2\pi}{T} = \frac{2\pi eB}{\hbar^2} \frac{d\varepsilon}{dA_k}; \qquad (10.2)$$

by comparison with $\omega_c = eB/m$ for free electrons we may define a cyclotron effective mass m_c as

$$m_c = \frac{\hbar^2}{2\pi} \frac{dA_k}{d\varepsilon}. \qquad (10.3)$$

It must, however, be remembered that in general m_c is quite a different quantity from the effective mass m^* defined in Eq. (9.20). Thus, by writing δA_k as $\oint |dk \times \delta k_\perp|$, Eq. (10.3) can be put in the form

$$m_c = \frac{\hbar}{2\pi} \oint \frac{|dk|}{v_F} = \hbar \overline{\left(\frac{k_F}{v_F}\right)},$$

where the bar indicates an average round the cross-section of the Fermi surface illustrated in Fig. 10.8. Taking the averaging process as understood we may write

$$\frac{1}{m_c} = \frac{1}{p} \frac{\partial\varepsilon}{\partial p},$$

which may be compared with Eq. (9.20) in the form

$$\frac{1}{m^*} = \frac{\partial^2\varepsilon}{\partial p^2}.$$

We thus see that m_c is related to a first derivative of the energy and m^* to a second derivative. They are identical only for an isotropic parabolic band, for which $\varepsilon = p^2/2m^*$.

Note also that the two masses give different criteria for electron-like or hole-like behaviour. m^* is negative if $(\partial^2\varepsilon/\partial k^2)$ is negative, whereas m_c is negative if $(\partial A_k/\partial\varepsilon)$ is negative, i.e. if the orbit in **k**-space shown in Fig. 10.8 has occupied states outside it. The latter criterion is physically more reasonable for those experiments in a magnetic field that usually reveal hole-like behaviour.

10.2.2   Cyclotron resonance in metals

Our treatment of cyclotron resonance is section 3.4.3 implicitly assumed that the microwave electric field was essentially uniform over the electron orbit. For semiconductors this is a reasonable assumption, since the small number of carriers leads to low conductivity and hence to a large microwave skin depth. In metals, on the other hand, the larger number of carriers gives a skin depth that is smaller than both the electron mean free path and the size of its orbit in the resonant magnetic field. This necessitates the use of a different geometry, shown in Fig. 10.9, for experiments on cyclotron resonance

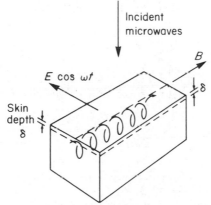

Fig. 10.9. Cyclotron resonance in metals. **B** is parallel to the sample surface and the electron is accelerated by E cos ωt each time it enters skin depth δ.

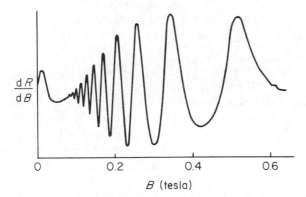

Fig. 10.10. Experimental values of the field derivative of the surface impedance of copper at 24 GHz.
{after A. F. Kip, D. N. Langenberg and T. W. Moore, *Phys. Rev.*, **124**, 359 (1961)}.

in metals; this geometry resembles quite closely the geometry of an ordinary cyclotron.

The essential feature of Fig. 10.9 is that the steady magnetic field **B** is parallel to a plane surface of the specimen, so that in performing their spiral orbits in accordance with Eq. (10.1) some of the electrons enter the skin depth where they can feel the microwave electric field, **E** cos ωt, once per revolution. The electrons therefore absorb energy from the microwaves if they see an electric field in the right direction when they enter the skin depth; since the skin depth is much smaller than the orbit diameter, resonance can occur for a frequency that is any harmonic of the cyclotron frequency. From Eq. (10.2) we therefore expect resonant absorption* when

$$\omega = n\frac{2\pi eB}{\hbar^2}\frac{d\varepsilon}{dA_k};$$

since the normal procedure is to work at a constant microwave frequency and vary the magnetic field we expect a power absorption that is periodic in $1/B$ with a period given by

$$\delta(1/B) = \frac{2\pi e}{\hbar^2\omega}\frac{d\varepsilon}{dA_k}. \tag{10.4}$$

Some measurements of the microwave surface resistance of copper showing this periodicity are illustrated in Fig. 10.10.

10.2.3   Quantization of electron orbits

Our discussion of electron motion in section 10.2.1 was essentially classical in that we assumed an electron wavepacket small compared with the size of the orbit. In reality, even under such quasi-classical conditions, the correspondence principle gives a Bohr quantization condition that has to be satisfied by the line integral of the momentum round the orbit:

$$\oint \mathbf{p} \cdot \mathbf{dr} = (n + \gamma)h = 2\pi(n + \gamma)\hbar, \tag{10.5}$$

where γ is a small constant, less than 1. We shall see shortly that this gives an energy difference $\hbar\omega_c$ between successive orbits, so that the radiation emitted in a transition between these orbits (for large n) has the frequency of the classical motion, in accordance with the correspondence principle. To evaluate the integral in Eq. (10.5) we note that the crystal momentum $\hbar\mathbf{k}$

---

* The resonance condition gives an impedance minimum at resonance (cf. Eq. (3.34)). For a metal this impedance is so low that the microwave source is essentially a current source, giving a minimum absorption at resonance. For a semiconductor the impedance is usually high enough for the microwave generator to behave as a voltage source, giving a maximum absorption at resonance.

corresponds to $m\mathbf{v}$ in the acceleration equation of a free particle, so that by analogy with Eq. (5.7) we write*

$$\mathbf{p} = \hbar\mathbf{k} + e\mathbf{A}. \tag{10.6}$$

We now calculate separately the contributions to Eq. (10.5) from the two terms in Eq. (10.6). The first term gives, by using Eq. (10.1) to relate changes in $\mathbf{k}$ to changes in $\mathbf{r}$,

$$\oint \hbar\mathbf{k} \cdot d\mathbf{r} = e \oint (\mathbf{r} \times \mathbf{B}) \cdot d\mathbf{r}$$

$$= -e \oint \mathbf{B} \cdot (\mathbf{r} \times d\mathbf{r}) = -2e\Phi, \tag{10.7}$$

where Φ is the magnetic flux through the orbit in $\mathbf{r}$-space. The contribution of the electromagnetic momentum $e\mathbf{A}$ is just

$$e \oint \mathbf{A} \cdot d\mathbf{r} = e\Phi, \tag{10.8}$$

so that by combining Eqs. (10.7) and (10.8) Eq. (10.5) becomes

$$\oint \mathbf{p} \cdot d\mathbf{r} = -e\Phi = (n + \gamma)h; \tag{10.9}$$

the flux through an electron orbit in $\mathbf{r}$-space is therefore quantized in units of h/e. This can be related to the Fermi surface in $\mathbf{k}$-space by noting that $\phi = BA_r$, where A_r is the area of the orbit in $\mathbf{r}$-space, and using our result that the orbit in $\mathbf{k}$-space is $eB/\hbar$ larger than the orbit in $\mathbf{r}$-space. Eq. (10.9) then becomes

$$A_k = \frac{2\pi eB}{\hbar}(n + \gamma). \tag{10.10}$$

The difference in area A_k for a change of quantum number by 1 is therefore $2\pi eB/\hbar$; substituting this in Eq. (10.2) we find that the corresponding energy change $\delta\varepsilon$ at the Fermi surface is

$$\delta\varepsilon - \hbar\omega_c. \tag{10.11}$$

This is just as it should be, and is, as we have mentioned, an example of the correspondence principle; the level spacing for large quantum numbers just corresponds to the classical orbital frequency. From a quantum point of view we can regard cyclotron resonance as an electron raising its energy by photon absorption.

---

* It is important to note that in a magnetic field $\mathbf{k}$ no longer has any simple relation to electron wavelength because the wavefunction is strongly perturbed by the field.

10.2.4   The de Haas–van Alphen effect

The results of section 10.2.3 show that when a magnetic field is applied in the z-direction our usual quantization of energy levels by translational momentum breaks down in the x–y plane. Essentially, the boundary conditions at the edge of the crystal are no longer relevant, because the magnetic field confines the electrons to localized orbits; instead we have an angular momentum quantization condition and the electrons are confined to orbits of quantized area in **k**-space. This situation is depicted for a spherical Fermi surface in Fig. 10.11; instead of being uniformly distributed throughout the Fermi sphere the electrons are confined to a series of cylinders. On the average the density of electron states in **k**-space is unchanged, but the quantization is coarser than in the absence of a field. We may use Eq. (10.10) to estimate the quantum number of the outermost occupied cylinder in Fig. 10.11. A typical cross-sectional area of a Fermi surface in **k**-space is $1\,\text{Å}^{-2} = 10^{20}\,\text{m}^{-2}$ and the largest field used in experiments is about $10\,\text{T}$; we thus have

$$n = \frac{\hbar A_k}{2\pi e B} \approx \frac{10^{-34} \times 10^{20}}{2\pi \times 10^{-19} \times 10} \sim 10^3.$$

Note that this quantum number is inversely proportional to field, so that as the field is increased the whole array of cylinders in Fig. 10.11(b) expands and the number of occupied cylinders within the Fermi sphere decreases; this leads to a periodic dependence of various quantities on magnetic field, as follows. As the cylinders expand quantities such as the free energy will in general change smoothly, except that as the outermost cylinder just reaches the Fermi-surface and becomes unoccupied its contribution will suddenly cease and there will be a discontinuity in the rate of change of free energy. These discontinuities will not be abrupt at finite temperatures, but should still be noticeable provided $kT \ll \hbar\omega_c$. Since magnetic moment is the rate of change of free energy with field, there will be corresponding discontinuities in magnetic moment. Consequently the magnetization (and also the susceptibility) varies periodically with $(1/B)$; from Eq. (10.10) the period is

$$\delta\left(\frac{1}{B}\right) = \frac{2\pi e}{\hbar A_k}, \tag{10.12}$$

where A_k is now the maximum (or minimum) cross-sectional area of the Fermi surface normal to the field.

It is interesting to compare this result with Eq. (10.4) for cyclotron resonance. We see that the cyclotron resonance period is longer, being governed by the smaller area $\hbar\omega_c(dA_k/d\varepsilon)$, which is the *difference* in the area between successive orbits. Strictly, just as A_k in Eq. (10.12) is an extremal cross-section of the Fermi surface, so is $(dA_k/d\varepsilon)$ in Eq. (10.4) an extremal value for

(*a*)   Occupied region of **k**-space without a magnetic field.

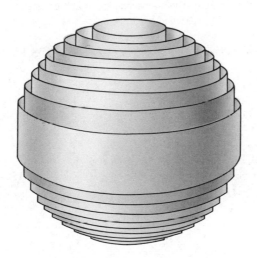

(*b*)   Occupied region of **k**-space with a magnetic field.

Fig. 10.11.

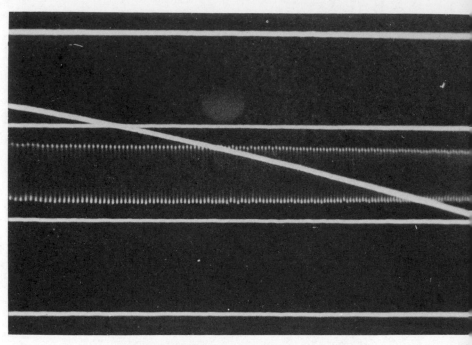

(*a*) Oscillogram of de Haas-van Alphen oscillations in copper, by the impulsive field method. The oblique line shows the field variation, and the horizontal lines indicate fields of 10.47, 10.70, 10.93 and 11.16 T; the picture is about 1 ms wide.

[after D. Shoenberg, *Phil. Trans. Roy. Soc. A*, **255**, 85 (1962)].

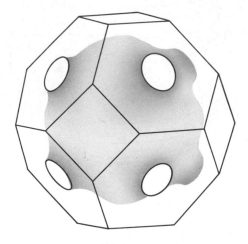

(*b*) Sketch of the Fermi surface of copper, showing contact with the hexagonal faces of the zone.

Fig. 10.12.

the various possible cross-sections of the Fermi surface normal to B. But for an ellipsoidal Fermi surface $(dA_k/d\varepsilon)$ is constant and this point does not arise. In general, however, different parts of the Fermi surface have different cyclotron frequencies, but the greatest contribution comes from those parts of the Fermi surface where the frequency is a maximum or a minimum, so that these are the frequencies actually observed. This is an example of a general result of combining a continuous spectrum of frequencies, and is known as the principle of stationary phase.

The de Haas–van Alphen effect was discovered in bismuth, which has small pockets of Fermi surface containing only about 10^{-5} electrons/atom, and consequently, from Eq. (10.12) gives a large and readily observable period. The much shorter periods for ~ 1 electron/atom mean that fields of order 10 T at temperatures of order 1 K are needed to observe the effect satisfactorily. Specimens are typically single crystal wires a few mm long mounted in a small pickup coil, often with a similar dummy coil connected in series opposition. Two basic methods of observation are used. In the impulsive field method a large condenser is discharged through a liquid nitrogen cooled solenoid to give a field that rises to 10 T and falls to zero again in a few milliseconds. While the field is changing there is an induced EMF proportional to

$$\frac{dM}{dt} = \frac{dM}{dB} \cdot \frac{dB}{dt}, *$$

so that an oscillatory signal proportional to the differential susceptibility (dM/dB) is obtained; an oscillogram for copper obtained by this method is shown in Fig. 10.12(a). The second basic method makes use of superconducting solenoids, with which it is possible to obtain steady fields of order 5 T of extremely high stability. Because of their zero resistance these solenoids, once energized, can carry a persistent current without external power supply. This steady field can be modulated at any convenient frequency by means of auxiliary coils, so that a signal at that frequency proportional to dM/dB is generated in the pickup coil.

The advantage of modulating a steady field at a fixed frequency is that phase sensitive detection techniques can be used to improve the signal to noise ratio. Integration times of several seconds, giving an effective detector bandwidth of a fraction of a Hz, can be used if necessary. On the other hand, the fact that an impulsive field experiment lasts only a few milliseconds means that the bandwidth is necessarily of the order of MHz.

A further refinement of detection technique is made possible by the non-linear relation between magnetization and field. If we expand the magnetization in a Taylor series about the steady field B_0, to which a small field

---

* The total diamagnetism of the metal is very weak, so that the difference between the applied field B_e and the average field B seen by the electrons in the metal is unimportant.

$B_1 \cos \omega t$ is being added, we have

$$M(t) = M(B_0) + (B_1 \cos \omega t)\left(\frac{\mathrm{d}M}{\mathrm{d}B}\right)_{B_0} + \tfrac{1}{2}(B_1 \cos \omega t)^2\left(\frac{\mathrm{d}^2 M}{\mathrm{d}B^2}\right)_{B_0} + \cdots,$$

from which

$$-\frac{\mathrm{d}M}{\mathrm{d}t} = -\omega B_1\left(\frac{\mathrm{d}M}{\mathrm{d}B}\right)_{B_0} \sin \omega t + \tfrac{1}{4}B_1^2\left(\frac{\mathrm{d}^2 M}{\mathrm{d}B^2}\right)_{B_0} \sin 2\omega t + \cdots, \qquad (10.13)$$

so there is an induced EMF at frequency 2ω which depends essentially on the non-linear nature of the de Haas–van Alphen effect ($\mathrm{d}^2 M/\mathrm{d}B^2 \neq 0$). Therefore, by employing phase sensitive detection at frequency 2ω, we can discriminate against all linear pickup effects at frequency ω; the modulation amplitude B_1 is best chosen to be comparable with the de Haas–van Alphen period.

The special value of the de Haas–van Alphen effect is that, for a complicated Fermi surface consisting of many pieces, it gives a separate periodic signal for each extremal area. The signal shown in Fig. 10.12(a) is approximately the frequency for a free-electron sphere containing one electron per atom, but quite different frequencies are also seen for copper (and also silver and gold). This leads to the conclusion that the free electron sphere is stretched

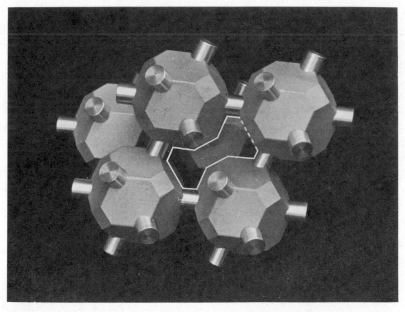

Fig. 10.13. Photograph of a schematic model of the Fermi surface of copper in the repeated zone scheme, showing the dog's bone orbit.

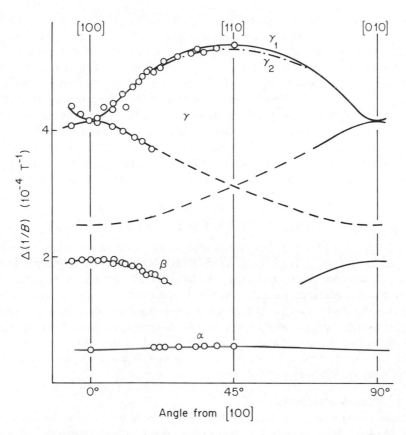

Fig. 10.14. Angular variation of the de Haas–van Alphen periods in lead.
{after A. V. Gold, *Phil. Trans. Roy. Soc. A*, **251**, 85 (1958)}.

towards all the zone faces, and actually makes contact with the hexagonal (111) faces, as shown in Fig. 10.12(*b*). Thus as well as the main 'belly' orbit, there is a 'neck' orbit of minimum area at the contact with the zone boundary. Fig. 10.13 shows a model* of part of the Fermi surface of copper in the repeated zone scheme; a hole orbit called the 'dog's bone' is indicated. This contact with the zone boundary makes the Fermi surface multiply connected in the repeated zone schemes, and various other hole orbits covering bits of several spheres are also possible.

---

*This model is constructed from modified rhombic dodecahedra (Fig. 1.17). A rhombic dodecahedron can be inscribed in the fcc Brillouin zone, with its vertices just touching the centres of the faces, and such a dodecahedron exactly half fills the zone. The Fermi surfaces of copper, silver, and gold are very closely related to this. They have large almost flat areas corresponding to the faces of the rhombic dodecahedron; the [100] vertices are rounded off within the zone; and the [111] vertices are rounded off across the zone boundary to form the necks.

The de Haas–van Alphen effect can also disentangle more complicated Fermi surfaces; thus, it has been used to show that lead has a Fermi surface essentially of the form shown in Fig. 10.6. The observed variation of period with field orientation, from [100] to [110], is shown in Fig. 10.14. The α oscillations correspond to the second zone holes in Fig. 10.6; the β oscillations come from a hole orbit on the square face of the zone arising from the third zone monster; and the γ oscillations come from the pockets in the fourth zone (these in general give three periods, as pockets in three different orientations are seen for any field direction). Over limited ranges of angle beats are seen on the γ oscillations, showing that two periods, γ_1 and γ_2, are present; the γ_2 period is due to orbits around limbs of the third zone monster.

★ 10.3 ARE ALKALI METALS FREE-ELECTRON LIKE?

The alkali metals crystallize in the body-centred cubic structure, so that their Brillouin zone is the highly symmetrical rhombic dodecahedron. This leads to the expectation that they will have Fermi surfaces very close to free-electron spheres, an expectation that is apparently borne out by experiments on the de Haas–van Alphen effect. Nevertheless, there have recently been theoretical suggestions, supported by some experimental anomalies, that potassium in particular differs qualitatively from the free-electron model. It is our purpose in this section to give some discussion of what is still an open question.

10.3.1 Possible instabilities

Although in this chapter we have considered Fermi surfaces of complicated topology that differ very much from free-electron spheres we have always assumed that the electrons behave essentially like a Fermi gas. In particular we have assumed the existence of a well defined Fermi surface in the vicinity of which the electron energy varies smoothly with $\mathbf{k}$. A question of central importance in the theory of metals is whether such a Fermi gas like state is the true ground state, stable against perturbations.

We shall meet an example of instability of the Fermi gas in Chapter 11; superconductivity arises when the electron gas is unstable against the formation of a particular type of electron pair. In this section we consider another type of instability, suggested by Overhauser (*Phys. Rev. B*, **3**, 3173 (1971)). Overhauser considers the effect of the Coulomb interaction between electrons, which contains terms of both exchange and correlation type, when there is a periodic perturbation to the electron density with a wavenumber slightly greater than $2k_F$. Such a perturbation introduces additional 'zone boundaries' as shown in Fig. 10.15. For a particular simplified model, Overhauser finds that the energy is minimized when the perturbation is such that the Fermi surface is distorted to the lemon shape shown in Fig. 10.15,

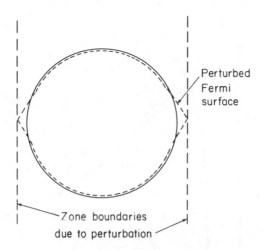

Fig. 10.15. Effect of a periodic perturbation
of wavenumber slightly greater than $2k_F$ on
a spherical Fermi surface.

which just makes contact with the extra zone boundaries. Overhauser considers the possibility of a phase difference between the density oscillations of spin-up and spin-down electrons, so that there are two principal types of instability: spin density waves, in which the oscillations are in antiphase; and charge density waves, in which the oscillations are in phase. We shall consider in Chapter 12 the evidence that the antiferromagnetism of chromium is due to a spin density wave, and that rare earth magnetism is accompanied by spin density waves in the conduction electrons. We now consider experimental evidence relevant to the suggestion of charge density waves in alkali metals, particularly potassium.

10.3.2   Optical absorption

The optical absorption of potassium, as deduced from the reflection coefficient of a clean metal surface, shows an absorption edge in the infrared at about 0.6 eV (2 μm) with a characteristically steep initial rise (Fig. 10.16). This unusual shape is exactly what would be expected for excitation across an energy gap if the density of occupied states were that for the conical regions of Fermi surface shown in Fig. 10.15. The curve in Fig. 10.16 is a theoretical curve calculated by Overhauser with the charge density wave energy gap as the only adjustable parameter; the agreement is impressive. However, the experimental evidence is not entirely clear, for experiments on evaporated films of potassium have not shown this absorption edge. This discrepancy could be explained by alignment of the wavevector of the charge density wave normal to the evaporated film. Although suggestions can be made why

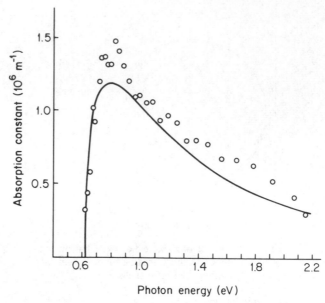

Fig. 10.16. Infrared absorption edge of potassium; the solid curve is Overhauser's theory.
{A. W. Overhauser, *Phys. Rev. Lett.*, **13**, 191 (1964)}.

such an alignment might occur there is no independent evidence of its occurrence.

10.3.3 Torque anomaly

An arrangement sometimes used to investigate magnetoresistance is shown in Fig. 10.17(a). A spherical single crystal is suspended from a torsion balance in a horizontal magnetic field, which is slowly rotated ($\lesssim 1$ rev/min.) in a horizontal plane. The changing magnetic flux through the sample produces eddy currents as shown in the figure, which interact with the field to produce a couple about the vertical suspension axis which can be measured; the currents, and hence the couple, are inversely proportional to the resistivity.

Fig. 10.17(b) shows the variations of this couple, during 360° rotation of the applied field, for a single crystal of potassium at 4.2 K. The striking feature of these curves is that they show an anisotropy of twofold symmetry, although a threefold [111] axis of the crystal is vertical. Moreover, this twofold anisotropy is observed for all crystal orientations; the crystal therefore cannot be truly of cubic symmetry. Since the main torque minima occur when the field is oriented such that part of the path of the eddy currents is in a [110] direction, the observation indicate that a particular [110] direction is a special direction

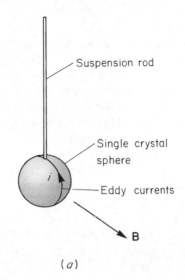

Suspension rod

Single crystal
sphere

Eddy currents

B

(a)

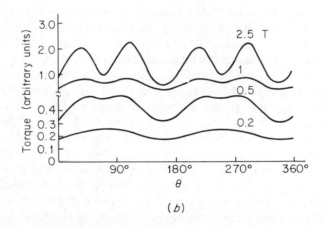

(b)

Fig. 10.17. The torque anomaly in potassium:
(a) Geometry of the experiment.
(b) Results with [111] vertical.
{J. A. Schaefer and J. A. Marcus, *Phys. Rev. Lett.*, **27**,
935 (1971)}.

in the crystal. If it is supposed that this special direction is the wavevector
of a charge density wave a semiquantitative explanation of the results can
be given. It must, however, be supposed that the charge density wave is
locked to a particular [110] direction by strains set up by differential thermal
contraction between the sample and its mounting.

10.3.4   Other observations

At first sight de Haas–van Alphen observations contradict the idea of a charge density wave in potassium, for, as we mentioned earlier, they indicate an equatorial area of Fermi surface that varies very little with crystal orientation. This result is not conclusive, however, because it is quite plausible that a charge density wave would orient parallel to a magnetic field as strong as is used to observe the de Haas–van Alphen effect. If this were the case no striking anisotropy would be observed.

There is indeed an anomaly in the de Haas–van Alphen observations that could be associated with a Fermi surface distortion of the type shown in Fig. 10.15. The volumes of the experimentally determined alkali Fermi surfaces differ significantly, by several tenths of a percent, from those required for one electron per atom; in all cases except sodium the experimental volume is too small. Fig. 10.15 shows that the distortion due to a charge density wave might be expected to produce just such a reduction in equatorial area, if the volume enclosed by the Fermi surface is maintained at the value for one electron per atom.

All observations we have discussed so far *could* be due to a charge density wave, but the possibility of other explanations cannot be ruled out. What is needed to settle the question is an effect that could *only* be due to a charge density wave. Overhauser has pointed out that the requirement of local electrical neutrality provides such an effect: an electron charge density wave must be neutralized by an appropriate periodic displacement of the positive ions from their usual positions. Such an effect would be detectable in a neutron diffraction experiment. The periodic modulation of the structure would produce satellite reflections, which might however be very weak because of fluctuations. More important, some of the main reflections would be weakened when the scattering wavevector $\mathbf{K} = (\mathbf{k} - \mathbf{k}')$ of the neutrons was parallel to the ionic displacement due to the charge density wave. Now, we have seen that the de Haas–van Alphen observations mean that a charge density wave, if present, must align itself with a strong magnetic field. This gives us a way to control the predicted reduction in neutron scattering intensity: as a strong magnetic field is rotated with respect to the crystal, the intensity of appropriate Bragg reflections should be modified. The importance of this proposal is that it gives a way to *disprove* the existence of charge density waves. If the effect is looked for and not seen, charge density waves cannot be present. Equally, if the effect is seen, charge density waves must be present.

PROBLEMS 10

10.1   In a cyclotron resonance experiment in potassium at 68 GHz three consecutive resonances were seen at magnetic fields of 0·74 T, 0·59 T, 0·49 T. Calculate the cyclotron effective mass of electrons in potassium.

10.2 The de Haas–van Alphen effect is studied in copper at a field of 10 T. What is the order of the maximum temperature which can be tolerated while still getting a good effect.

If the impurity density n and electron collision time τ are related by $n\tau = 10^{14}$ m^{-3} s, up to what density of impurities can the effect still be readily observed?

10.3 What is the pattern on the front cover of this book?

CHAPTER

Superconductivity

11.1 CLASSICAL SUPERCONDUCTIVITY

Superconductivity is not a classical phenomenon; like ferromagnetism, it is a state in which metallic electrons display macroscopic properties qualitatively different from those of an ideal Fermi gas, and it is an essentially quantum mechanical consequence of interactions between electrons that are ignored in the independent particle model. Nevertheless, we may apply the word classical to it in the present section for two reasons. First, we shall be concerned with phenomena in which actual quantization conditions are not macroscopically observable; and second, the phenomena we discuss in this section were almost all discovered in what might be called the classical period of superconductivity, prior to the second world war. We shall discuss macroscopic quantum effects in sections 11.2 and 11.3.

11.1.1 Zero resistance and the Meissner effect

Superconductivity was discovered by Kamerlingh Onnes in 1911, three years after his first liquefaction of helium. He was investigating the resistance of metals in the low temperature region this made available, and had chosen mercury for study because it could readily be purified by distillation, and should therefore show very small resistance. This expectation was dramatically fulfilled (Fig. 11.1(a)), but it was soon established that there was no essential connection with high purity; quite substantial amounts of added

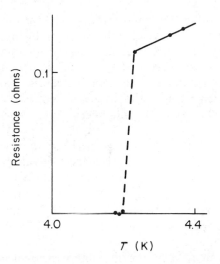

(*a*)   Superconducting transition of Hg (after Onnes, 1911).

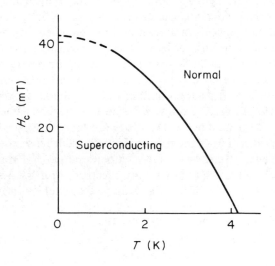

(*b*)   Critical field curve of Hg.

Fig. 11.1.

impurity had very little effect on the transition to a state of unmeasureably small resistance, which Onnes called the superconducting state. In his original experiments the transition occurred in a temperature range of about 10^{-2} K, but transition widths as small as 10^{-3} K can be obtained in well-annealed single crystals of a metal such as tin. Many metals and alloys have now been shown to become superconducting at a well-defined transition temperature T_c below about 20 K. (see D. Shoenberg, *Superconductivity*, Cambridge, 1952, Appendix I).

At first it was hoped that superconductivity would give a means of obtaining high magnetic fields with dissipationless solenoids, but it was soon found that the property of superconductivity was destroyed by a quite modest magnetic field, as shown in Fig. 11.1(*b*). It is only in the last decade that materials retaining zero resistance in usefully high magnetic fields have been discovered; the properties of these materials will be discussed in sections 11.2.2 and 11.2.3.

A material of infinite conductivity cannot support an electric field, and consequently the magnetic field within it cannot change with time $\partial \mathbf{B}/\partial t = 0$; superficial eddy currents are induced to prevent any change of field. For simplicity we consider the magnetic properties of a long superconducting rod parallel to the applied field $\mathbf{B}_e$, so that the demagnetizing factor (see Appendix C) is small and $\mathbf{B}_e \approx \mathbf{H}$.* We then have the situation illustrated in Fig. 11.2(*a*); magnetic field is excluded until the critical magnetic field $\mathbf{H}_c$† is reached, when it suddenly penetrates. When the field is decreased we would expect the magnetic flux in the sample at $\mathbf{H}_c$ to be trapped, so that the sequence in Fig. 11.2(*b*) is followed, leaving the sample with a permanent magnetic moment. Some trapped flux is indeed usually observed, but it was not until 1933 that Meissner and Ochsenfeld made quantitative investigations of the magnetic field near a superconductor. They then discovered that in general sequence (*b*) of Fig. 11.2 does not occur, but rather sequence (*a*) is reversed. On reducing the field below $\mathbf{H}_c$ flux is suddenly expelled; for a well annealed sample less than 1 % of the flux at $\mathbf{H}_c$ is trapped. This expulsion of field is the Meissner effect. It shows that a superconductor is more than a perfect conductor, and it also shows that the transition from superconductor to normal metal is reversible: for a given field and temperature there is a unique equilibrium state. A superconductor can therefore be regarded as a *perfect diamagnetic* in which $\mathbf{B} = 0, \mathbf{M} = -\mathbf{H}/\mu_0$. Fig. 11.3 shows that an annealed sample of a superconductor departs very little from this ideal, displaying only very slight magnetic hysteresis.

---

* For the definition of **H**, see Appendix E.

† The notation $\mathbf{H}_c$ is used because in a sample of finite demagnetizing factor it is the field $\mathbf{H} = \mathbf{B} - \mu_0\mathbf{M}$ that becomes critical. Magnetization curves of superconductors are usually plotted as $\mathbf{M}(\mathbf{B}_e)$, and such curves depend on sample shape, but the $\mathbf{M}(\mathbf{H})$ curve is the same for any ellipsoidal sample.

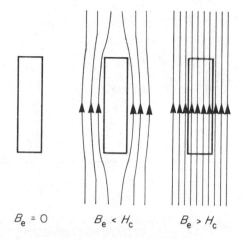

$B_e = 0$ $B_e < H_c$ $B_e > H_c$

(*a*) A perfect conductor in increasing magnetic fields.

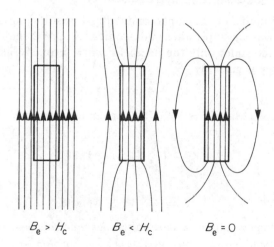

$B_e > H_c$ $B_e < H_c$ $B_e = 0$

(*b*) A perfect conductor in decreasing magnetic fields.

Fig. 11.2. For a *superconductor* sequence (*b*) is *not* followed,
but sequence (*a*) is reversed.

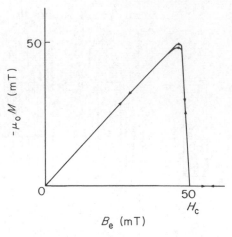

Fig. 11.3. Magnetization curve of a long rod
of superconducting lead (annealed).

11.1.2 Thermodynamics of superconductivity

The reversibility of the superconducting transition in a magnetic field means that we can apply thermodynamics to it. We shall ignore volume changes and consider only the magnetic work term, so that the change in internal energy is given by

$$dU = T\,dS + \mathbf{H}\cdot d\mathbf{M} \tag{11.1}$$

(see for example, Mandl,[2] p. 109). It is convenient to define a magnetic Gibbs function by

$$G = U - TS - \mathbf{M}\cdot\mathbf{H}, \tag{11.2}$$

so that

$$dG = -S\,dT - \mathbf{M}\cdot d\mathbf{H}, \tag{11.3}$$

where all extensive quantities are taken *per unit volume*. Because T and $\mathbf{H}$ are the independent variables in Eq. (11.3), G is the free energy that is a minimum for thermal equilibrium at constant temperature and applied field. For the superconducting state we have, by integrating Eq. (11.3) at constant temperature

$$G_s(\mathbf{H}, T) = G_s(0, T) - \int_0^{\mathbf{H}} \mathbf{M}\cdot d\mathbf{H}$$

$$= G_s(0, T) + \frac{H^2}{2\mu_0}, \tag{11.4}$$

since $\mathbf{M} = -\mathbf{H}/\mu_0$. At the critical field $\mathbf{H}_c$ where the normal and super-conducting states are in equilibrium, their Gibbs functions must be equal

$$G_n(T) = G_s(\mathbf{H}_c, T);$$

we ignore any weak magnetism in the normal state and assume that G_n is independent of field. With Eq. (11.4) this gives

$$G_n(T) - G_s(0, T) = \frac{H_c^2}{2\mu_0}. \tag{11.5}$$

Eq. (11.5) shows clearly the energetics of the superconducting transition. In zero field the superconducting state is lower in free energy by $(H_c^2/2\mu_0)$ per unit volume; for a typical critical field of 10–100 mT this is about 10^3J/m^3. The electrons have a lower ground state than a normal Fermi gas, but the Meissner effect shows that it is an essential property of this state that magnetic field is excluded. Therefore, in a field, the system has to choose between gaining energy by being superconducting and gaining energy by letting the field in. For small fields the superconducting state is energetically preferable, and for large fields the normal state is energetically preferable; the critical field is determined by Eq. (11.5).

It is also instructive to determine entropies from $S = -(\partial G/\partial T)_{\mathbf{H}}$. Eq. (11.5) gives

$$S_n - S_s = -\frac{H_c}{\mu_0} \frac{dH_c}{dT}. \tag{11.6}$$

Since the critical field curve of Fig. 11.1(b) has a negative slope, $S_n \geqslant S_s$, the superconducting state is therefore a *more ordered state* than the normal state. Eq. (11.6) also shows that the fact that dH_c/dT tends to zero at absolute zero is a consequence of the requirement of the third law of thermodynamics that $S_n \to S_s$ as $T \to 0$.

11.1.3   The London equation

We saw in section 5.3 that an unperturbed electron wavefunction led to strong diamagnetism in the sense that an applied magnetic field would be excluded from the interior of the region occupied by the electron, except for a surface layer about 300 Å thick. The fact that ordinary atomic wave-functions are small in extent compared with 300 Å provides an explanation of the small diamagnetic susceptibilities ordinarily observed. The arguments of section 5.3 therefore suggest that the origin of the Meissner effect may be in electronic wavefunctions of macroscopic extent not readily perturbed by a magnetic field.

The key equation is Eq. (5.12), which we rewrite here for a general current carrier of charge q,

$$\mathbf{j} = \frac{nq}{m}(\hbar\nabla\theta - q\mathbf{A}).$$ (11.7)

This was originally postulated phenomenologically by London (with a general gradient term instead of $\hbar\nabla\theta$) in order to account for the Meissner effect. But it is instructive to note here that Eq. (5.12) was derived for a single particle wavefunction of the form $\psi = n^{1/2}e^{i\theta}$. If an equation such as Eq. (11.7) is to apply to our many electron system, it means that the de Broglie waves describing the carriers of charge q must all be *coherent* in the sense that they have a common phase θ. This is quite analogous to the coherence of photons in laser light, and under such conditions (many particles per cubic wavelength) the phase of the wavefunction becomes an observable quantity. For photons it is easy for many to have the same wavefunction, because of Bose–Einstein statistics, but for electrons this is prevented by the exclusion principle. However, there is nothing to prevent *pairs* of electrons (which behave as bosons) having a common wavefunction, and this is indeed what the microscopic theory of superconductivity due to Bardeen, Cooper and Schrieffer* (B.C.S.) shows to happen: an attractive interaction between electrons leads to the formation of weakly-bound electron pairs *all of which have the same centre of mass momentum.*

The consequences of the $\hbar\nabla\theta$ term in Eq. (11.7) and the evidence that $q = 2e$ we shall consider in section 11.2. Our present purpose is just to derive the Meissner effect, for which only the curl of Eq. (11.7) is required:

$$\text{curl } \mathbf{j} = -\frac{nq^2}{m}\mathbf{B}.$$ (11.8)

This is the equation originally postulated by London; we see that the $\hbar\nabla\theta$ term has disappeared from it. With Maxwell's equations for static fields

$$\text{curl } \mathbf{B} = \mu_0\mathbf{j}$$

and

$$\text{div } \mathbf{B} = 0.$$

Eq. (11.8) yields

$$\lambda^2\nabla^2\mathbf{B} = \mathbf{B},$$ (11.9)

in which $\lambda^2 = (m/\mu_0 nq^2)$, as in section 5.3.

It is important to note that we are now taking the screening currents explicitly into account rather than considering them as a diamagnetic

* For an elementary account, see Rose-Innes and Rhoderick.[22]

moment, as in section 11.1.2. Either point of view is possible, but it is important to be consistent. On the viewpoint of section 11.1.2, which is useful for macroscopic problems (on a scale $\gg \lambda$), we have for a superconductor in a static magnetic field

$$\mathbf{B} = \mathbf{H} + \mu_0\mathbf{M} = 0 \text{ inside}$$

$$\mathbf{j} = 0, \qquad \mathbf{M} = -\mathbf{H}/\mu_0 \neq 0.$$

If all currents are explicitly taken into account, as in the present section, we have

$$\mathbf{B} = \mathbf{H} = 0 \text{ inside}$$

$$\mathbf{j} \neq 0, \qquad \mathbf{M} = 0.$$

As we saw in section 5.3, the consequence of Eq. (11.9) is that $\mathbf{B}$ decays exponentially into the interior of the sample in a characteristic distance λ. Since $\lambda \sim 300 \text{ Å}$, it is an extremely good approximation for a macroscopic sample to say that $\mathbf{B} = 0$ inside. The finite thickness of the surface layer containing the screening currents is a manifestation of electron inertia. This may be seen most clearly by noting that the acceleration equation

$$m\frac{d\mathbf{v}}{dt} = q\mathbf{E},$$

together with $\mathbf{j} = nq\mathbf{v}$ and Maxwell's equations, leads to the time derivative of Eq. (11.8); in the limit $m \to 0$, $\lambda \to 0$, since $\lambda = (m/\mu_0 nq^2)^{1/2}$. Superconductivity is more than zero resistance, however, because Eq. (11.8) itself, rather than its time derivative, is needed to describe the Meissner effect. In fact, Eqs. (11.7) and (11.8) are not exact in that the relation between $\mathbf{j}$ and $\mathbf{A}$ is not a local one—$\mathbf{j}$ at any point is actually related to the average of $\mathbf{A}$ over a small region round the point. But this subtlety need not concern us here; Eq. (11.8) gives the correct qualitative behaviour.

For samples with one or more dimensions of the order of λ (evaporated films or colloids) there is considerable field penetration and magnetization measurements on such samples can be used to measure λ. Such measurements give the expected order of magnitude, but the detailed results are part of the evidence for a non-local relation between $\mathbf{j}$ and $\mathbf{A}$, referred to above.

11.1.4   High frequency resistance

Although the DC resistance transition of a superconductor is sharp, at radiofrequencies it becomes steadily less abrupt, though the resistance still decreases to zero at absolute zero. There is always an inductive component of surface impedance due to electron inertia (the penetration depth λ is an inductive skin depth), but in general there is also a resistive component of surface impedance. In the millimetre wave region power absorption at absolute zero sets in, and finally in the infrared region the superconducting transition is not detectable. A typical set of results is shown in Fig. 11.4(*a*).

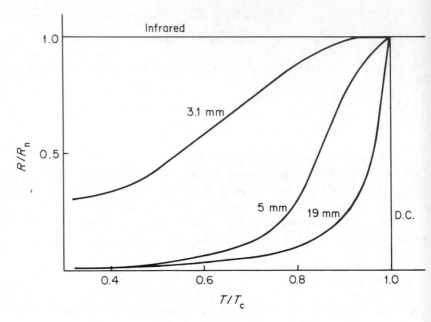

(*a*)   Microwave resistance of Al, for the indicated wavelengths.

[after M.A.Biondi and M.P.Garfunkel, *Phys. Rev. Lett.*, **2**, 143 (1959)].

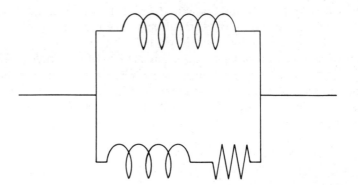

(*b*)   Radiofrequency equivalent circuit of a superconductor.

Fig. 11.4.

At the lower frequencies the situation may be qualitatively represented by the equivalent circuit shown in Fig. 11.4(*b*). The superconducting electrons are represented by a pure inductance, but in parallel with these are some normal electrons represented by a resistance and an inductance in series. For DC, the normal electrons are short circuited by the superconducting electrons. These normal electrons result from the dissociation of superconducting electron pairs by thermal agitation; at T_c all pairs have become dissociated and the metal reverts to the normal state.

The absorption at absolute zero cannot be explained in this way; it is due to the direct dissociation of electron pairs by absorption of microwave photons. The photon energy at which absorption begins is a measure of the binding energy of an electron pair; it is found to be 3.5–4.5 $k_B T_c$, about 10^{-22} J in a typical case. Since we have seen in section 11.1.2 that the total binding energy of the superconducting state relative to the normal state is about 10^3 J/m^3, it follows that there are only about 10^{25} electron pairs/m^3, so that effectively only about 1 in 10^4 of the electrons are paired. This is in accordance with the B.C.S. theory, which gives a binding energy only for electrons within about $k_B T_c$ of the Fermi energy; $(T_c/T_F) \sim 10^{-4}$.

11.2   TIME-INDEPENDENT QUANTIZATION

11.2.1   The flux quantum

If we consider Eq. (11.7) in the interior of a superconductor, where $\mathbf{j} = 0$, it becomes

$$\hbar \nabla \theta = q\mathbf{A} \qquad (11.10)$$

Since θ is defined as the phase of a wavefunction, it is not necessarily a single valued function of position; the wavefunction itself must be single valued but this only requires θ to be defined within 2π. Thus, if we integrate $\nabla \theta$ along a closed path the result need not in general be zero, it can be any multiple of 2π. Integrating Eq. (11.10) round a closed path in the superconductor we have

$$2\pi n\hbar = q \oint \mathbf{A} \cdot d\mathbf{l};$$

or in other words the magnetic flux through our closed path is given by

$$\Phi = \iint \mathbf{B} \cdot d\mathbf{S} = \oint \mathbf{A} \cdot d\mathbf{l} = n\frac{h}{q}, \qquad (11.11)$$

and is thus quantized. For a block of superconductor the Meissner effect means that $n = 0$, but if we have a ring-shaped sample we may have $n \neq 0$ for a path of integration going round the ring: flux may be trapped in the ring without any field penetrating the metal. Such trapped flux has been known for a long time, and its constancy with time gives the best experimental evidence for strictly zero resistance in the superconducting state. But to detect its quantization is more difficult, since (h/e) is only 4×10^{-15} Wb $(= 4 \times 10^{-7}$ gauss cm^2).

An experiment to measure the flux quantum is illustrated schematically in Fig. 11.5. The specimen is a fine copper wire a few mm long and about 10 μm diameter with an electroplated film of tin on its centre section (remember that copper is an insulator for superconductors!); the reason for this specimen shape is that one flux quantum then corresponds to a reasonable field of order $10 \, \mu T$ (0.1 gauss) trapped inside the tin cylinder. In an experiment the sample is placed in a magnetic field of this order and cooled through the transition temperature; it is then vibrated between two search coils connected in series opposition so as to give an AC signal proportional to its magnetic moment. A series of such experiments gives results which are shown in Fig. 11.6(a); There is a magnetic moment, but this is made just about as small as possible by allowing one or more flux quanta to pass through the annular superconducting region.

If the external field is now removed and the magnetic moment measured again it is found that this quantized flux remains trapped by the specimen, as shown in Fig. 11.6(b). The higher quanta in Fig. 11.6 become less well defined because of an experimental difficulty: there may be a flaw in the evaporated tin film part way along its length through which one or more flux quanta can pass.

Apart from the existence of the flux quantum, the important result of this experiment is its magnitude, $h/2e$. This is strong evidence that the charge carriers in a superconductor are indeed electron pairs.

11.2.2   Type II superconductors

We now consider rather more closely the reason for the Meissner effect, in the light of Eqs. (11.10) and (11.11). Why can we not have a closed path in the interior of a superconductor for which

$$\Phi = \oint \mathbf{A} \cdot d\mathbf{l} = \frac{h}{q} \oint \nabla \theta \cdot d\mathbf{l} \neq 0 ?$$

Continuity of the function $\theta(\mathbf{r})$ surely requires that this integral must be the same for all paths into which our original path can be continuously deformed. The problem arises when we try to shrink the path to zero (which cannot always be done for a ring while remaining inside the superconductor). As we shrink the path the variations in θ become more and more rapid until we reach a singularity where $\nabla \theta$ is infinite; to keep $\nabla \psi$ finite this must be a nodal point in ψ for which $|\psi| = 0$. A little thought shows that this singular point

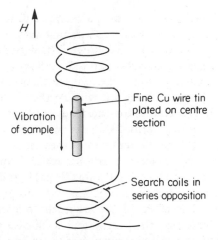

Fig. 11.5. Arrangement of experiment
to measure flux quanta.

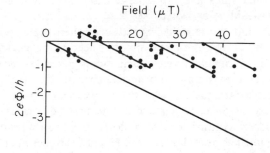

(*a*)   Magnetic moment of sample after cooling in a field.

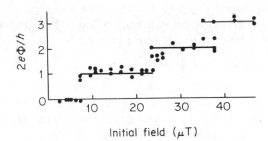

(*b*)   Trapped flux after removal of field.

Fig. 11.6. Quantized flux steps in a plated tin cylinder.
{after B. S. Deaver and W. M. Fairbank, *Phys. Rev. Lett.*, **7**,
43 (1961)}.

must lie on a nodal line of ψ; we have a quantized flux line, which can be thought of as a filament of non-superconducting material in the metal.

The structure of such a flux line is shown schematically in Fig. 11.7. The density of superconducting electron pairs $|\psi|^2$ falls to zero on the axis, and there is a circulating current round the axis which contains the flux. Such a flux line has kinetic energy associated with it; there is the classical kinetic energy of the screening currents arising from gradients in the phase of ψ, and there is also a quantum term, like zero point energy, arising from gradients in the amplitude of ψ. Adjustment of these two energies to minimize the total determines a characteristic length ξ which is the size of the region within which $|\psi|^2$ is reduced to a small value (Fig. 11.7(a)). However, because of Eq. (11.9), the magnetic field of a flux line is spread over a region of order λ in size (Fig. 11.7(c)). Therefore, by the argument following Eq. (11.5), there is a gain of energy by letting the magnetic field in over an area of order λ^2 and a loss of energy because the metal is effectively not superconducting over an area of order ξ^2. It is therefore energetically preferable to form flux lines, and a complete Meissner effect does not occur, if $\lambda > \xi$.* On the other hand, if $\lambda < \xi$ flux lines are not energetically favoured, and a complete Meissner effect would be expected.

It so happens that the length ξ is reduced when the electron mean-free path is reduced by alloying, for reasons that need not concern us here. This gave rise to a belief, for many years, that incomplete Meissner effect was in some way associated with alloy inhomogeneity and flux trapping. Fig. 11.8 shows that this is not the case, however; an *annealed* lead alloy has an almost *reversible* magnetization curve in which flux first penetrates at a field H_{c1} less than the thermodynamic critical field H_c of pure lead, and penetration is not complete until a much larger field H_{c2}. In the region $H_{c1} < H < H_{c2}$ the sample is threaded by a close packed array of single quantum flux lines, as has been shown by electron microscopy.† Materials showing this gradual flux penetration are known as type II superconductors, whereas those showing a Meissner effect up to H_c are known as type I. The fact that if $H_{c1} \ll H_c$, $H_{c2} \gg H_c$, is a consequence of Eq. (11.4): the area under the reversible magnetization curve is always $H_c^2/2\mu_0$.

★ **11.2.3 Superconducting magnets**

The basic fact behind the possibility of making useful superconducting magnets is that a type II superconductor remains superconducting in a field very much larger than its thermodynamic critical field H_c. Alloys such as Nb_3Sn remain superconducting in fields over 10 T.

---

* By use of Eq. (11.5), this argument leads to the conclusion that H_{c1} in Fig. 11.8 is given by $(H_{c1}/H_c) \sim (\xi/\lambda)$, which is correct in order of magnitude.

† See, for example, U. Essman and H. Traüble, *Scientific American*, **224** (Mar.) 74 (1971).

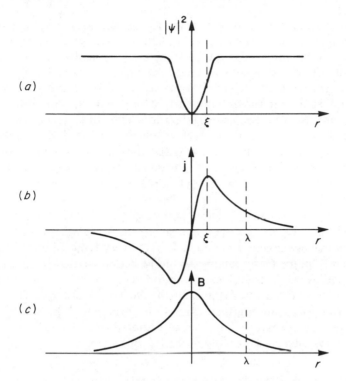

Fig. 11.7. Schematic cross-sections of a quantized flux line.

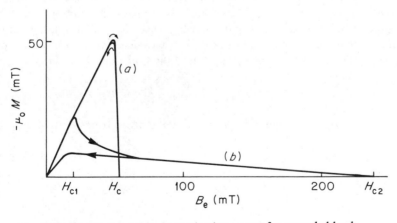

Fig. 11.8. Almost reversible magnetization curves for annealed lead:
(a) Pure lead (type I).
(b) Lead made type II by alloying with 8.23% In.
{after J. P. Livingston, *Phys. Rev.*, **129**, 1943 (1963)}.

There is a snag, however. A ring of type I superconductor traps a magnetic flux because the Meissner effect does not allow flux to cross it. But flux can penetrate, and therefore cross, a type II material for fields between H_{c1} and H_{c2}. For this reason a ring of type II material with a reversible magnetization curve (as shown in Fig. 11.8) cannot trap a magnetic field, and correspondingly a solenoid of such material will not carry a persistent current. Some mechanism is required to prevent the free migration of flux lines, giving an *irreversible* magnetization curve, before such a solenoid will sustain a persistent current. This is usually done by making the alloy inhomogeneous, either by precipitation or work hardening. By this means regions where the flux line energy is low are produced, which act as pinning centres for the lines, and thus prevent flux migration.

When a superconducting solenoid is operating near its maximum field a small region may revert to the normal state, which has a high resistivity. If precautions are not taken this can result in the very large amount of stored energy being dumped as heat in the liquid helium bath, with disastrous consequences. A very satisfactory precaution in practice is to coat the superconductor with a thick layer of copper; in the event of a small region becoming normal the copper cladding then carries the current with little dissipation and prevents the normal region from spreading disastrously; consequently the field decays only slowly, without drastic consequences.

11.3   TIME-DEPENDENT QUANTIZATION

The phase variations of the superconducting wavefunctions considered in section 11.2 represent a spatial quantization analogous to the time independent Schrödinger equation. We should also expect the common phase of the superconducting electron pairs to vary with time in an analogous way to the phase of an ordinary single particle wavefunction. The relevant energy is the chemical potential μ of a pair, so that the time dependence of ψ is given by

$$\psi \propto e^{-i\mu t/\hbar}. \tag{11.12}$$

More generally, if the chemical potential depends on time we have

$$\psi \propto e^{+i\phi(t)},$$

in which

$$\hbar \frac{\partial \phi}{\partial t} = -\mu. \tag{11.13}$$

Ordinarily, because a superconductor cannot sustain potential differences, μ is uniform and there are no observable consequences of Eq. (11.12). It was, however, noticed by Josephson that since a voltage difference can be applied to two pieces of superconductor separated by a thin insulating layer, Eqs. (11.12) or (11.13) should have interesting observable consequences if the probability of an electron pair tunnelling through the insulating barrier is not zero.

11.3.1   The Josephson junction

Consider two pieces of superconductor separated by an insulating barrier, as in Fig. 11.9. The superconducting wavefunction ψ can be considered as made up of two parts: one originating in region 1 and decaying exponentially into the insulator; and one originating in region 2 and decaying exponentially into the insulator. The absolute values of these two contributions must be equal because the electron concentrations on the two sides are equal, so the only freedom we have is to choose their phases ϕ_1 and ϕ_2. If the junction is thick enough that the exponential originating in region 1 has practically decayed to zero by the time it reaches 2, ϕ_1 and ϕ_2 are the phases of the *total* wavefunction in regions 1 and 2 respectively. The total wavefunction in the insulating region is thus

$$\psi = e^{[i\phi_1 - K(x+a)]} + e^{[i\phi_2 + K(x-a)]}$$

$$= e^{-Ka}(e^{[i\phi_1 - Kx]} + e^{[i\phi_2 + Kx]}). \tag{11.14}$$

The current j in the junction may now be calculated from Eq. (5.11) for charge q in the absence of a magnetic field

$$j = \frac{\hbar q}{2mi}[\psi^*\nabla\psi - \psi\nabla\psi^*]. \tag{11.15}$$

Eq. (11.14) gives

$$\psi^*\nabla\psi = e^{-2Ka}[-K\,e^{i(\phi_1-\phi_2)} + Ke^{i(\phi_2-\phi_1)} - K\,e^{-2Kx} + K\,e^{2Kx}],$$

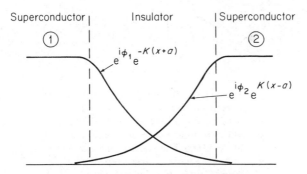

Fig. 11.9. Schematic contributions to the wavefunction of superconducting carriers in an insulating barrier.

with which Eq. (11.15) gives

$$j = \frac{2\hbar q}{m} K \, e^{-2Ka} \sin(\phi_2 - \phi_1)$$

$$= j_0 \sin \delta, \tag{11.16}$$

in which j_0 is a constant characteristic of the junction and $\delta = (\phi_2 - \phi_1)$ is the phase difference across the junction.

If there is a potential difference V across the junction the difference of chemical potential μ is qV so that from Eq. (11.13) we also have

$$\delta = qVt/\hbar + \delta_0 \tag{11.17}$$

with which Eq. (11.16) becomes

$$j = j_0 \sin[(qVt/\hbar) + \delta_0]. \tag{11.18}$$

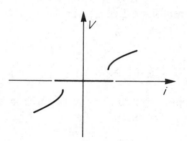

(*a*)   Without microwaves.

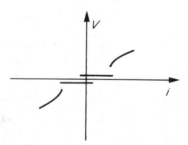

(*b*)   With microwaves.

Fig. 11.10. Stable part of the current-voltage characteristic of a Joshephson junction with and without microwave irradiation at 9.3 GHz. The steps in (*b*) are at $V = \pm \hbar\omega/2e$; more steps are seen at higher microwave power. {after S. Shapiro, *Phys. Rev. Lett.*, **11**, 80 (1963)}.

This is the basic equation of a Josephson junction; it is interesting to notice that this result may alternatively be derived by treating the regions 1 and 2 as weakly coupled quantum mechanical systems and applying Eqs. (1.7) (see Feynman,[1] section 21-9).

Eq. (11.18) indicates that this type of junction should have very remarkable properties. For $V = 0$ any current between $\pm j_0$ can flow by adjustment of δ_0; $j = \pm j_0$ for $\delta_0 = \pm \pi/2$. But as soon as a voltage is applied this current stops—there is only an oscillatory current at angular frequency $qV/\hbar$.

If the junction is now irradiated with microwaves at frequency ω an additional voltage $v \cos \omega t$ at this frequency appears across it so that Eqs. (11.13) and (11.16) now give

$$ j = j_0 \sin \left[\frac{q}{\hbar} \left(Vt + \frac{v}{\omega} \sin \omega t \right) + \delta_0 \right]. \tag{11.19} $$

This is a frequency modulated current and contains components at angular frequencies

$$ \frac{qV}{\hbar} \pm n\omega $$

where n is any integer. Thus a DC current (zero frequency) now flows if

$$ V = \frac{n\hbar\omega}{q}. \tag{11.20} $$

This predicted behaviour is somewhat blurred by the fact that an actual junction shows a tunnel current of single electrons as well as pairs. Nevertheless, Fig. 11.10(b), which is the current–voltage characteristic of a microwave irradiated Josephson junction, does show very well defined voltage steps in agreement with Eq. (11.18) with $q = 2e$; the zero-voltage tunnel current without microwave irradiation, is shown for comparison in Fig. 11.10(a). These voltage steps are so precise that they have been used to determine $\hbar/e$ to a precision that is useful in deciding best values of the fundamental constants.[*]

11.3.2   Quantum interferometers

Since Josephson junctions are sensitive to the phase difference of the superconducting wavefunctions across them, an interferometer may be constructed by connecting two such junctions in parallel as in Fig. 11.11(a). Eq. (11.11) for the phase difference in going round a superconducting ring

* W. H. Parker, D. N. Langenberg, A. Denenstein and B. N. Taylor, *Phys. Rev.*, **177**, 639 (1969).

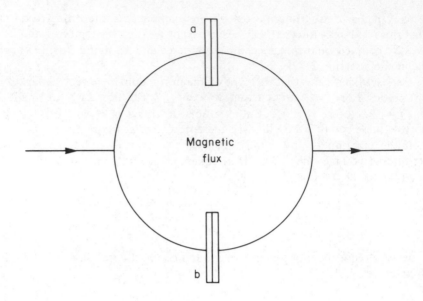

(*a*)   Two   Josephson   junctions   in   parallel.

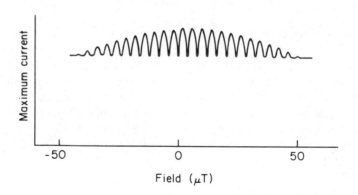

(*b*)   Maximum   current   passed   by   a   pair   of   evaporated   film
        Josephson   junctions   as   a   function   of   magnetic   field.
        [after   R.C. Jaklevic,   J. Lambe,   J.E. Mercereau,   and   A.H. Silver,
        Phys. Rev., **140A**, 1628 (1965)].

Fig. 11.11.

is now replaced, for the ring containing two junctions, by

$$\frac{q}{\hbar} \oint \mathbf{A} \cdot d\mathbf{l} + \delta_a - \delta_b = 2\pi n, \qquad (11.21)$$

where δ_a and δ_b are the phase differences across the junctions. The maximum current is passed by the two junctions in parallel if $\delta_a = \delta_b = \pi/2$, from Eq. (11.16). But this condition can be satisfied only if the flux through the loop is quantized according to Eq. (11.11), for other values of flux the maximum current is reduced. Therefore, if the maximum current is measured as a function of magnetic field an interference pattern is obtained, as in Fig. 11.11(b). Experiments have been done with junctions separated by distances of order 1 cm, giving impressive evidence that superconducting electron pairs are phase coherent over this sort of distance.

Because of the smallness of the flux quantum, a pair of junctions as in Fig. 11.11(a) embracing an area of 1 cm^2 would change from maximum to minimum critical current for a change in field of only 10^{-11} T (10^{-7} gauss). Josephson junctions can therefore be applied to measure magnetic fields with very great precision; they may also be used to construct very sensitive ammeters by sensing the field due to the current.

CHAPTER

Magnetic order

12.1 ARE MOMENTS LOCALIZED?

In Chapter 5 we considered magnetism in terms of definite magnetic moments localized on the atoms of a crystal. In this chapter we shall first consider some experimental evidence that moments are not always localized in this way and that more exotic types of magnetic ordering than those we have so far considered can occur. In the optional section 12.2 we show that the Weiss molecular field model is not strongly dependent on the assumption of localized moments, and that it can give rise to other types of ordering. The most important practical example of non-uniform magnetic ordering, the domain structure in a ferromagnet, we consider briefly in section 12.3.

In deciding whether magnetic moments are localized two pieces of evidence are particularly important: does the moment per atom correspond to anything we might reasonably expect from our knowledge of atomic structure; and does the magnetic order–disorder transition have associated with it the entropy change $R \ln (2S + 1)$ that one would expect for N localized spins S?

12.1.1 Saturation magnetization in the iron group

To consider the question of localization in the iron group of ferromagnetic metals we calculate the number of Bohr magnetons per atom corresponding to the experimental saturation magnetization. The results are given in

Table 12.1, and we notice at once that they are not integers. Also shown in Table 12.1 are the magnetic moments per ion for these elements in paramagnetic salts, deduced from the Curie law susceptibility; it can be seen that these moments are considerably larger.

Table 12.1. Magnetic moments of iron group metals and ions (in Bohr magnetons per atom)

	Saturation magnetization	S	Curie law moment	$2\sqrt{S(S + 1)}$	$g\sqrt{J(J + 1)}$
Fe	2.22	Fe^{3+} $\frac{5}{2}$	5.82	5.91	5.91
		Fe^{2+} 2	5.36	4.90	6.71
Co	1.72	$\frac{3}{2}$	4.90	3.88	6.63
Ni	0.61	1	3.12	2.83	5.59

Let us first try to understand the paramagnetic salt results. These should give a moment per ion, in Bohr magnetons, of $g\sqrt{J(J + 1)}$, where g is the Landé splitting factor. In fact, as the table shows, the results are much closer to the value $2\sqrt{S(S + 1)}$ that one might expect if only the spin moment coupled to the external magnetic field. This is a well-understood effect known as quenching of orbital angular momentum. It arises because orbital angular momentum, unlike spin, is associated with an aspherical charge cloud which is strongly influenced by crystalline electric fields. The symmetry of the environment is usually such that states with an angular variation as $\cos l_z \phi$ and $\sin l_z \phi$ have different energies. Since these states are mixtures of equal parts of $e^{il_z \phi}$ and $e^{-il_z \phi}$ they have zero expectation value of l_z and therefore no magnetic moment. Consequently, provided the states $\cos l_z \phi$ and $\sin l_z \phi$ are not degenerate, first order perturbation by a magnetic field produces no orbital magnetization, and the orbital moment is quenched.

The paramagnetic salt results suggest that we should try to understand the ferromagnetism of iron, cobalt, and nickel in terms of a spin-only moment. Confirmatory evidence for this idea comes from gyromagnetic experiments on these metals. It is possible to measure the small angular momentum impulsively generated when an iron rod is suddenly magnetized, and hence the gyromagnetic ration $\gamma (= ge/2m)$; the experiments give g within one or two percent of the value 2 for spin only.

On this basis we might expect the saturation magnetization per atom to be $2S$ Bohr magnetons.* Table 12.1 shows that the observed values are much less than this, as well as non-integral. We deduce that the magnetic entities are

---

* Remember that the saturation magnetization gives maximum *resolved component* of the moment, not the magnitude of the moment.

electron spins, but they are not localized on atoms in any simple way. This leads us to visualize the $3d$ magnetic electrons, like the $4s$ conduction electrons, as a band of itinerant states of the type we considered in Chapter 4. The simplest form of this band picture, for iron, is shown in Fig. 12.1; we show the spin-up and spin-down densities of states for each band separately,

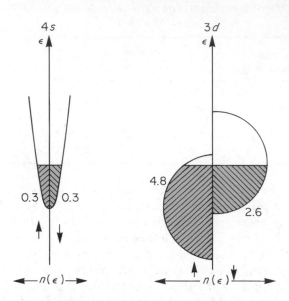

Fig. 12.1. Simplified band picture of ferromagnetism in iron. The vertical arrows indicate spin directions, and the numbers against the curves show the number of electrons per atom in each band.

as we did when calculating conduction electron paramagnetism (Fig. 5.2). We suppose that the Weiss molecular field acts on the $3d$ electrons to lower the energy of spin-up relative to spin-down, and thus produces a magnetization of the $3d$ band via the Pauli paramagnetism (compare Figs. 5.2 and 12.1). Because the Pauli susceptibility is finite at $T = 0$, we now obtain a spontaneous magnetization related to the Weiss molecular field, in contrast to the total spin alignment at $T = 0$ that we obtained previously with the Curie law susceptibility of localized moments. The effect of delocalizing the magnetic electrons is thus, by introducing Fermi degeneracy, to reduce the spontaneous magnetization at $T = 0$.

The above model is oversimplified in that there is probably some s electron polarization too; indeed it is quite possible that the s electrons mediate the exchange interaction between d electrons that is the basis of the Weiss molecular field model.

12.1.2   The antiferromagnetism of chromium

Chromium crystallizes in the body-centred cubic structure. The first neutron scattering experiments, by the powder method on a polycrystalline sample, showed that (100) reflections were present below the Néel temperature of 310 K, although these reflections should be absent for the bcc structure (see section 6.3.2.). This suggested that the magnetic structure was effectively simple cubic, with the body-centring atoms having spins opposed to the others. However, the intensity of the (100) reflection corresponded to a moment of only about $0.4\mu_B$ per atom, and the specific heat anomaly associated with the transition involved an entropy change of only about $0.002R$, more typical of a Fermi gas. Even these early observations were therefore rather suggestive of delocalized magnetic electrons.

Later neutron scattering experiments on single crystals have shown that the magnetic structure is stranger than the simple antiferromagnetic pattern described above, and quite incompatible with localized magnetic electrons. The (100) magnetic reflection in fact consists of a group of six closely spaced reflections, each displaced from the (100) reciprocal lattice point by small amounts along [100] directions. This corresponds to a slow sinusoidal modulation of the structure previously described, as illustrated schematically in Fig. 12.2. This structure may alternatively be described as a standing spin

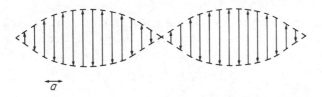

Fig. 12.2. The antiferromagnetic structure of chromium
in a [100] direction.

density wave of wavenumber slightly less that $2\pi/a$ (wavelength slightly greater than a, compare Fig. 2.8). The direction of the arrows in Fig. 12.2 is not necessarily meant to indicate the polarization of the spin-density wave; in fact, the neutron scattering intensities rather suggest transverse polarization above 115 K, with a transition to longitudinal polarization below that temperature.

Overhauser has suggested that an electron gas may be unstable to perturbations of wavenumber slightly greater than $2k_F$.* The Fermi surface of chromium is certainly not spherical, but it is thought to contain, *inter alia*, electron and hole surfaces that can be brought into near coincidence over a

* For a discussion of his suggestion of possible charge density waves in potassium, see section 10.3.

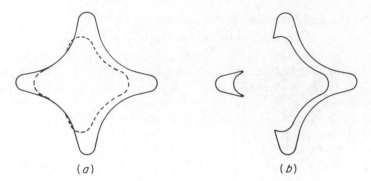

Fig. 12.3. The schematic electron Fermi surface of Cr is shown at (*a*), together with the hole Fermi surface (broken curve) displaced by the wavevector of the antiferromagnetic spin density wave. The effect of a periodic perturbation due to the spin-density wave on these Fermi surfaces is shown at (*b*).
{based on W. M. Lomer, *Proc. Phys. Soc.*, **80**, 489 (1962)}.

considerable area by relative displacement in **k**-space by a wavevector slightly less than $2\pi/a$, as shown in Fig. 12.3(*a*). This means that a perturbation of this wavenumber would couple these two electron energy bands together, reduce the energy of the occupied states, and reform the Fermi surface as shown in Fig. 12.3(*b*).* There is some experimental evidence that an effect of this type does indeed occur in chromium, for the electrical resistivity is observed to rise sharply just below the Néel point, as would be the case if the free area of Fermi surface were reduced in the antiferromagnetic state.

12.1.3   Rare earth magnetism

A plane-polarized spin-density wave, such as we considered in the last section, is clearly incompatible with localized moments at absolute zero, for the magnitude of the magnetic moment per atom varies sinusoidally through the crystal. However, if we consider instead a *circularly* polarized transverse standing wave, the magnitude of the moment per atom is constant; such a structure is not incompatible with localized moments, and is typical of the rare earth metals, which we now discuss.

The saturation magnetization of ferromagnetic rare earth metals corresponds closely to $gJ\mu_{\mathrm{B}}$ with unquenched orbital angular momentum.† We

---

*The true situation is more complicated than that shown in Fig. 12.3, which shows the perturbation due to a single wavenumber $k = (2\pi/a - \varepsilon)$, because even a single plane polarized spin density wave contains wavenumbers $+k$ and $-k$. Also, it is not obvious that a spin-density wave is the most energetically favourable mode of perturbation.

† Small departures, for example in gadolinium, give evidence for significant polarization of the unlocalized conduction electrons.

may understand this result by noting that the wavefunctions of the $4f$ shell, which give rise to rare earth magnetism, are buried much deeper within the ion than are the $3d$ wavefunctions responsible for the magnetism of the iron group. Consequently, crystalline electric fields have a much smaller effect on $4f$ electrons, so that the orbital angular momentum is unquenched, and the $4f$ wavefunctions probably do not overlap sufficiently to produce a band of mobile states. The experimental and theoretical evidence thus favours localized free-ion moments as the basis of rare earth magnetism.

Another consequence of the small overlap of $4f$ wavefunction centred on adjacent atoms is that direct exchange interaction is almost certainly inadequate to account for the observed magnetic ordering in these metals. It is highly probable that the relevant exchange interaction is a two stage process: ion A interacts with the $6s$ conduction band which in turn interacts with ion B. This means that the magnetic ordering may well be accompanied by some form of standing spin density wave in the $6s$ conduction band, and the alignment of the rare earth ionic moments will be correlated with this.

The rare earth metals all crystallize in either the hexagonal close-packed structure, or in very similar structures of heagonal symmetry with a more complicated stacking sequence of close-packed layers. Extensive neutron scattering experiments on single crystals of these metals have revealed a great variety of magnetic ordering arrangements. Typical is dysprosium, which is ferromagnetic below 85 K with the easy direction of magnetization in the basal plane (i.e., in a close-packed layer). Between 85 K and 179 K the ordering remains ferromagnetic in a close-packed layer, but as we go from one layer to the next the direction of magnetization turns through an angle of order 30° about the c-axis; this angle changes with temperature. This spiral ordering may be described as a transverse circularly polarized standing spin density wave, with a wavenumber incommensurate with the lattice periodicity. Above 179 K dysprosium is paramagnetic.

Most rare earths show this spiral ordering under some conditions. Even more complicated arrangements can occur, in which spiral ordering is accompanied by a component of magnetization along the c-axis which may also show periodic behaviour.

12.2 MOLECULAR FIELD THEORY IN A CONTINUUM

In section 5.4.1 we formulated the Weiss molecular field model in terms of magnetic moments localized on atoms. In this section we wish to reformulate the model in such a way as to show that it can encompass the phenomena we have just discussed, itinerant magnetic carriers and magnetic ordering of the spin density wave type. The model will remain oversimplified—for example, we shall continue to ignore crystalline anistropy—but we shall obtain some important qualitative results.

In the elementary form of the Weiss model we consider only a uniformly magnetized sample, and relate the molecular field $\mathbf{B}_{eff}$ to this magnetization $\mathbf{M}$, although we know that $\mathbf{B}_{eff}$ at an atom really depends mainly on the magnetization of its near neighbours. We now make the model appropriate to itinerant electrons by considering a continuous distribution of magnetization density, and for simplicity we ignore variations on an atomic scale. The idea of short range exchange interactions we incorporate by putting

$$\mathbf{B}_{eff}(\mathbf{r}) \propto \int d^3r' J(\mathbf{r} - \mathbf{r}')\mathbf{M}(\mathbf{r}'), \tag{12.1}$$

where $J(\mathbf{r} - \mathbf{r}')$ is an exchange integral. For our purposes it is convenient to transform the non-local relation (12.1) into a local one. To do this we expand $\mathbf{M}(\mathbf{r}')$ in a Taylor series about the point $\mathbf{r}$, and for simplicity confine our attention to situations in which $\mathbf{M}$ varies only in the z direction. The terms involving odd derivatives of $\mathbf{M}(\mathbf{r})$ then integrate to zero, and the other integrals are merely numerical coefficients of the even derivatives of $\mathbf{M}(\mathbf{r})$. We thus have

$$\mathbf{B}_{eff} = \lambda_0 \mathbf{M} + \lambda_2 \frac{\partial^2 \mathbf{M}}{\partial z^2} + \lambda_4 \frac{\partial^4 \mathbf{M}}{\partial z^4} + \cdots, \tag{12.2}$$

where the λs are constants; note that $\lambda_0 = \mu_0 \lambda$ in the notation of section 5.4. We now combine Eq. (12.2) with the gyroscopic equation of motion of the magnetization and obtain

$$\frac{\partial \mathbf{M}}{\partial t} = \gamma \mathbf{M} \times \mathbf{B}_{eff}$$

$$= \gamma \lambda_2 \mathbf{M} \times \frac{\partial^2 \mathbf{M}}{\partial z^2} + \gamma \lambda_4 \mathbf{M} \times \frac{\partial^4 \mathbf{M}}{\partial z^4} + \cdots. \tag{12.3}$$

In what follows we shall ignore the higher order terms; note that if only the λ_2 term is retained, Eq. (5.30) is essentially the linearized finite-difference form of Eq. (12.3).

12.2.1   Ferromagnetic spin waves

We first rederive, for the continuum model, the $\omega(k)$ relation for spin wave excitations of a ferromagnetic ground state that we obtained in section 5.4.3. We linearize Eq. (12.3) by setting $\mathbf{M} = \mathbf{M}_0 + \mathbf{m}$, where $\mathbf{m}$ ($\ll \mathbf{M}_0$) is perpendicular to $\mathbf{M}_0$.* We take $\mathbf{m}$ as in the x–y plane, which need not be normal to the direction z in which $\mathbf{M}$ varies. To first order in $\mathbf{m}$ Eqs. (12.3)

---

* In other words, we assume the saturation magnetization $\mathbf{M}_0$ is tilted through a small angle, but fixed in magnitude.

then become

$$\frac{\partial m_y}{\gamma t} = \gamma\lambda_2 M_0 \frac{\partial^2 m_x}{\partial z^2} + \gamma\lambda_4 M_0 \frac{\partial^4 m_x}{\partial z^4}$$

$$-\frac{\partial m_x}{\partial t} = \gamma\lambda_2 M_0 \frac{\partial^2 m_y}{\partial z^2} + \gamma\lambda_4 M_0 \frac{\partial^4 m_y}{\partial z^4}.$$

$$(12.4)$$

With the new variable $m_+ = m_x + i m_y$ Eqs. (12.4) reduce to the single equation

$$-i\frac{\partial m_+}{\partial t} = \gamma\lambda_2 M_0 \frac{\partial^2 m_+}{\partial z^2} + \gamma\lambda_4 M_0 \frac{\partial^4 m_+}{\partial z^4},$$ $$(12.5)$$

which has a wavelike solution $m_+ \propto e^{i(kz - \omega t)}$ if

$$\omega = \gamma\lambda_2 M_0 k^2 - \gamma\lambda_4 M_0 k^4,$$ $$(12.6)$$

which for small k is of the same form as Eq. (5.34). We thus reproduce our earlier result for $\lambda_2 > 0, \lambda_4 > 0$ (Fig. 12.4(a)). But it is interesting to consider also what would happen in the case $\lambda_2 < 0, \lambda_4 < 0$, illustrated in Fig. 12.4(b), for which the spin-wave energy is negative for a range of k values. This means that the energy of the system can be reduced below that of the supposed ground state by adding spin waves! Clearly, we have assumed the wrong ground state, and one might conjecture that the true ground state would be periodic with a wavenumber near that giving the minimum spin wave energy.

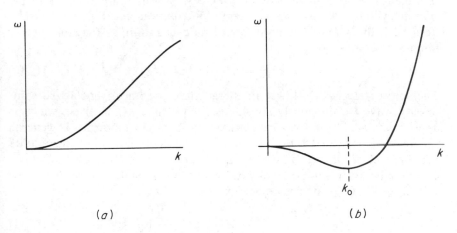

(a)   (b)

Fig. 12.4. Spin wave energy spectra in a continuum:
(a) for $\lambda_2, \lambda_4 > 0$;
(b) for $\lambda_2, \lambda_4 < 0$.

12.2.2 Periodic ground state

With the case $\lambda_2, \lambda_4 < 0$ particularly in mind we now consider the possibility of a standing spin density wave ground state, $\mathbf{M} = \mathbf{M}_0 \cos kz$. Note that, since the derivatives of $\mathbf{M}$ are parallel to $\mathbf{M}$ for this state, the right hand side of Eq. (12.3) vanishes and we do have a possible stationary state. We may calculate the exchange energy density of this state as $-\frac{1}{2}\mathbf{M} \cdot \mathbf{B}_{\text{eff}}$,[*] which is

$$-\tfrac{1}{2}\mathbf{M}_0 \cos kz \cdot \mathbf{M}_0 \left[\lambda_0 \cos kz + \lambda_2 \frac{\partial^2}{\partial z^2}(\cos kz) + \lambda_4 \frac{\partial^4}{\partial z^4}(\cos kz) \right]$$

$$= \tfrac{1}{2}(-\lambda_0 + k^2\lambda_2 - k^4\lambda_4)\mathbf{M}_0^2 \cos^2 kz, \tag{12.7}$$

which gives an average energy density

$$U = \tfrac{1}{4}M_0^2(-\lambda_0 + k^2\lambda_2 - k^4\lambda_4). \tag{12.8}$$

As expected, we see that for $\lambda_2 > 0$ this is a minimum for $k = 0$, the ferromagnetic state. For $\lambda_2, \lambda_4 < 0$ it is a minimum for $k = k_0$ (Fig. 12.4(b)), which, by differentiating Eq. (12.8) with respect to k and equating to zero, is given by

$$k_0^2 = \lambda_2/2\lambda_4; \tag{12.9}$$

our conjecture of a periodic ground state is thus confirmed.

We now consider the possibility of a spiral ground state given by

$$M_z = 0, \qquad M_x = M_0 \cos kz, \qquad M_y = M_0 \sin kz.$$

It is readily verified by direct substitution that this distribution of magnetization makes the right-hand side of Eq. (12.3) vanish, and is thus a possible stationary state. An exchange energy contribution like Eq. (12.8) results from both the x and y components of magnetization, so that the total is doubled:

$$U = \tfrac{1}{2}M_0^2(-\lambda_0 + k^2\lambda_2 - k^4\lambda_4). \tag{12.10}$$

The spiral state is thus lower in energy than the linearly polarized spin-density wave we previously considered, if we take M_0 as fixed, which is appropriate at $T = 0$ if we are considering localized moments. In general, of course, $\mathbf{M}_0$ is related to $\mathbf{B}_{\text{eff}}$ by a nonlinear susceptibility, whose magnitude may depend on the type of ordering, and crystal fields may also affect the energy balance; it is not then obvious which type of state is lower in energy. At k_0 our spiral state has an energy

$$U_{\text{min}} = -\tfrac{1}{2}M_0^2\left(\lambda_0 + \frac{\lambda_2^2}{4|\lambda_4|} \right). \tag{12.11}$$

---

[*] The factor $\tfrac{1}{2}$ compensates for counting each spin twice, once as a magnetic moment and once as a source of $\mathbf{B}_{\text{eff}}$.

12.2.3 Excitations of the spiral ground state

To find the spin-wave excitations of the spiral ground state we follow our previous procedure of adding a small magnetization **m** perpendicular to the ground state magnetization $\mathbf{M}_0$, and linearizing Eq. (12.3) in **m**. Specifically, we now take

$$M_z = m_z, \qquad M_x = M_0 \cos k_0 z + m_x, \qquad M_y = M_0 \sin k_0 y + m_y$$

with k_0 given by Eq. (12.9). We deal first with the equations for the x and y components of magnetization, which are simplest. Substitution in Eq. (12.3) gives

$$\frac{\partial m_x}{\partial t} = \gamma \lambda_2 \left[M_0 \sin k_0 z \frac{\partial^2 m_z}{\partial z^2} - m_z(-k_0^2 M_0 \sin k_0 z) \right]$$
$$+ \gamma \lambda_4 \left[M_0 \sin k_0 z \frac{\partial^4 m_z}{\partial z^4} - m_z(k_0^4 M_0 \sin k_0 z) \right], \qquad (12.12)$$

$$\frac{\partial m_y}{\partial t} = \gamma \lambda_2 \left[m_z(-k_0^2 M_0 \cos k_0 z) - M_0 \cos k_0 z \frac{\partial^2 m_z}{\partial z^2} \right]$$
$$+ \gamma \lambda_4 \left[m_z(k_0^4 M_0 \cos k_0 z) - M_0 \cos k_0 z \frac{\partial^4 m_z}{\partial z^4} \right]. \qquad (12.13)$$

The sines and cosines may clearly be eliminated from these equations if we take

$$m_x = -m_w \sin k_0 z, \qquad m_y = m_w \cos k_0 z, \qquad (12.14)$$

which also accords with our idea that **m** is perpendicular to $\mathbf{M}_0$. With the substitution (12.14) equations (12.12) and (12.13) both yield:

$$-\frac{\partial m_w}{\partial t} = \gamma \lambda_2 M_0 \left(\frac{\partial^2}{\partial z^2} + k_0^2 \right) m_z + \gamma \lambda_4 M_0 \left(\frac{\partial^4}{\partial z^4} - k_0^4 \right) m_z. \qquad (12.15)$$

We now consider the equation for m_z, which we have so far shirked; it is, by substitution in Eq. (12.3),

$$\frac{\partial m_z}{\partial t} = \gamma \lambda_2 \left[M_0 \cos k_0 z \frac{\partial^2 m_y}{\partial z^2} - m_y(-k_0^2 M_0 \cos k_0 z) \right.$$
$$+ m_x(-k_0^2 M_0 \sin k_0 z) - M_0 \sin k_0 z \frac{\partial^2 m_x}{\partial z^2} \right]$$
$$+ \gamma \lambda_4 \left[M_0 \cos k_0 z \frac{\partial^4 m_y}{\partial z^4} - m_y(k_0^4 M_0 \cos k_0 z) \right.$$
$$+ m_x(k_0^4 M_0 \sin k_0 z) - M_0 \sin k_0 z \frac{\partial^2 m_x}{\partial z^2} \right]. \qquad (12.16)$$

To express Eq. (12.16) in terms of m_w rather than m_x and m_y requires some rather lengthy but straightforward manipulation which we merely summarize. By evaluating expressions like

$$\frac{\partial^4 m_y}{\partial z^4} = \frac{\partial^4}{\partial z^4}(m_w \cos k_0 z)$$

$$= \cos k_0 z \frac{\partial^4 m_w}{\partial z^4} - 4k_0 \sin k_0 z \frac{\partial^3 m_w}{\partial z^3} - 6k_0^2 \cos k_0 z \frac{\partial^2 m_w}{\partial z^2}$$

$$+ 4k_0^3 \sin k_0 z \frac{\partial m_w}{\partial z} + k_0^4 m_w \cos k_0 z,$$

we can prove useful identities

$$\cos k_0 z \left(\frac{\partial^2}{\partial z^2} + k_0^2\right) m_y - \sin k_0 z \left(\frac{\partial^2}{\partial z^2} + k_0^2\right) m_x = \frac{\partial^2 m_w}{\partial z^2}$$

and

$$\cos k_0 z \left(\frac{\partial^4}{\partial z^4} - k_0^4\right) m_y - \sin k_0 z \left(\frac{\partial^4}{\partial z^4} - k_0^4\right) m_x = \frac{\partial^4 m_w}{\partial z^4} - 6k_0^2 \frac{\partial^2 m_w}{\partial z^2},$$

which by substitution in Eq. (12.16) yield

$$\frac{\partial m_z}{\partial t} = \gamma \lambda_2 M_0 \frac{\partial^2 m_w}{\partial z^2} + \gamma \lambda_4 M_0 \left(\frac{\partial^4 m_w}{\partial z^4} - 6k_0^2 \frac{\partial^2 m_w}{\partial z^2}\right). \tag{12.17}$$

Eqs. (12.15) and (12.17) are the coupled equations we have to solve. If we remember that k_0^2 is given by Eq. (12.9) they become

$$\frac{\partial m_z}{\partial t} = -2\gamma \lambda_2 M_0 \frac{\partial^2 m_w}{\partial z^2} + \gamma \lambda_4 M_0 \frac{\partial^4 m_w}{\partial z^4}$$

$$-\frac{\partial m_w}{\partial t} = \gamma \frac{\lambda_2^2}{4\lambda_4} M_0 m_z + \gamma \lambda_2 M_0 \frac{\partial^2 m_z}{\partial z^2} + \gamma \lambda_4 M_0 \frac{\partial^4 m_z}{\partial z^4}. \tag{12.18}$$

We try a solution of the form

$$m_w = e^{i(qz - \omega t)}, \qquad m_z = \alpha\, e^{i(qz - \omega t)},$$

and by substitution in Eqs. (12.18) obtain a pair of simultaneous equations for $\omega(q)$ and $\alpha(q)$:

$$-\alpha i\omega = 2\gamma \lambda_2 M_0 q^2 + \gamma \lambda_4 M_0 q^4$$

$$i\omega = \alpha\left[\gamma \frac{\lambda_2^2}{4\lambda_4} M_0 - \gamma \lambda_2 M_0 q^2 + \gamma \lambda_4 M_0 q^4\right]. \tag{12.19}$$

These give

$$\alpha = \frac{2\lambda_2 q^2 + \lambda_4 q^4}{-i\omega/\gamma M_0} = \frac{i\omega/\gamma M_0}{(\lambda_2^2/4\lambda_4) - \lambda_2 q^2 + \lambda_4 q^4}, \tag{12.20}$$

from which, by use of Eq. (12.9),

$$\omega^2 = (\gamma \lambda_2 M_0)^2 \left(2q^2 + \frac{q^4}{2k_0^2} \right) \left(\tfrac{1}{2} k_0^2 - q^2 + \frac{q^4}{2k_0^2} \right). \tag{12.21}$$

Note that Eq. (12.21) gives zero spin wave energy ($\omega = 0$) for $q = 0$ and also $q = k_0$. For $q = 0$, $\alpha = 0$ and the oscillation is purely in m_w; the mode therefore corresponds simply to rotation of the spiral as a whole about the z axis, which clearly requires no energy. For $q = k_0$, $\alpha = \infty$, so that the oscillation is purely in m_z. At wavenumber k_0 such an oscillation corresponds to tilting the spins on one side of the spiral in the $+z$ direction, and tilting those on the other side of the spiral in the $-z$ direction; in other words the plane in which $\mathbf{M}$ lies is tilted. That this is a 'soft mode' is an artefact of our model; we have not included any crystalline anistropy to determine the plane in which $\mathbf{M}$ lies in the spiral ground state.

Of more physical interest is the behaviour of the spin-wave energy for small q; from Eq. (12.21)

$$\omega \approx \gamma M_0 |\lambda_2| k_0 q, \tag{12.22}$$

so that ω is initially linear in q, in contrast to the quadratic behaviour for ferromagnetic spin waves, Eq. (12.6). We can see that this linear behaviour originates in the presence of a term in m_z (without differential operators) in the second of Eqs. (12.18); this in turn comes from an x–y component of ground-state magnetization that varies in the z direction. Our arguments therefore suggest that the spin wave energy is likely to be linear in q for any magnetically ordered structure in which the ground state magnetization is non-uniform. It is certainly true that the spin wave spectrum is linear for small q in simple antiferromagnetic structures.

12.3   FERROMAGNETIC DOMAINS

The equations we have used in sections 5.4.3 and 12.2 to discuss the space and time dependence of magnetization are clearly incomplete. For example, we have ignored the anisotropy energies resulting from the orienting effects of crystalline electric fields, and we have also assumed that the exchange interaction is isotropic; nor have we considered any elastic energy resulting from magnetostriction. Nevertheless, if our models were improved by taking these effects into account, we could in principle use the same equations that we have used to discuss spin waves to treat any problem of space or time dependent magnetization.

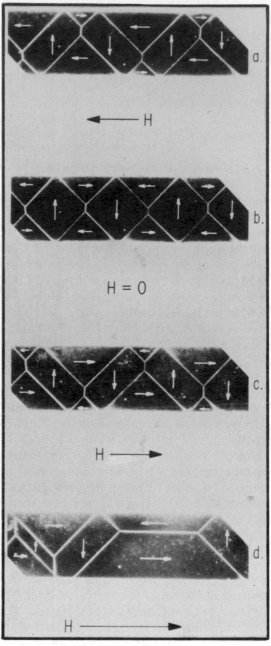

Fig. 12.5. Photographs showing reversible domain wall
motion in a 50 μm whisker from (*a*) to (*b*) to (*c*), with
an irreversible jump from (*c*) to (*d*).
{R. W. de Blois and C. D. Graham, *J. Appl. Phys.*, **29**,
931 (1958)}.

In this context it is important to note that Eqs. (5.28) and (12.3) are *non-linear* partial differential equations, which we have solved only by linearizing them for suitable special circumstances. It is a most important property of nonlinear differential equations that the principle of superposition does not apply to their solutions, with the consequence that a solution for given boundary conditions is not necessarily unique: for example, many metastable stationary states may be possible.

As an example of the sort of complexity that can result from nonlinear partial differential equations, consider hydrodynamics, where we have an everyday acquaintance with the nature of the solutions. For the same boundary conditions we can have laminar flow, a regular pattern of eddies, or turbulence—though not all these types of solution may be stable. It is thus not usually possible to solve a nonlinear partial differential equation without some physical idea of what the solution is like. In the case of magnetism, experiment gives us a very clear indication of the type of static solution, through the observation of **ferromagnetic domains** (Fig. 12.5): it is observed that a ferromagnetic crystal can be divided into **domains** where the magnetization is essentially uniform separated by relatively narrow **Bloch walls** where the magnetization direction changes rather rapidly. This observation enables us to avoid a full solution of the equations, and consider merely the balance of energy between the various regions.

12.3.1 Bloch wall energy

In a domain boundary the magnetization direction slowly changes; for a boundary at which the magnetization direction changes by 180° a spiral change of direction, as illustrated in Fig. 12.6, avoids free poles at the boundary and the consequent magnetostatic energy. In fact, the magnetization direction will spiral most rapidly past those directions for which the crystalline anisotropy energy is least favourable, but we may obtain a rough estimate of the wall energy by supposing that M_0 rotates uniformly through a thickness t.

Fig. 12.6. Schematic diagram of a Bloch wall.

In a uniformly magnetized sample the exchange energy density (see Eq. (5.19) or (5.39)) is

$$E_{ex} = -W = -\tfrac{1}{2}\mu_0\lambda M_0^2. \tag{12.23}$$

If we suppose that in the wall successive layers of spins are rotated by an angle ε with respect to each other, the inter-layer contribution to E_{ex} is multiplied by a factor $\cos\varepsilon \approx (1 - \tfrac{1}{2}\varepsilon^2)$ for nearest neighbour interactions. The exchange energy density in the wall region is therefore raised by an amount

$$\Delta E_{ex} = \tfrac{1}{2}\alpha W\varepsilon^2, \tag{12.24}$$

in which the numerical factor α ($\tfrac{1}{3}$ for a simple cubic lattice with nearest neighbour interactions) takes account of the fact that the exchange interaction within atomic layers parallel to the wall is unmodified. If the atomic spacing is a, we have $\varepsilon = (\pi a/t)$ so that

$$\Delta E_{ex} = \frac{\alpha W\pi^2 a^2}{2t^2}, \tag{12.25}$$

giving surface energy per unit area of wall of

$$\Delta\sigma_{ex} = t\Delta E_{ex} = \frac{\alpha W\pi^2 a^2}{2t}. \tag{12.26}$$

We can see that this expression decreases without limit as t increases, so if this were the only contribution no definite wall would exist. However, about half the spins in the wall are pointing in directions opposed by crystalline anisotropy, and thus contribute an anisotropy energy K per unit volume. Anisotropy thus gives a surface energy

$$\sigma_{anis} \approx \tfrac{1}{2}Kt$$

so that the total surface energy is

$$\sigma = \sigma_{ex} + \sigma_{anis} = \frac{\alpha W\pi^2 a^2}{2t} + \tfrac{1}{2}Kt; \tag{12.27}$$

this is minimized for

$$t = \left(\frac{\alpha W}{K}\right)^{1/2}\pi a. \tag{12.28}$$

The exchange energy is usually much larger than the anisotropy energy, so that the wall is typically several hundred atoms thick. For this thickness

Eq. (12.27) gives

$$\sigma = (\alpha W K)^{1/2} \pi a$$

$$= \left(\frac{\alpha \mu_0 \lambda K}{2}\right)^{1/2} M_0 \pi a. \tag{12.29}$$

The continuum approach we used in section 12.2 gives an alternative approach to ΔE_{ex} which shows that the energy required to twist the magnetization direction is really related to λ_2 rather than $\lambda_0 = \mu_0 \lambda$. Thus, by inspection of Fig. 12.6, the total exchange energy density in the wall is given by Eq. (12.10) for $k = (\pi/t)$, so that

$$U = \tfrac{1}{2}M_0^2 \left(-\lambda_0 + \left(\frac{\pi}{t}\right)^2 \lambda_2\right),$$

if we set $\lambda_4 = 0$. Thus,

$$\Delta E_{ex} = \frac{\pi^2 \lambda_2 M_0^2}{2t^2} \tag{12.30}$$

instead of Eq. (12.25); we see that $\tfrac{1}{2}\alpha\lambda_0 M_0^2$ is replaced by $\lambda_2 M_0^2/a^2$. Of course, if these quantities are compared for the same detailed model, they must be equal; and in general we expect

$$\lambda_2 \sim \lambda_0 a^2$$

on dimensional grounds, since atomic dimensions are the only relevant length scale in the problem.

From Eq. (12.30) we find the alternative result

$$\sigma = \pi M_0 (\lambda_2 K)^{1/2}, \tag{12.31}$$

which directly relates wall energy to spin-wave energy.

12.3.2   The energy balance

If a crystal of iron is magnetized as a single domain, Fig. 12.7(a), there is a considerable contribution to the total energy from the energy density of the magnetic field outside the crystal, $B^2/2\mu_0$; this energy is of order $\mu_0 M_0^2 V$, where V is the volume of the crystal. It is to reduce this energy that all but the smallest samples find it preferable to break up into domains of opposite magnetization, as in Fig. 12.7(b), separated by Bloch walls. The scale of the domains is determined by a balance between Bloch wall energy and free field energy, adjusted to minimize the total.

In a crystal with only one easy direction of magnetization the actual domain pattern is as in Fig. 12.7(b), but in cubic crystals with several equivalent easy direction closure domains form as in Fig. 12.7(c), so as to

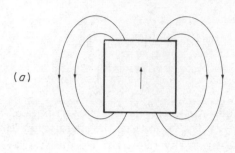

(a)

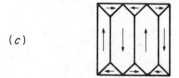

(b)

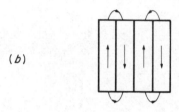

(c)

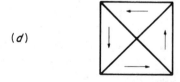

(d)

Fig. 12.7. Reduction of field energy by
domain formation.

further reduce the free field energy. Having formed closure domains, one might wonder why a reduction in wall energy, without producing free fields, as in Fig. 12.7(*d*), is not energetically desirable. The reason is **magneto-striction**; a magnetized crystal tends to expand or contract along the magnetization direction. These distortions are incompatible for the domains in Fig. 12.7(*d*), resulting in a positive elastic stress energy. This elastic stress energy is reduced by having only small closure domains as in Fig. 12.7(*c*). Thus, the scale of domains in Fig. 12.7(*c*) is determined primarily by a balance between Bloch wall energy and magnetoelastic energy.

12.3.3   The magnetization process

The magnetization process is illustrated in its simplest form, for a single-crystal iron whisker, by the domain patterns shown in Fig. 12.5. For the small fields shown in Figs. 12.5(*a*)–(*c*) magnetization occurs reversibly by the movement of domain walls, but for larger fields (Fig. 12.5(*d*)) irreversible changes in magnetization occur by the removal of unfavourable domains. The processes are essentially the same, though less distinct, in bulk polycrystalline material. Domain boundary motion may then not be quite reversible, because of inhomogeneities. Indeed, an important technique in making high coercive force materials for permanent magnets is the deliberate introduction of inhomogencities to inhibit domain boundary motion—for example by producing a two-phase alloy system.

In polycrystalline materials there is in high fields a final slow approach to saturation as the domain magnetization rotates from the nearest easy direction to the direction of the applied field.

CHAPTER

13

Disordered solids

13.1 THE PROBLEM

So far in this book we have studied the physics of infinite perfect single crystals and called it solid state physics. We have, reluctantly, allowed a few imperfections and boundaries, when forced to do so, but that is as near as we have got to the real world. The reason for this is that crystalline solids have been the most extensively studied experimentally, and are the easiest to study theoretically because of the simplifying consequences of lattice periodicity. But we should not overlook the fact that the real world contains disordered solids, such as glasses and plastics, and liquids, newtonian and non-newtonian. The physics of condensed matter should encompass an understanding of these materials also. In this final chapter we shall take a brief look outside our conventional blinkers at some of the problems that arise in trying to understand disordered solids.

The extensive use we have made of the consequences of lattice periodicity is a weakness as well as a strength. For in Chapter 4 we used the ideas of Brillouin zones, energy gaps, and full bands to distinguish between conductors and insulators. All this powerful theoretical apparatus leaves us powerless to face the simple question 'Why is glass an insulator?' Yet, to a chemist, the question is trivially simple. Both crystalline and fused quartz consist of a three dimensional network of SiO_4 tetrahedra in which all the corner oxygen atoms are shared between two silicons; the whole is a structure of saturated covalent bonds and should therefore be an insulator. The only difference between ordered and disordered SiO_2 is one of **topological disorder**; the local order is essentially the same, but the length and configuration of a closed Si—O—Si—O . . . ring is random in the glassy structure. Yet this rather trivial difference, which one would not expect to lead to gross changes in properties, is a complete barrier to the use of the theoretical methods we have developed.

The essential simplifying feature in a perfect crystal is that, as we saw in Chapter 6, the limited translational invariance of a crystal enables us to label each energy eigenstate with a **k**-vector, in exactly the same way that full translational invariance makes momentum a good quantum number. In a disordered structure **k** is no longer a good quantum number, and a satisfactory alternative label has not yet been found. Considerable theoretical attention has been given, in one dimensional models, to the effect of random atom spacing and random potential well depth; this is equivalent to randomising the coefficients in a set of equations such as Eqs. (2.2) or (3.1). But this may not even be the main problem; for this type of disorder can be continuously deformed into an ordered structure, whereas topological disorder cannot. Topological disorder, which cannot occur in one dimension, amounts to altering which pairs of equations are coupled together in a set such as Eqs. (2.2) or (3.1); this is a discontinuous change.

13.2   ELECTRONIC STRUCTURE

In order to understand in a more general way how forbidden gaps arise in the electronic energy levels of a solid, it is convenient to begin by considering a little more closely the solution we obtained to Eqs. (3.1); this was (Eq. (3.2))

$$E = E_0 - 2A \cos ka \qquad (13.1)$$

for a wavelike solution $e^{i(kx - \omega t)}$, for which $E = \hbar\omega$. Inspection of Eq. (13.1) shows that E is bounded within an allowed band $E_0 \pm 2A$ because we insist on k being real. Our reason for doing this in Chapter 3 was that a complex k would give a probability amplitude growing indefinitely in some direction in an infinite crystal. As a step towards considering a disordered solid, let us consider what happens near an impurity or other defect in our linear chain. A suitable defect will enable us to match exponentials that decay away on both sides of the impurity, so that in this situation we may consider complex k, which will give a state localized near the defect (see Feynman,[1] section 13-7). Remembering that

$$\cos (k + i\alpha)a = \cos ka \cosh \alpha a - i \sin ka \sinh \alpha a$$

we see that a real E can be obtained in Eq. (13.1) with $k = i\alpha$, in which case

$$E = E_0 - 2A \cosh \alpha a, \tag{13.2}$$

or with $k = (\pi/a) + i\alpha$, in which case

$$E = E_0 + 2A \cosh \alpha a. \tag{13.3}$$

Note that in each case the real part of k remains at the value, 0 or (π/a), appropriate to the minimum or maximum of the band. We have noticed already, in Chapter 4, that energy gaps arise when the electron wavelength is synchronous with the lattice periodicity. Eqs. (13.2) and (13.3) show that exponentially decaying states of the same wavelength exist in the energy gap; these may be combined at a suitable lattice defect to give a localized state.

It is interesting to look at the donor and acceptor levels we considered in Chapter 3 from the present viewpoint. For $\alpha a \ll 1$ Eqs. (13.2) and (13.3) give localized states with energies given by

$$E = E_0 \pm (2A + \Delta E),$$

where

$$\Delta E = A(\alpha a)^2$$
$$= \frac{\hbar^2 \alpha^2}{2m^*}, \tag{13.4}$$

with m^* defined by Eq. (3.3). This may be compared with the treatment by analogy with a hydrogen atom in Chapter 3, which gives

$$\Delta E = \frac{\hbar^2}{2m^* r_1^2} \tag{13.5}$$

with r_1 given by Eq. (3.5). Since r_1 is, like $(1/\alpha)$, the exponential decay distance of the localized state, Eqs. (13.4) and (13.5) are equivalent.

We thus see that there is a range of energy in which the electron wavelength is synchronized with the lattice periodicity, but only localized states centred on lattice defects are possible. We may express this synchronization without explicit reference to wavelength or lattice periodicity, by saying that the nodes of the electron wavefunction tend to lock to the atomic positions when the number of nodes is commensurate with the number of atoms. Expressed in this form, we might expect the result to apply to disordered structures also. Since a disordered structure can be thought of as a crystal with very many defects, suitable centres for the formation of localized states will always be available in such a structure. We therefore conclude that, in a disordered structure as in a crystal, the tendency of wavefunction nodes to synchronize with atomic positions will lead to ranges of energy in which electron states propagating right through the crystal are not possible. In these energy gaps the allowed electron states will be localized; such localized states can occur only at lattice defects in a crystal, but we expect many more of them in a disordered structure, depending on the degree of disorder. This conclusion, that in a disordered structure there are energy ranges in which only localized states are allowed, is often expressed by saying that we have a **mobility gap** rather than an energy gap.

Experimental evidence for the existence of these localized states is provided by measurements of the resistivity of amorphous germanium prepared by vacuum deposition on a cooled substrate, Fig. 13.1. These measurements do not show the straight line on a $\ln \rho$ vs. $(1/T)$ plot that is characteristic of crystalline germanium. The reason for this is that the ordinary intrinsic conduction is short circuited by hopping of electrons from one localized state to another. Mott pointed out that there are two competing factors in such a process. An electron can tunnel easily to a neighbouring site, but this state will probably have a rather different energy so that an activation energy is required. If the electron is prepared to tunnel further it is more likely to

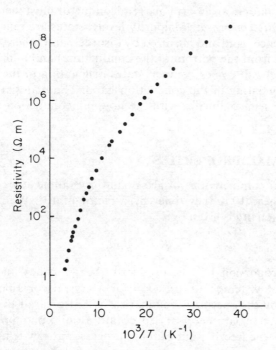

Fig. 13.1. Temperature dependence of the resistivity
of amorphous germanium.
{A. H. Clark, *Phys. Rev.*, **154**, 750 (1967)}.

find a state requiring only a low activation energy, but on the other hand the tunnelling process is less probable. The net result is that as the sample is cooled, and activation energy becomes less readily available, the electron tends to tunnel further, and the mean activation energy falls, in agreement with the results shown in Fig. 13.1. Mott finds that this balance of tunnelling and activation leads to the result

$$\rho \propto \exp\left[(A/T)^{1/4}\right] \tag{13.6}$$

which is in quite good agreement with experiment.

It is interesting to note that if a sample of amorphous germanium is crystallized by annealing it, the resistivity *falls* dramatically to values indicative of an impurity concentration of order 10^{16} cm^{-3}; the reduction of resistivity by impurities is thus an effect of crystalline order. This is easy to understand from a chemical point of view. We saw in Chapter 3 that the formation of donor and acceptor states depended on the impurity going into a substitutional position, where it had to form *four* bonds, leaving a spare electron (or hole). But in an irregular structure it is easy for the impurity to form the five or three bonds natural to it, leaving no spare electron or hole. The localized states in an amorphous semiconductor must therefore have the character of *fully compensated* impurity levels: states are removed from the top of the valence band and localized by disorder, and other states are similarly removed from the bottom of the conduction band and localized; but the Fermi level still comes between states originating in the valence band and those originating in the conduction band. The material thus behaves as an *intrinsic* semiconductor, with the hopping conduction mechanism in parallel.

13.3 THERMAL PROPERTIES

The thermal conductivities of glassy and crystalline silica are shown in in Fig. 13.2; near 10 K they differ by a factor of 10^4. We recall that the thermal conductivity is given by

$$K = \tfrac{1}{3}C_v c l, \tag{13.7}$$

and is thus proportional to the product of heat capacity and phonon free path; the T^3 low temperature region for a crystal corresponds to a free path limited by the specimen boundaries. We also recall that in a crystal Umklapp processes are necessary to obtain a finite phonon thermal conductivity; it is the freezing out of these processes at low temperatures that gives rise to the sharp maximum in the region of 10 K. We saw in Chapter 8 that what was really essential to obtain a finite thermal conductivity was the thermal equilibration of high energy phonons near the zone boundary.

Now in a glassy substance neither Umklapp nor zone boundaries are meaningful ideas. As with electrons, though, we should expect an energy range between the acoustic and optic modes where only localized modes of lattice vibration are possible. These local phonons will fulfil the function of Umklapp processes in tending to remove 'crystal momentum' (i.e., total wavevector) from the assembly of low energy phonons. However, from what has been said, we would expect these local modes to have an energy comparable to the Debye cutoff energy, so that equilibration with them should freeze out in much the same way as for Umklapp processes. Local phonons are therefore unlikely to be responsible for the large difference in Fig. 13.2.

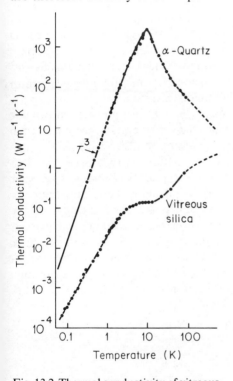

Fig. 13.2. Thermal conductivity of vitreous and crystalline SiO_2.
{R. C. Zeller and R. O. Pohl, *Phys. Rev.*, **B 4**, 2029 (1971)}.

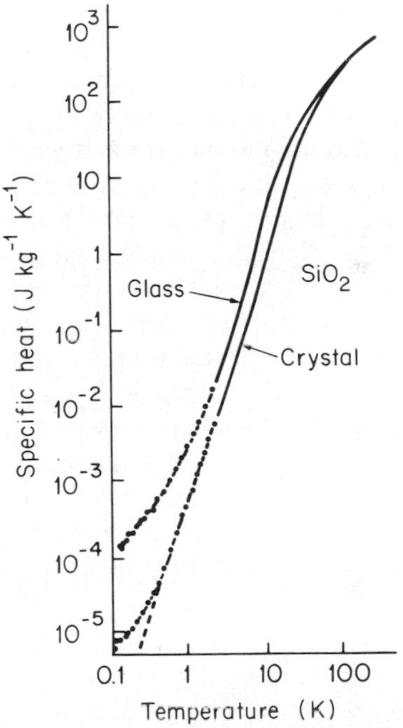

Fig. 13.3. Specific heat of vitreous and crystalline SiO_2.
{R. C. Zeller and R. O. Pohl, *Phys. Rev.*, **B 4**, 2029 (1971)}.

Another process that can limit phonon free paths in crystals is scattering by lattice defects. Since a glass is in some sense 'all defects' we might expect this effect to be strong. Phonon wavelengths are typically much larger than atomic sizes, and we therefore expect Rayleigh scattering $\propto \lambda^{-4}$; since $\lambda \propto (1/T)$ for a typical thermal phonon, we expect $l \propto T^{-4}$; with $C_v \propto T^3$

for $T \ll \Theta_D$ we thus expect $K \propto (1/T)$. The plateau in K around 10 K, which is observed for a wide variety of glasses, may be an indication of this effect. At higher temperatures the free path becomes comparable with the atomic spacing and ceases to fall, and at still higher temperatures the specific heat reaches the classical limit. Thus, at the highest temperatures, both glass and crystal thermal conductivity tend to a value given by the classical specific heat and a mean free path equal to the atomic spacing,

$$K \sim \frac{Rca}{V} \tag{13.8}$$

where a is the atomic spacing and V the molar volume.

Below 10 K the thermal conductivity does not increase as $(1/T)$, but falls approximately as $T^{1.8}$; this temperature dependence, with a strikingly similar absolute magnitude, is found for many substances, both glasses and plastics. It is presumably due to some additional, rather ubiquitous, phonon scattering mechanism, but what this is remains a mystery.

A possible clue is provided by specific heat measurements, shown for SiO_2 in Fig. 13.3. Below 1 K the specific heat of the glass becomes much larger than that of the corresponding crystal and tends to vary as T rather than T^3; similar behaviour has been observed for many glasses. Fig. 13.3 shows a trace of a similar anomaly in crystalline quartz also, but it is an order of magnitude smaller than in the glass and is probably due to impurities.

These additional low temperature specific heats are indicative of the presence of another possible mode of excitation, besides phonons, in glasses. It may be that these excitations will also serve to limit the phonon free path, and thus explain the behaviour of the thermal conductivity. But that possibility is conjectural until we have some idea of what these extra excitations in glasses might be; and at the moment we do not.

Coupled probability amplitudes

In this section we derive the coupled probability amplitude equations (1.7) from the Schrödinger equation (1.2). To do this we express our wavefunction $\Psi(\mathbf{r}, t)$ as a series in some set of functions $\psi_l(\mathbf{r})$ (which might, for example, be atomic eigenfunctions) by means of Eq. (1.8):

$$\Psi(\mathbf{r}, t) = \sum_l a_l(t)\psi_l(\mathbf{r}). \qquad (A.1)$$

Such an expansion is always possible provided the ψ_l form what is called a **complete set** of functions. An example of such a complete set is the sine and cosine functions; in this case Eq. (A.1) is the expansion of Ψ as a Fourier series (or, in the limit, a Fourier integral). The expansion coefficients a_l in Eq. (A.1) are time dependent because Ψ is time dependent. We shall not bother here with the formal question of what constitutes a complete set, but shall just assume for the derivation of Eq. (1.7) that we have a set of functions $\psi(\mathbf{r})$ such that the expansion (A.1) is both possible and unique. Later in this appendix we discuss how far the functions $\psi_l(\mathbf{r})$ that we actually make use of in this book satisfy this condition.

If we assume that both $\Psi(\mathbf{r}, t)$ and the $\psi_l(\mathbf{r})$ are normalized, the probability of finding the system in the state ψ_n at time t is $|c_n(t)|^2$, where the probability amplitude $c_n(t)$ is given by

$$c_n(t) = \int \psi_n^*(\mathbf{r})\Psi(\mathbf{r}, t)\, d^3\mathbf{r}$$

$$= \sum_l a_l(t) \int \psi_n^*(\mathbf{r})\psi_l(\mathbf{r})\, d^3\mathbf{r}, \qquad (A.2)$$

by use of Eq. (A.1). If the functions ψ_l are orthogonal as well as normalized the integral in Eq. (A.2) is zero for $l \neq n$ and one otherwise, so that $c_n(t) = a_n(t)$. We shall want to use atomic wavefunctions centred on adjacent lattice sites, which are not orthogonal to each other, and we shall therefore continue the argument without assuming orthogonality. A geometrical analogy may help to clarify Eq. (A.2). Eq. (A.1) is like writing a vector in terms of its components in a multidimensional space. In this analogy the a_n correspond to the components and the c_n to the projections on the coordinate axes; these are equal only if the coordinate axes are orthogonal (mutually perpendicular).*

We now substitute the expansion (A.1) in the Schrödinger equation (1.2) to obtain

$$i\hbar \sum_l \frac{da_l}{dt}\psi_l(\mathbf{r}) = \sum_l a_l H\psi_l(\mathbf{r}). \qquad (A.3)$$

If Eq. (A.3) is multiplied on the left by ψ_n^* and integrated throughout all space we obtain, by use of Eq. (A.2), an expression for dc_n/dt:

$$i\hbar\frac{dc_n}{dt} = i\hbar \sum_l \frac{da_l}{dt} \int \psi_n^*(\mathbf{r})\psi_l(\mathbf{r})\, d^3\mathbf{r}$$

$$= \sum_l a_l \int \psi_n^*(\mathbf{r})H\psi_l(\mathbf{r})\, d^3\mathbf{r}$$

$$= \sum_l a_l \int \psi_l(\mathbf{r})H\psi_n^*(\mathbf{r})\, d^3\mathbf{r}, \qquad (A.4)$$

where the last step is an example of a general quantum mechanical result. It is obviously true for the potential part of H; to see that it is true also for the operator ∇^2, note that successive integration by parts gives

$$\int \psi_1\nabla^2\psi_2\, d^3\mathbf{r} = -\int (\nabla\psi_1)\cdot(\nabla\psi_2)\, d^3\mathbf{r}$$

$$= \int (\nabla^2\psi_1)\psi_2\, d^3\mathbf{r},$$

provided only that ψ and its derivatives vanish at infinity so that the surface integrals vanish.

* Do not worry that the total probability $\sum_n|c_n|^2 \neq 1$ when the ψ_n are not orthogonal; this is just because the possibilities are not mutually exclusive in this case.

Notice now that $H\psi_n^*(\mathbf{r})$ is just another function of position, so that it can be expanded in a way analogous to Eq. (A.1):

$$H\psi_n^*(\mathbf{r}) = \sum_m E_{nm}\psi_m^*(\mathbf{r}).$$
(A.5)

The only differences are that the expansion coefficients E_{nm} are independent of time because $H\psi_n^*(\mathbf{r})$ is independent of time, and we have chosen to expand in terms of the complete set of functions ψ_m^*, rather than the set ψ_m. Substitution of the expansion (A.5) in Eq. (A.4) now gives

$$i\hbar\frac{dc_n}{dt} = \sum_l \sum_m E_{nm}a_l \int \psi_l(\mathbf{r})\psi_m^*(\mathbf{r})\,d^3\mathbf{r}$$
$$= \sum_m E_{nm}c_m,$$
(A.6)

by use of Eq. (A.2); this is Eq. (1.7). By comparing Eqs. (A.4) and (A.6) we see that the coefficients E_{nm} may be calculated by solving the set of simultaneous equations:

$$\sum_m E_{nm} \int \psi_m^*(\mathbf{r})\psi_l(\mathbf{r})\,d^3\mathbf{r} = \int \psi_n^*(\mathbf{r})H\psi_l(\mathbf{r})\,d^3\mathbf{r}.$$
(A.7)

In the special case when the $\psi_l(\mathbf{r})$ are all mutually orthogonal, Eq. (A.7) reduces to

$$E_{nl} = \int \psi_n^*(\mathbf{r})H\psi_l(\mathbf{r})\,d^3\mathbf{r} = H_{nl},$$

so that in this case the E_{nl} are what are usually called the *matrix elements* of the operator H, and our Eq. (1.7) takes precisely the form used by Feynman. In general, if the non-orthogonality is small, the E_{nl} are quite close to these values.

We now consider how to obtain a complete set of wavefunctions $\psi_n(\mathbf{r})$ so that the expansions (A.1) and (A.5) can be made. All the energy eigenstates, bound and unbound, for an electron in the field of a single positive ion constitute a complete set, by a general quantum mechanical theorem. But this is not a useful complete set, because very many terms would be required to describe an electron bound to another ion in a remote part of the crystal. We could try to meet this difficulty by considering all the wavefunctions centred on each ion in the crystal; but this enlarged set is clearly overcomplete, in that some of the functions can be expanded in terms of others. This has the unfortunate consequence that the expansions (A.1) and (A.5) are not unique.

In this book we use a limited set of wavefunctions, consisting of only the lowest bound state centred on each ion. This set is actually incomplete, so that the expansions (A.1) and (A.5) can only be made with very limited accuracy. However, this simplification does enable us to obtain qualitatively correct results with the minimum of algebra. To understand the approximation involved, consider the application of Eq. (A.5) to the problem of H_2^+, considered in section 1.2.2. The hamiltonian for the problem is

$$H = -\frac{\hbar^2}{2m}\nabla^2 + V_a + V_b, \tag{A.8}$$

where V_a and V_b are coulomb potentials centred at R_a and R_b respectively. Since $\psi_a(\mathbf{r})$ is the ground state of the atom centred at R_a, of energy E_0, we have

$$H\psi_a = E_0\psi_a + V_b\psi_a. \tag{A.9}$$

Note that here ψ_a is real, so that $\psi_a^* = \psi_a$.

The terms in Eq. (A.9) are illustrated in Fig. A.1. Our approximation consists in supposing that $V_b\psi_a$ can be expressed in terms of ψ_a and ψ_b. This

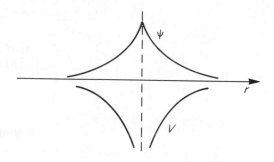

(*a*)   Ground state wavefunction and potential
for an isolated hydrogen atom.

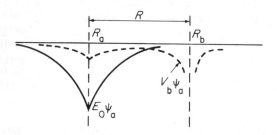

(*b*)   The terms in Eq. (A.9).

Fig. A.1.

is roughly possible, because $V_b \psi_a$ consists of peaks near R_a and R_b, but with only ψ_a and ψ_b we are limited to specifying the *size* of these peaks,* and are unable to fit their *shape* with this truncated form of Eq. (A.5). If we accept this limitation we can write

$$V_b \psi_a \approx -(\alpha \psi_a + \beta \psi_b),$$

with α and β positive quantities, so that Eq. (A.9) becomes

$$H \psi_a \approx (E_0 - \alpha)\psi_a - \beta \psi_b. \tag{A.10}$$

Thus, in Eqs. (1.9) we have $E = (E_0 - \alpha)$ and $A = \beta$. A little thought suffices to show that when the two protons are far apart

$$-\alpha \approx V_b(R_a) = -e^2/4\pi\varepsilon_0 R;$$

this just cancels the internuclear repulsion, and thus accounts for the symmetrical splitting of bonding and antibonding levels at large R, shown in Fig. 1.2(b).

---

* For example, by requiring that our fitted peaks have the same integral over all space as the true peaks.

The exchange integral

We wish to evaluate the exchange interaction between two electrons, which might be on the same atom, on neighbouring atoms in a crystal, or occupying Bloch states of a crystal lattice. A problem sufficiently general to cover these cases is the interaction of two electrons where the unperturbed state of one, ψ_a, is an eigenstate of the hamiltonian

$$H_a = -\frac{\hbar^2}{2m}\nabla^2 + V_a, \qquad (B.1)$$

and the unperturbed state of the other, ψ_b, is an eigenstate of the hamiltonian

$$H_b = -\frac{\hbar^2}{2m}\nabla^2 + V_b, \qquad (B.2)$$

The hamiltonian for the two electrons, including their mutual coulomb repulsion, is therefore

$$H = -\frac{\hbar^2}{2m}(\nabla_1^2 + \nabla_2^2) + \frac{e^2}{4\pi\varepsilon_0|\mathbf{r}_1 - \mathbf{r}_2|} + V(\mathbf{r}_1) + V(\mathbf{r}_2), \qquad (B.3)$$

where ∇_1 and ∇_2 are the operations of differentiations with respect to $\mathbf{r}_1$ and $\mathbf{r}_2$ respectively. The total potential $V(\mathbf{r})$ in Eq. (B.3) is $V_a(\mathbf{r}) + V_b(\mathbf{r})$ for electrons on neighbouring atoms; but for electrons on the same atom, or Bloch electrons, $V_a(\mathbf{r}) = V_b(\mathbf{r}) = V(\mathbf{r})$.

We now use the hamiltonian (B.3) to evaluate the expectation value of the energy,

$$\langle E \rangle = \frac{\iint d^3\mathbf{r}_1\, d^3\mathbf{r}_2\, \Psi^*(\mathbf{r}_1,\mathbf{r}_2)H\Psi(\mathbf{r}_1,\mathbf{r}_2)}{\iint d^3\mathbf{r}_1\, d^3\mathbf{r}_2\, \Psi^*(\mathbf{r}_1,\mathbf{r}_2)\Psi(\mathbf{r}_1,\mathbf{r}_2)}, \tag{B.4}$$

for the properly symmetrized (or antisymmetrized) trial wavefunction

$$\Psi(\mathbf{r}_1,\mathbf{r}_2) = \psi_a(\mathbf{r}_1)\psi_b(\mathbf{r}_2) \pm \psi_b(\mathbf{r}_1)\psi_a(\mathbf{r}_2). \tag{B.5}$$

The denominator of Eq. (B.4) is merely a normalization factor, so we have

$$\begin{aligned}
\langle E \rangle \propto \iint d^3\mathbf{r}_1\, d^3\mathbf{r}_2 [&\psi_a^*(\mathbf{r}_1)\psi_b^*(\mathbf{r}_2)H\psi_a(\mathbf{r}_1)\psi_b(\mathbf{r}_2) \\
&+\psi_b^*(\mathbf{r}_1)\psi_a^*(\mathbf{r}_2)H\psi_b(\mathbf{r}_1)\psi_a(\mathbf{r}_2) \\
&\pm\psi_a^*(\mathbf{r}_1)\psi_b^*(\mathbf{r}_2)H\psi_b(\mathbf{r}_1)\psi_a(\mathbf{r}_2) \\
&\pm\psi_b^*(\mathbf{r}_1)\psi_a^*(\mathbf{r}_2)H\psi_a(\mathbf{r}_1)\psi_b(\mathbf{r}_2)] \\
= 2 \iint d^3\mathbf{r}_1\, d^3\mathbf{r}_2 [&\psi_a^*(\mathbf{r}_1)\psi_b^*(\mathbf{r}_2)H\psi_a(\mathbf{r}_1)\psi_b(\mathbf{r}_2) \\
&\pm\psi_a^*(\mathbf{r}_1)\psi_b^*(\mathbf{r}_2)H\psi_b(\mathbf{r}_1)\psi_a(\mathbf{r}_2)],
\end{aligned} \tag{B.6}$$

where the last line follows by exchanging the labels 1 and 2 in the second and fourth terms of the integrand. It is the second term in Eq. (B.6) which changes sign according to the symmetry of Ψ, and is called the exchange integral:

$$I = \iint d^3\mathbf{r}_1\, d^3\mathbf{r}_2\, \psi_a^*(\mathbf{r}_1)\psi_b^*(\mathbf{r}_2)H\psi_b(\mathbf{r}_1)\psi_a(\mathbf{r}_2). \tag{B.7}$$

Remembering that

$$\left(-\frac{\hbar^2}{2m}\nabla^2 + V_a\right)\psi_a = E_a\psi_a, \tag{B.8}$$

and

$$\left(-\frac{\hbar^2}{2m}\nabla^2 + V_b\right)\psi_b = E_b\psi_b, \tag{B.9}$$

we obtain from Eq. (B.7)

$$\begin{aligned}
I = \iint d^3\mathbf{r}_1\, d^3\mathbf{r}_2\, \psi_a^*(\mathbf{r}_1)\psi_b^*(\mathbf{r}_2)\Bigg[E_a + E_b + &\frac{e^2}{4\pi\varepsilon_0|\mathbf{r}_1 - \mathbf{r}_2|} \\
+ V(\mathbf{r}_1) - V_b(\mathbf{r}_1) + V(\mathbf{r}_2) - V_a(\mathbf{r}_2) &\Bigg]\psi_b(\mathbf{r}_1)\psi_a(\mathbf{r}_2).
\end{aligned} \tag{B.10}$$

The E_a and E_b terms in Eq. (B.10) combine with the corresponding parts of the first term in Eq. (B.6) to separate out from Eq. (B.4) a term $(E_a + E_b)$, the unperturbed energy of the system. We are left with an exchange interaction proportional to

$$
I' = \iint d^3\mathbf{r}_1\, d^3\mathbf{r}_2\, \psi_a^*(\mathbf{r}_1)\psi_b^*(\mathbf{r}_2)\frac{e^2}{4\pi\varepsilon_0|\mathbf{r}_1 - \mathbf{r}_2|}\psi_b(\mathbf{r}_1)\psi_a(\mathbf{r}_2)
$$

$$
+ \int d^3\mathbf{r}_2\, \psi_b^*(\mathbf{r}_2)\psi_a(\mathbf{r}_2) \int d^3\mathbf{r}_1 [V(\mathbf{r}_1) - V_b(\mathbf{r}_1)]\psi_a^*(\mathbf{r}_1)\psi_b(\mathbf{r}_1)
$$

$$
+ \int d^3\mathbf{r}_1\, \psi_a^*(\mathbf{r}_1)\psi_b(\mathbf{r}_2) \int d^3\mathbf{r}_2 [V(\mathbf{r}_2) - V_a(\mathbf{r}_2)]\psi_b^*(\mathbf{r}_2)\psi_a(\mathbf{r}_2). \tag{B.11}
$$

Note, however, that if we are considering two electrons on the same atom ψ_a and ψ_b are eigenstates of the *same* potential, so that $V_a(\mathbf{r}) = V_b(\mathbf{r}) = V(\mathbf{r})$, and only the first term in Eq. (B.11) remains. This term has the form of the coulomb self energy for a so-called **exchange charge density** defined by

$$
\rho(\mathbf{r}) = \psi_a(\mathbf{r})\psi_b(\mathbf{r}), \tag{B.12}
$$

at any rate for real ψ_a, ψ_b. This term is therefore always positive, and this serves to explain Hund's rule that spins on the same atom tend to align parallel, since this corresponds to an antisymmetric space wavefunction (negative sign in Eq. (B.5), see Heitler,[9] Chapter IX).

Similarly, the second and third terms in Eq. (B.11) are like the potential energy of the exchange charge (B.12) in an attractive atomic potential; they are therefore negative. In the H_2 molecule these negative terms dominate, to give binding for the symmetric space wavefunction (positive sign in Eq. (B.5)), which goes with an antisymmetric spin wavefunction, corresponding to antiparallel spin.

APPENDIX

Local fields and demagnetizing fields

The local magnetic field in a crystal, like the local magnetization density, varies rapidly on an atomic scale within a unit cell. However, in calculating the magnetic field at a given point, it is clear that the variations of magnetization within *distant* unit cells are unimportant; only the average magnetization of such unit cells is important.

It is therefore convenient to divide the calculation of the local magnetic field at a point in a magnetized crystal into two parts. The effect of the 'near' region, within a radius r, we calculate from the detailed microscopic distribution of atomic dipoles; the radius r is chosen large compared with atomic dimensions, but small compared with the size of the sample. The 'far' region, beyond the radius r, we may treat macroscopically, by means of a uniform intensity of magnetization M. The magnetic field of this far region is just that of the current distribution equivalent to the magnetization: a surface current of magnitude M per unit length in the direction of **M**, around the boundaries of the magnetized region (see Purcell,[4] Chapter 10). These current distributions are sketched for several sample shapes in Fig. C.1; the small spherical cavity in the samples is the boundary separating the 'near' and 'far' regions defined above.

We may thus write the local magnetic field as

$$\mathbf{B}_{loc} = \mathbf{B}_e + \mathbf{B}_1 + \mathbf{B}_2 + \mathbf{B}_3, \tag{C.1}$$

where $\mathbf{B}_e$ is the externally applied field;

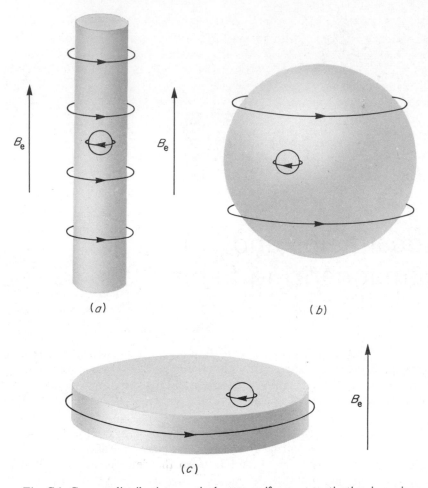

Fig. C.1. Current distributions equivalent to uniform magnetization in various shapes of specimen. The small spherical cavities are the regions within which the contributions of individual dipoles are considered from a microscopic viewpoint.

$\mathbf{B}_1$ is the field due to a surface current M per unit length on the outer surface of the sample;

$\mathbf{B}_2$ is the field due to a surface current M per unit length in the opposite sense on the inner boundary of the 'far' region; and

$\mathbf{B}_3$ is the field due to atomic dipoles within the 'near' region.

A particular reason for using a spherical boundary to separate the near and far regions is that in this case $\mathbf{B}_3$ vanishes identically for dipoles arranged at random (as in a gas), or for uncorrelated dipoles arranged with cubic symmetry about the point under consideration (as in many paramagnetic salts). We shall not prove this result, but you may convince yourself of its truth for the random case by integrating the contribution from dipoles in a spherical shell. In this Appendix we shall evaluate local fields on the assumption that $\mathbf{B}_3 = 0$; if $\mathbf{B}_3 \neq 0$, this contribution must be added to the field we calculate.

Note that the macroscopic field $\mathbf{B}$, defined by

$$\text{div } \mathbf{B} = 0, \qquad \text{curl } \mathbf{B} = \mu_0 \mathbf{j},$$

where $\mathbf{j}$ is the total current density including currents causing magnetization, averaged over a small region containing many atoms, is given by

$$\mathbf{B} = \mathbf{B}_e + \mathbf{B}_1. \tag{C.2}$$

For a proof that $\mathbf{B}$ so defined is the space average of the microscopic magnetic field, see Purcell,[4] Chapter 10. Thus, $\mathbf{B}$ is equal to the *average value of $\mathbf{B}_{loc}$ over all points in the unit cell* of a crystal; but it is in general *not* equal to $\mathbf{B}_{loc}$ at the site of any ion.

Now consider the various contributions to Eq. (C.1) for the three sample shapes illustrated in Fig. C.1. For a long thin rod (Fig. C.1(a)), $\mathbf{B}_1$ is the field of a long solenoid, so that $\mathbf{B}_1 = \mu_0 \mathbf{M}$. Since the field at the centre of a current loop is universely proportional to the radius, $\mathbf{B}_1 \to 0$ in the opposite limit of a very thin disc (Fig. C.1(c)). To evaluate the other contributions to $\mathbf{B}_{loc}$ in Fig. C.1 we need to calculate the field at the centre of a spherical current sheet M per unit length.* If the magnetization is along the z-axis in Fig. C.2, the element of surface shown, subtending an angle $d\theta$ at the centre carries a current $Mr \, d\theta \sin \theta$, in order to give a density M per unit distance in the z-direction. The field due to a short element dl of this ring is

$$dB = \frac{\mu_0}{4\pi} \frac{Mr \, d\theta \sin \theta}{r^2} \, dl$$

in the direction shown in Fig. C.2. When contributions from the whole current ring are added the components of B perpendicular to the z-axis cancel, giving a field along the z-axis equal to

$$\frac{\mu_0}{4\pi} \frac{Mr \, d\theta \sin^2 \theta}{r^2} 2\pi r \sin \theta$$

$$= \tfrac{1}{2} \mu_0 M \sin^3 \theta \, d\theta.$$

This may now be integrated over the sphere to give a total field

$$\int_0^\pi \tfrac{1}{2} \mu_0 M \sin^3 \theta \, d\theta = \int_1^{-1} -\tfrac{1}{2} \mu_0 M (1 - c^2) \, dc$$

$$= \tfrac{1}{2} \mu_0 M \left[c - \frac{c^3}{3} \right]_{-1}^{1} = \tfrac{2}{3} \mu_0 M, \tag{C.3}$$

* This field is in fact uniform over the interior of the sphere, but we shall not prove this.

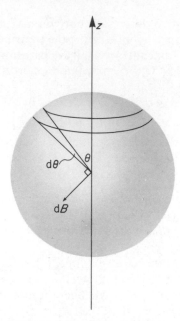

Fig. C.2. Calculation of field at
the centre of a spherical current
sheet.

where we have used the substitution $c = \cos\theta$. We thus have $\mathbf{B}_2 = -\frac{2}{3}\mu_0\mathbf{M}$ in all cases, since the current round the sphere of radius r is opposite to the current round the outside of the sample.

The expression (C.3) enables us to evaluate Eqs. (C.1) and (C.2) for various sample shapes. For the long rod (Fig. C.1(a)) we have

$$\mathbf{B}_{\text{loc}} = \mathbf{B}_e + \mu_0\mathbf{M} - \tfrac{2}{3}\mu_0\mathbf{M}$$

$$= \mathbf{B}_e + \tfrac{1}{3}\mu_0\mathbf{M} \tag{C.4}$$

and

$$\mathbf{B} = \mathbf{B}_e + \mu_0\mathbf{M}, \tag{C.5}$$

so that

$$\mathbf{H} = \mathbf{B}_e. \tag{C.6}$$

It is convenient to express magnetic measurements on ferromagnetic materials as a function of $\mathbf{H}$, since this is the parameter of a magnetic circuit that an electrical engineer can control. Such measurements are therefore usually made on long rod samples, for Eq. (C.6) shows that in this case $\mathbf{H}$ is just the applied field.

For a spherical sample (Fig. C.1(b)) we have

$$\mathbf{B}_{loc} = \mathbf{B}_e + \tfrac{2}{3}\mu_0\mathbf{M} - \tfrac{2}{3}\mu_0\mathbf{M}$$

$$= \mathbf{B}_e \tag{C.7}$$

$$\mathbf{B} = \mathbf{B}_e + \tfrac{2}{3}\mu_0\mathbf{M} \tag{C.8}$$

$$\mathbf{H} = \mathbf{B}_e - \tfrac{1}{3}\mu_0\mathbf{M}. \tag{C.9}$$

In measurements on paramagnetic salts, the magnetic field at an ion is the interesting independent variable, because it is the argument of theoretical expressions for the magnetization. Such measurements are therefore either made on (or corrected to) a spherical sample, so that the local field is just the applied field (provided that $\mathbf{B}_3 = 0$).

For completeness we note that in the opposite extreme of a disc shaped sample (Fig. C.1(e)), since $\mathbf{B}_1 = 0$,

$$\mathbf{B}_{loc} = \mathbf{B}_e - \tfrac{2}{3}\mu_0\mathbf{M} \tag{C.10}$$

$$\mathbf{B} = \mathbf{B}_e \tag{C.11}$$

$$\mathbf{H} = \mathbf{B}_e - \mu_0\mathbf{M}. \tag{C.12}$$

Note that Eq. (5.38) comes from using Eq. (C.11) for fields normal to a metal foil, together with Eq. (C.5) for fields in the plane of the foil.

Another concept often encountered in magnetic measurements is the **demagnetizing field,** defined as $\mathbf{H} - \mathbf{B}_e$. It is the field distribution calculated from the fictitious magnetic poles at the ends of the sample, and arises in the traditional form of magnetostatics using fictitious poles and $\mathbf{H}$ as the independent variable. The demagnetizing field is the correction to be added to the applied field to obtain $\mathbf{H}$; it is zero for a long rod. The **demagnetizing factor** is the ratio of demagnetizing field to $\mu_0 M$ ($\tfrac{1}{3}$ for a sphere, 1 for a disc).*

Although magnetic poles are fictitious, they provide a useful analogy with the corresponding calculation of local electric fields from distributions of polarization charge. We shall not give details of the electric case, since we do not need it in this book; we content ourselves with a brief comparison, for the sake of completeness.

Fig. C.3 compares pictorially the calculation of magnetic and electric local fields, for a long rod sample. The depolarizing field due to charge on the ends of such a sample is negligible, and the only correction to $\mathbf{E}_e$ is the field in the spherical cavity, $\tfrac{1}{3}\mathbf{P}/\varepsilon_0$. We thus have

$$\mathbf{E}_{loc} = \mathbf{E}_e + \tfrac{1}{3}\mathbf{P}/\varepsilon_0, \tag{C.13}$$

analogously to Eq. (C.4). Similar results apply in the other cases, with the

* In c.g.s. electromagnetic units it is usual to include the factor $\mu_0 = 4\pi$ in the definition of the demagnetizing factor.

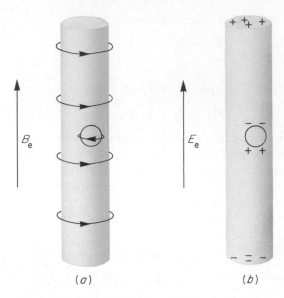

Fig. C.3. Comparison of local field calculations in
(a) the magnetic case, (b) the electric case.

correspondence

$$\mathbf{B}_e \rightarrow \mathbf{E}_e, \qquad \mathbf{B}_{loc} \rightarrow \mathbf{E}_{loc},$$

$$\mathbf{M} \rightarrow \mathbf{P}, \qquad \mu_0 \rightarrow 1/\varepsilon_0, \qquad\qquad (C.14)*$$

$$\mathbf{B} \rightarrow \mathbf{D}, \qquad \mathbf{H} \rightarrow \mathbf{E}.$$

The important point to note about this correspondence is that the relation between applied field and local field is independent of the nature of the dipoles. This is reasonable, because magnetic dipoles (due to current loops) and electric dipoles (due to separated charges) do not differ in their *external* field distribution, and our calculations have not involved any consideration of the field in the *interior* of a dipole. Such considerations could arise only in the calculation of $\mathbf{B}_3$, or its electrical analogue, and even then only if the dipoles were not well localized on lattice sites.†

The difference of the two types of dipole does however show up in the macroscopic average fields; it is because of this that the spatial average of the microscopic magnetic field is $\mathbf{B}$, but the spatial average of the microscopic electric field is $\mathbf{E}$, not $\mathbf{D}$; for a proof of this see Purcell,[4] Chapter 10.

---

* However, it is customary in SI units (see Appendix E) to measure $\mathbf{H}$ in dipole moment units, not, as we do, in field units. It is likewise customary in SI units to measure $\mathbf{D}$ in dipole moment units rather than field units. This convention complicates the correspondence (C.14); in conventional SI units our $\mathbf{H}$ must be written $\mu_0\mathbf{H}$ and our $\mathbf{D}$ must be written $\mathbf{D}/\varepsilon_0$.

† For example, in calculating the local magnetic field in a metal, the *interior* of a spinning electron dipole is relevant.

Distribution function with constant total momentum

We wish to calculate the thermal equilibrium distribution of particles over energy levels ε_i, each of which has a momentum (or quasi-momentum) $\mathbf{p}_i$ associated with it. We are interested in particular in an isolated system for which total energy and momentum are conserved, but we shall suppose that particles are not conserved since it is the application to phonons that concerns us. We thus have to find a distribution function subject to the constraints of constant energy and momentum, instead of the more usual constraints of constant energy and particle number.

We shall adopt the usual elementary method, that of finding the most probable distribution; as usual, we shall suppose that this represents a very sharp maximum in probability, and thus corresponds to thermal equilibrium. As in Mandl,[2] section 11.2, we consider the number N_i of particles in a very narrow range of energy containing G_i states of energy ε_i and momentum $\mathbf{p}_i$; the purpose of this grouping of energy levels is to ensure that N_i and G_i are large numbers, which facilitates the mathematics.

We now define Ω_i as the number of ways in which N_i particles may be arranged on G_i states. Then the total probability of a given distribution is given by

$$\Omega = \prod_i \Omega_i \tag{D.1}$$

this is to be maximized subject to the constraints

$$E = \sum_i N_i \varepsilon_i = \text{constant},$$
(D.2)

and

$$\mathbf{P} = \sum_i N_i \mathbf{p}_i = \text{constant}.$$
(D.3)

It is convenient to maximize $\ln \Omega$ rather than Ω, since this converts the product in Eq. (D.1) into a sum; the condition for a maximum is then

$$\delta(\ln \Omega) = \sum_i \frac{\partial \ln \Omega_i}{\partial N_i} \delta N_i = 0,$$
(D.4)

subject to the constraints

$$\delta E = \sum_i \varepsilon_i \, \delta N_i = 0$$
(D.5)

and

$$\delta \mathbf{P} = \sum_i \mathbf{p}_i \, \delta N_i = 0.$$
(D.6)

We solve Eqs. (D.4)–(D.6) by the standard method of Lagrange undetermined multipliers. From these equations it is clearly true that

$$\sum_i \delta N_i \left[\frac{\partial \ln \Omega_i}{\partial N_i} - \beta \varepsilon_i + \gamma \cdot \mathbf{p}_i \right] = 0$$
(D.7)

for *any* value of the constants β, γ (note that γ is really three constants, since it is a vector). Because of the constraints (D.5) and (D.6) not all the δN_i are independent; two of them—δN_1 and δN_2, say—must be chosen to ensure that (D.5) and (D.6) are satisfied. If we also choose β and γ so that

$$\frac{\partial \ln \Omega_1}{\partial N_1} - \beta \varepsilon_1 + \gamma \cdot \mathbf{p}_1 = 0$$

$$\frac{\partial \ln \Omega_2}{\partial N_2} - \beta \varepsilon_2 + \gamma \cdot \mathbf{p}_2 = 0,$$
(D.8)

the first two terms in the sum (D.7) vanish so that we have

$$\sum_{i>2} \delta N_i \left[\frac{\partial \ln \Omega_i}{\partial N_i} - \beta \varepsilon_i + \gamma \cdot \mathbf{p}_i \right] = 0,$$
(D.9)

where *all* the δN_i (for $i > 2$) are independent. Eq. (D.9) is the equation for an unrestricted maximum, and can be satisfied only if

$$\frac{\partial \ln \Omega_i}{\partial N_i} = \beta \varepsilon_i - \gamma \cdot \mathbf{p}_i$$
(D.10)

for all $i > 2$. But we have already chosen β and γ to satisfy Eqs. (D.8); the condition for maximum Ω is therefore that Eq. (D.10) is true for *all* i. β and γ are to be chosen to make the total energy and momentum come out right; by standard arguments (for example, by consideration of thermal equilibrium with a Boltzmann distribution) it may be shown that $\beta = (1/k_B T)$.

To complete the derivation we simply note that for bosons (such as phonons)

$$\Omega_i = \frac{(N_i + G_i - 1)!}{N_i!(G_i - 1)!} \tag{D.11}$$

so that for $N_i, G_i \gg 1$, by the use of Stirling's approximation for $N!$,

$$\frac{\partial \ln \Omega_i}{\partial N_i} = \ln \left(\frac{G_i + N_i}{N_i} \right) \tag{D.12}$$

(Mandl,[2] Eq. (11.53)). With Eq. (D.10) we thus obtain

$$N_i = \frac{G_i}{e^{(\beta \varepsilon_i - \gamma \cdot \mathbf{p}_i)} - 1}, \tag{D.13}$$

which is Eq. (8.12), since $n = N/G$.

We may finally note that other methods of deriving distribution functions may be similarly adapted to give the distribution with constant total momentum. Thus, in the method of Mandl,[2] section 11.1, we obtain a suitably modified form of his Eq. (11.3) by considering instead the function

$$\mathscr{Z}'(T, V, \gamma) \equiv \sum_{\mathbf{P}=0}^{\infty} Z(T, V, \mathbf{P}) e^{\gamma \cdot \mathbf{P}}; \tag{D.14}$$

from this starting point the argument proceeds in a quite similar way to that for constant particle number.

APPENDIX

Units in electromagnetism

'... It does not seem possible at present to set up a system of units which will satisfy the electrical engineer and the physicist alike. With regard to Maxwell's theory, the difference between the physicist and the electrician is not a matter of notation merely, but of principle. The technical view adheres much more strictly than current physics does to the original form of the Faraday–Maxwell theory. The engineer looks upon vectors **E** and **D**—even in a vacuum—as magnitudes of quite different kinds, related to one another more or less like tension and extension in the theory of elasticity. From this point of view it must of course seem a very questionable procedure, in an exposition of fundamental principles, to put the factor of proportionality K, in the equation **D** $= K$**E**, equal to 1 for empty space, thus artificially attributing to **D** and **E** the same dimensions. On the other hand, the distinction in principle between **D** and **E**, which is closely connected with the mechanical theory of the aether, has been absolutely abandoned in modern physics, the electromagnetic conditions at any point in empty space being now regarded as completely defined when we are given *one* electric vector **E** and *one* magnetic vector **B** (or **H**). The numerical identity of **E** and **D** (for empty space) in the Gaussian system of units is not, for the physicist, the result of an arbitrary definition, but an expression of the fact that **E** and **D** are actually the same thing. The introduction by the engineer of a dielectric constant and permeability not equal to 1 in a vacuum seems to the physicist to be merely an artifice, by means of which formulae are reduced to a shape which is convenient for practical calculations.'

R. Becker

[M. Abraham and R. Becker, *Electricity and Magnetism*, Blackie, London (1937); from the preface to the eighth German edition (1930).]

E.1 PREAMBLE

Unfortunately, the dilemma so clearly explained by Becker is still with us forty years later, and what is worse, the engineering viewpoint has been given exclusive international approval in the current form of SI units. Because of this dilemma atomic and solid state physics has hitherto been almost entirely written in c.g.s. Gaussian units. In this book I have adhered to SI usage as closely as is consistent with the physicist's point of view described by Becker. This I have done by using MKSA units with the electromagnetic system of defining equations, as described in detail below.

E.2 DEFINING EQUATIONS

It is important to distinguish defining equations from units. The *equations* embody a set of conventions about how various quantities are defined, and three systems are in common use: electromagnetic, as used in this book; SI; and Gaussian. The main defining equations are listed in Tables E.1 and E.2,

Table E.1. Electric defining equations

Defined quantity	SI	Electromagnetic	Gaussian
Charge		$\mathbf{F} = \dfrac{qq'}{4\pi\varepsilon_0 r^2}\hat{\mathbf{r}}$	
Electric field		$\mathbf{F} = q\mathbf{E}$	
Field due to a charge		$\operatorname{div}\mathbf{E} = \rho/\varepsilon_0$	
Electrostatic potential		$\mathbf{E} = -\operatorname{grad}\phi$	
Dipole moment		$\mathbf{p} = q\mathbf{d}$	
Polarization		$\operatorname{div}\mathbf{P} = -\rho_{\mathbf{P}}$	
D field	$\mathbf{D} = \varepsilon_0\mathbf{E} + \mathbf{P}$	$\mathbf{D} = \mathbf{E} + \mathbf{P}/\varepsilon_0$	
Electric susceptibility	$\mathbf{P} = \chi_e\varepsilon_0\mathbf{E}$	$\mathbf{P} = \chi_e\mathbf{E}$	

Table E.2. Magnetic defining equations

Defined quantity	SI	Electromagnetic	Gaussian
Current	$$dF = \dfrac{\mu_0 i i'}{4\pi r^2}[dl \times (dl' \times \hat{r})]$$		
Magnetic field	$\mathbf{F} = q\mathbf{v} \times \mathbf{B}$ $d\mathbf{F} = i\,d\mathbf{l} \times \mathbf{B}$		$\mathbf{F} = q\mathbf{v} \times \mathbf{B}/c$ $d\mathbf{F} = i\,d\mathbf{l} \times \mathbf{B}/c$
Field due to a current	curl $\mathbf{B} = \mu_0 \mathbf{j}$		curl $\mathbf{B} = \mu_0 c\mathbf{j}$ $= \mathbf{j}/\varepsilon_0 c$
Vector potential	$\mathbf{B} = $ curl $\mathbf{A}$		
Magnetic moment	$\mathbf{m} = i\,d\mathbf{S}$		$\mathbf{m} = i\,d\mathbf{S}/c$
Magnetization	curl $\mathbf{M} = \mathbf{j_M}$		curl $\mathbf{M} = \mathbf{j_M}/c$
H field	$\mathbf{H} = \dfrac{\mathbf{B}}{\mu_0} - \mathbf{M}$	$\mathbf{H} = \mathbf{B} - \mu_0\mathbf{M}$	$\mathbf{H} = \mathbf{B} - \mu_0 c^2 \mathbf{M}$ $= \mathbf{B} - \mathbf{M}/\varepsilon_0$
Magnetic susceptibility	$\mathbf{M} = \chi_m \mathbf{H}$		

from which it can be seen that SI or Gaussian equations may be obtained from electromagnetic equations by making the following changes everywhere:

to convert to SI

$$\mathbf{D} \Rightarrow \mathbf{D}/\varepsilon_0, \qquad \chi_e \Rightarrow \varepsilon_0 \chi_e,$$

$$\mathbf{H} \Rightarrow \mu_0 \mathbf{H}, \qquad \chi_m \Rightarrow \chi_m/\mu_0 ;$$

to convert to Gaussian

$$\mathbf{B} \Rightarrow \mathbf{B}/c, \qquad \mathbf{H} \Rightarrow \mathbf{H}/c,$$

$$\mathbf{M} \Rightarrow c\mathbf{M}, \qquad \chi_m \Rightarrow c^2 \chi_m.$$

E.3 UNITS

Once a system of equations has been selected, the *units* are determined by the units of mass, length, and time, and the choice of the constants μ_0 and ε_0 ; in all systems $\mu_0 \varepsilon_0 c^2 = 1$.

c.g.s. units

Electromagnetic equations with $\mu_0 = 4\pi$ give c.g.s. e.m.u.. Gaussian equations with $\varepsilon_0 = 1/4\pi$ give c.g.s. Gaussian units; note that the factors of c in the defining equations give magnetic quantities in e.m.u. and electric quantities in e.s.u.

MSKA units

In MKS units the fourth fundamental unit is always chosen as the ampere. From the definition of the ampere $\mu_0 = 4\pi \times 10^{-7}$ H/m. MKSA units may conveniently be used with either electromagnetic or SI equations; MKSA Gaussian units (in which magnetic fields are in V/m) are *not* used.

E.4   RATIONALIZATION

This is a famous red herring. It is the name given to the convention, which I have used, whereby 4π appears explicitly in Coulomb's law but not in Maxwell's equations. But it is a misnomer to call any currently used *units* rationalized; the 4π reappears in μ_0 and ε_0.

Bibliography

GENERAL

1. R. P. Feynman, *Lectures on Physics*, Vol. III, Addison-Wesley, Reading, Mass., 1965. (A full exposition of the approach to quantum mechanics we adopt in this book. Chapters 13, 14, 15 and 21 are particularly relevant.)
2. F. Mandl, *Statistical Physics*, Wiley, London, 1970. (Contains the statistical mechanics background required in this book.)
3. B. H. Flowers and E. Mendoza, *Properties of Matter*, Wiley, London, 1970. (A good exposition of the basic atomistic view of matter we presuppose, particularly kinetic theory of gases.)
4. E. M. Purcell, *Electricity and Magnetism*, McGraw-Hill, New York, 1965. (The essential background in electromagnetism. The discussion of fields in matter in Chapters 9 and 10 is particularly good.)
5. C. Kittel, *Introduction to Solid State Physics* (3rd Edn.), Wiley, New York, 1966. (An excellent book full of facts.)
6. J. M. Ziman, *Principles of the Theory of Solids*, Cambridge, 1964. (More theoretical and less factual than Kittel, beginning postgraduate level.)
7. C. Kittel, *Quantum Theory of Solids*, Wiley, New York, 1963. (Not on any account to be confused with Kittel[5], it assumes that all experimental facts are known. A compendium of useful calculations, mainly postgraduate level.)

CHAPTER 1

8. F. C. Phillips, *Introduction to Crystallography*, Longmans, London, 1949. (An excellent account of crystal and lattice geometry.)

9. W. Heitler, *Wave Mechanics* (2nd Edn.), Oxford, 1956. (A clearly elementary account which gives a good grounding for the quantum mechanics in this book.)
10. R. C. Evans, *Crystal Chemistry* (2nd Edn.), Cambridge, 1966. (An excellent account of the relation between chemical binding and crystal structure.)

CHAPTER 3

11. M. J. Morant, *Introduction to Semiconductor Devices* (2nd Edn.), Harrap, London, 1970. (Gives further information on semiconductor devices at a suitable level.)
Feynman,[1] Chapter 14, is also relevant here.

CHAPTER 6

12. L. Brillouin, *Wave Propagation in Periodic Structures*, Dover, New York, 1953. (An excellent readable account by the inventor of the subject.)
13. F. G. Smith and J. H. Thomson, *Optics*, Wiley, London, 1971. (Gives the necessary background in diffraction theory.)
14. W. A. Harrison, *Pseudopotentials in the Theory of Metals*, Benjamin, New York, 1966. (A largely postgraduate book, useful for pictures of zones and free electron like Fermi surfaces.)

CHAPTER 7

15. G. E. Bacon, *Neutron Diffraction* (2nd Edn.), Oxford, 1962. (A textbook of neutron crystallography, excluding inelastic scattering.)
16. P. A. Egelstaff (Ed.), *Thermal Neutron Scattering*, Academic Press, London, 1965. (A collection of articles on inelastic scattering.)

CHAPTER 8

19. R. E. Peierls, *Quantum Theory of Solids*, Oxford, 1956. (Includes an excellent clear account of basic ideas about phonons.)
20. C. T. Lane, *Superfluid Physics*, McGraw-Hill, New York, 1962. (A good elementary account of superfluid helium.)
21. B. Bertman and D. J. Sandiford, *Scientific American*, **222** (May) 92, (1970). (A good accessible account of collective phonon flow effects.)

CHAPTER 10

Harrison[14] is particularly useful for diagrams showing the construction of multiply connected Fermi surfaces.

CHAPTER 11

22. A. C. Rose-Innes and E. H. Rhoderick, *Introduction to Superconductivity*, Pergamon, Oxford, 1969. (Gives a good, elementary, and up to date account.)
Feynman,[1] Chapter 21, gives an excellent briefer account.

CHAPTER 12

23. F. Brailsford, *Physical Principles of Magnetism*, Van Nostrand, New York, (1966). (A specialist text giving details of real magnetic materials.)

Solutions to problems

1.1 The corners of a tetrahedron are in the directions $[1, 1, 1], [-1, -1, 1], [1, -1, -1]$, $[-1, 1, -1]$, and the sets of coefficients (a_x, a_y, a_z) must be proportional to these vectors. For normalization:

$$\int |\psi|^2 \, dV = a_x^2 + a_y^2 + a_z^2 = 1$$

since the p states are orthonormal.

The required sets of coefficients are therefore $(3^{-1/2}, 3^{-1/2}, 3^{-1/2})$, $(-3^{-1/2}, -3^{-1/2}, 3^{-1/2})$, $(3^{-1/2}, -3^{-1/2}, -3^{-1/2})$, $(-3^{-1/2}, 3^{-1/2}, -3^{-1/2})$.

1.2 For example

$$\phi_{111} = bs + \frac{c}{\sqrt{3}}(p_x + p_y + p_z).$$

$$\int |\phi_{111}|^2 \, dV = b^2 + c^2 = 1 \quad \text{for normalization.}$$

$$\int \phi_{111}\phi_{-1-11} \, dV = \int b^2 s^2 \, dV + \int \frac{c^2}{3}(p_x + p_y + p_z)(-p_x - p_y + p_z) \, dV$$

$$= b^2 - \tfrac{1}{3}c^2 = 0 \quad \text{for orthogonality.}$$

Therefore, $b = \tfrac{1}{2}$, $c = \sqrt{3}/2$, or

$$\phi_{111} = \tfrac{1}{2}(s + p_x + p_y + p_z),$$

and similarly for the other ϕs.

1.3 Vectors (β, γ) at $120°$ are $(1, 0), (-\frac{1}{2}, \sqrt{3}/2), (-\frac{1}{2}, -\sqrt{3}/2)$. Consider states

$$\chi_1 = \alpha s + p_x$$

$$\chi_2 = \alpha s - \tfrac{1}{2}p_x + \frac{\sqrt{3}}{2}p_y;$$

α may be taken the same in both cases because the wavefunctions differ only in orientation.

$$\int \chi_1 \chi_2 \, dV = \alpha^2 - \tfrac{1}{2} = 0 \quad \text{for orthogonality,}$$

so that $\alpha = 1/\sqrt{2}$. The normalization integral for χ_1 is thus

$$\int \left| \frac{1}{\sqrt{2}} s + p_x \right|^2 dV = \tfrac{3}{2},$$

so for normalization our coefficients must be multiplied by $\sqrt{2/3}$, so that the required values of (α, β, γ) are $(1/\sqrt{3}, \sqrt{2}/\sqrt{3}, 0)$, $(1/\sqrt{3}, -1/\sqrt{6}, 1/\sqrt{2})$, $(1/\sqrt{3}, -1/\sqrt{6}, -1/\sqrt{2})$. The other orthogonality and normalization integrals may be checked similarly.

1.4 Compare Fig. 1.14.

1.5 The plane

$$\frac{x}{\alpha} + \frac{y}{\beta} + \frac{z}{\gamma} = 1$$

has intercepts on the axes

$$x = \alpha, \qquad y = \beta, \qquad z = \gamma.$$

The plane $(h \ k \ l)$ nearest the origin therefore has the equation

$$hx + ky + lz = a,$$

where a is the lattice spacing, and the plane parallel to this through the origin is

$$hx + ky + lz = 0$$

An arbitrary vector in the latter plane is

$$(x, y, -(hx + ky)/l);$$

the scalar product of this with $[h, k, l]$ is

$$hx + ky - (hx + ky) = 0.$$

The vector $[h \ k \ l]$ is therefore normal to the plane $(h \ k \ l)$. Note that this proof depends on the use of orthogonal axes with equal length scales.

1.6 d is the normal distance from the origin to the plane $hx + ky + lz = a$. The direction cosines of the normal to this plane are $d \div (a/h)$, etc. Therefore

$$\frac{h^2 d^2}{a^2} + \frac{k^2 d^2}{a^2} + \frac{l^2 d^2}{a^2} = 1$$

or

$$d = a/(h^2 + k^2 + l^2)^{1/2}.$$

1.7

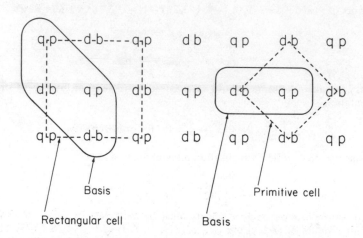

1.8
(a)

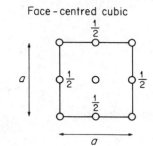

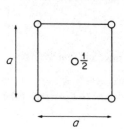

Face-centred cubic

Body-centred cubic

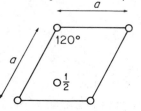

Hexagonal closed-packed

Diamond

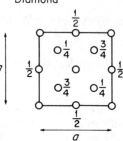

(b) fcc (000), $(\frac{1}{2}\frac{1}{2}0)$, $(0\frac{1}{2}\frac{1}{2})$, $(\frac{1}{2}0\frac{1}{2})$
  bcc (000), $(\frac{1}{2}\frac{1}{2}\frac{1}{2})$
  hcp (000), $(\frac{2}{3}\frac{1}{3}\frac{1}{2})$
  diamond (000), $(\frac{1}{2}\frac{1}{2}0)$, $(0\frac{1}{2}\frac{1}{2})$, $(\frac{1}{2}0\frac{1}{2})$, $(\frac{1}{4}\frac{1}{4}\frac{1}{4})$, $(\frac{3}{4}\frac{3}{4}\frac{1}{4})$, $(\frac{1}{4}\frac{3}{4}\frac{3}{4})$, $(\frac{3}{4}\frac{1}{4}\frac{3}{4})$.

(c) Structure	Sphere diameter	Number in cell	Fraction occupied
fcc	$a/\sqrt{2}$	4	0.740
bcc	$\sqrt{3}a/2$	2	0.680
hcp	—	—	as fcc
diamond	$\sqrt{3}a/4$	8	0.340

1.9 Note that if $\mathbf{c}' = 3\mathbf{k}$, then

$$\mathbf{c} = \tfrac{1}{2}(\mathbf{a} + \mathbf{b} + \mathbf{c}'),$$

which is the body centering position of a cubic unit cell defined by $\mathbf{a}, \mathbf{b}, \mathbf{c}'$. The Bravais lattice is therefore body centred cubic with a conventional unit cell of volume 27 Å^3. The primitive cell is half this volume, 13.5 Å^3, since there are two atoms in the conventional cell.

The most densely packed planes are {110}.

1.10 For fcc see Fig. 1.16(b), and for bcc Fig. 6.10 gives a clue.

For fcc the directions of $\mathbf{a}, \mathbf{b}, \mathbf{c}$ are $[110], [011], [101]$.

$$\cos \alpha = \frac{\mathbf{a} \cdot \mathbf{b}}{\mathbf{a} \cdot \mathbf{a}} = \frac{1}{2}$$

$$\alpha = 60°.$$

For bcc the directions of $\mathbf{a}, \mathbf{b}, \mathbf{c}$ are $[\bar{1}11], [1\bar{1}1], [11\bar{1}],$

$$\cos \alpha = \frac{\mathbf{a} \cdot \mathbf{b}}{\mathbf{a} \cdot \mathbf{a}} = \frac{-1}{3}$$

$$\alpha = 109° \, 27'.$$

2.1 With $u_n = A \exp i(kna - \omega t)$ the momentum of the nth atom is

$$M\dot{u}_n = -i\omega M A \exp i(kna - \omega t),$$

and the total momentum of the chain is obtained by summing the atomic momenta:

$$P(k) = -i\omega M A \, e^{-i\omega t} \sum_{n=1}^{N} e^{ikna}$$

$$= -i\omega M A \, e^{-i\omega t} \sum_{n=1}^{N} \exp 2\pi i\left(\frac{pn}{N}\right)$$

with periodic boundary conditions, from Eq. (2.4). By summing the geometric progression

$$P(k) = -i\omega M A \, e^{-i\omega t} \left\{ \frac{1 - \exp(2\pi i p)}{1 - \exp(2\pi i p/N)} \right\} = 0 \qquad \text{for } p \neq 0.$$

For $p = 0$ ($k = 0$) all the atoms are moving with velocity $-i\omega A$, and $P(0)$ is just the momentum of the whole chain with this velocity.

2.2 From Eqs. (2.27) and (2.31) the exact result is

$$C = \frac{2R}{\pi} \int_0^{2(K/M)^{1/2}} \left(\frac{4K}{M} - \omega^2\right)^{-1/2} \left(\frac{\hbar\omega}{k_B T}\right)^2 \frac{\exp(\hbar\omega/k_B T)}{[\exp(\hbar\omega/k_B T) - 1]^2} \, d\omega.$$

With

$$\Theta = \frac{\pi \hbar}{k_B} \left(\frac{K}{M}\right)^{1/2}$$

and $x = \hbar\omega/k_B T$ this becomes

$$C = 2R \int_0^{2\Theta/\pi T} \left[\left(\frac{2\Theta}{T}\right)^2 - (\pi x)^2\right]^{-1/2} \frac{x^2 e^x}{(e^x - 1)^2} \, dx,$$

and with a Debye cutoff frequency $\pi(K/M)^{1/2}$ and constant density of states this becomes

$$C = \frac{RT}{\Theta} \int_0^{\Theta/T} \frac{x^2 e^x \, dx}{(e^x - 1)^2}.$$

Because the average frequency is lower in the exact density of states, this gives a higher specific heat than the Debye approximation at all temperatures. The factor $\sim e^{-x}$ in the integrand means that $x \sim 1$ makes the most important contribution. When $T \ll \Theta$ the two densities of states are equivalent and the integrals may be taken to infinity, so that

$$C = \frac{RT}{\Theta} \int_0^\infty \frac{x^2 e^x \, dx}{(e^x - 1)^2}.$$

2.3 The equations of motion are

$$M\ddot{u}_{2n} = K_1 u_{2n+1} - (K_1 + K_2)u_{2n} + K_2 u_{2n-1}$$

$$M\ddot{u}_{2n-1} = K_2 u_{2n} - (K_1 + K_2)u_{2n-1} + K_1 u_{2n-2}.$$

These may be solved as in section 2.3 by taking

$$u_{2n} = A \exp i(kna - \omega t)$$

and

$$u_{2n-1} = \alpha A \exp i(kna - \omega t);$$

note that the repeat distance a is the distance from an atom to the next but one. The separation between the $(2n - 1)$th and $(2n)$th atoms is irrelevant because any phase factor arising from it is absorbed in the constant α (which depends on k).

The solution proceeds just as in section 2.3 and the dispersion curves are as in Fig. 2.6, except that points B and C are at wavenumber (π/a) with frequencies $(2K_1/M)^{1/2}$ and $(2K_2/M)^{1/2}$, and point A has frequency $[2(K_1 + K_2)/M]^{1/2}$.

2.4 For extension along [100] NaCl may be considered as a set of linear chains in parallel and the lateral forces ignored. With this assumption Young's modulus is related to the atomic spacing $d = 2.8$ Å by

$$\text{Young's modulus } Y = \frac{K/d^2}{1/d} = \frac{K}{d}.$$

If we equate the reststrahl frequency to that of the $k = 0$ optical mode

$$\omega = \frac{2\pi c}{\lambda} = \left[\frac{2K(M + m)}{Mm}\right]^{1/2},$$

and with masses 23 and $37 \times 1.66 \times 10^{-27}$ kg

$$\lambda = 2\pi c \left[\frac{Mm}{2Yd(M + m)} \right]^{1/2}$$

$$= 6\pi \times 10^8 \left[\frac{23 \times 37 \times 1.66 \times 10^{-27}}{2 \times 5 \times 10^{10} \times 2.8 \times 10^{-10}(23 + 37)} \right]^{1/2} m$$

$$= 55 \ \mu m.$$

2.5 In two dimensions the density of states is, for area A,

$$g(k) \, dk = \frac{A}{2\pi} k \, dk,$$

so that with $\omega^2 = (\sigma k^3/\rho)$

$$g(\omega) = g(k)\frac{dk}{d\omega} = \frac{A}{3\pi} \left(\frac{\rho}{\sigma} \right)^{2/3} \omega^{1/3}.$$

With Eq. 2.25 for the average energy of a harmonic oscillator this gives the energy per unit area

$$E = E_0 + \frac{1}{3\pi} \left(\frac{\rho}{\sigma} \right)^{2/3} h \left(\frac{k_B T}{h} \right)^{7/3} \int_0^{\omega = \omega_c} \frac{x^{4/3} \, dx}{e^x - 1},$$

where E_0 is zero point energy, $x = \hbar\omega/k_B T$, and ω_c is a suitable cutoff frequency. This cutoff is rather hard to determine since we do not know how many degrees of freedom are associated with the surface, though we expect it to be of the order of the number of atoms in unit area of surface. However, at low enough temperatures this does not matter, for we can take the integral to infinity in this limit, giving

$$E = E_0 + aT^{7/3}.$$

The surface specific heat is then

$$C = \frac{dE}{dT} = \frac{7}{3}aT^{4/3},$$

and the entropy is

$$S = \int_0^T \frac{C \, dT}{T} = \frac{7}{4}aT^{4/3}.$$

Since σ is the surface free energy

$$\sigma(T) - \sigma(0) = E - E_0 - TS = -\tfrac{3}{4}aT^{7/3}.$$

2.6 For $\hbar\omega \ll k_B T$ Eq. (2.26) gives

$$n(\omega) \approx k_B T/\hbar\omega,$$

so that average energy $= \hbar\omega n(\omega) \approx k_B T$. We may consider roughly that modes are excited for $\hbar\omega \lesssim k_B T$, or for wavenumber

$$k \lesssim \frac{k_B T}{\hbar v} = k_0,$$

where v is the velocity of sound. If we ignore the difference between transverse and longitudinal sound, Eq. (2.39) gives

$$\left(\frac{k_B \Theta_D}{\hbar v} \right)^3 = \frac{6\pi^2 N}{V} = k_c^3$$

where k_c is the Debye cutoff wavenumber. Since modes are uniformly distributed in k-space the fraction of modes excited is $(k_0/k_c)^3$ and the number is thus

$$3N\left(\frac{k_0}{k_c}\right)^3 = 3N\left(\frac{T}{\Theta_D}\right)^3.$$

If we make the approximation that modes with $k < k_0$ have energy $k_B T$ and modes with $k > k_0$ are entirely unexcited the internal energy is

$$U = 3Nk_B T\left(\frac{T}{\Theta_D}\right)^3,$$

so that

$$C_v = 12Nk_B\left(\frac{T}{\Theta_D}\right)^3.$$

This is the right functional form, but about 20 times smaller than the exact result. A good numerical answer cannot be expected, because $C_v \propto k_0^3$, and k_0 is a rather arbitrary quantity.

2.7 From the previous problem

$$\Theta_D = \frac{hv}{k_B}\left(\frac{6\pi^2 N}{V}\right)^{1/3},$$

and we may take

$$v^2 = (Y/\rho)$$

and

$$\frac{N}{V} = \frac{\rho}{m},$$

where Y = Young's modulus, ρ = density, and $m = 12 \times 1.66 \times 10^{-27}$ kg is the mass of a carbon atom.

Hence $\Theta_D \approx 2700$ K.

To sketch the graph, remember the general shape of the Debye curve, and in particular that the specific heat is about half classical when $T \approx \frac{1}{3}\Theta_D$.

3.1 The sample will cease to show intrinsic behaviour when the impurity concentration becomes of the same order as the intrinsic carrier concentration, i.e.

$$N_D \gtrsim N_c \exp\left[-E_G/2k_B T\right].$$

With N_c from Eq. (3.10) this becomes

$$e^{5800/T} \gtrsim 430 T^{3/2}$$

or

$$T \lesssim 400 \text{ K}.$$

3.2 With the approximation $\hbar = 10^{-34}$ joule sec
   (a) $m^* = 5 \times 10^{-32}$ kg
   (b) $\mathbf{k} = -10^9 \hat{\mathbf{k}}_x \, \text{m}^{-1}$
   (c) $\mathbf{v} = -2 \times 10^6 \hat{\mathbf{k}}_x \, \text{m s}^{-1}$
   (d) $\varepsilon = +10^{-19}$ joule.

3.3 The figure shows experimental results for these concentrations

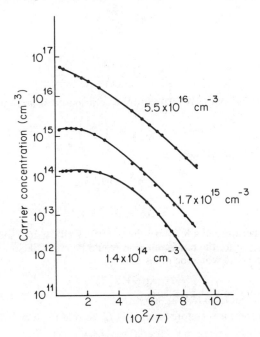

Note:
  (i) Intrinsic behaviour at high temperatures (not illustrated);
  (ii) $n \sim N_D$ at intermediate temperatures;
  (iii) low temperature behaviour given by Eq. (3.18) or Eq. (3.20). For the lowest donor concentration there are sufficient acceptors to pin down the Fermi level at the donor level, and Eq. (3.18) is quite well obeyed. For higher donor concentrations there is a transition to Eq. (3.20).

3.4 (a)
$$E_D = \frac{m_e}{m \varepsilon_r^2} \times 13.6 \, \text{eV}$$

$$= 6.6 \times 10^{-4} \, \text{eV}.$$

The acceptor energy is different if $m_h \neq m_e$.

(b) $r = \dfrac{\varepsilon_r m}{m_e} \times 0.53 \, \text{Å} = 650 \, \text{Å} = 6.5 \times 10^{-8} \, \text{m}.$

(c) Overlap is significant when $N_D \sim \dfrac{1}{(2r)^3}$

$$\sim 10^{21} \, \text{m}^{-3} = 10^{15} \, \text{cm}^{-3}.$$

At about this concentration an impurity band of mobile states is formed, see section 4.1.2.

3.5 For a single type of carrier the Hall effect gives the sign and magnitude of the carrier concentration. Measurement of the conductivity then enables the mobility to be determined from Eq. (3.28). The effective mass is determined from cyclotron resonance.

The condition for observation of cyclotron resonance is

$$\omega_c \tau = \frac{eB\tau}{m^*} \gg 1,$$

i.e.

$$B \gg \frac{m^*}{e\tau}.$$

Also the mean free path l is related to the collision cross-section A by

$$l = \frac{1}{N_D A} = \tau \left(\frac{k_B T}{m^*}\right)^{1/2}$$

so that the condition becomes

$$\frac{B}{T^{1/2}} \gg \frac{N_D A}{e}(k_B m^*)^{1/2}$$

$$\approx 0.6 \times 10^{-3}\,\text{T}\,\text{K}^{-1/2}.$$

Thus at a temperature of 4 K even a field of 0.01 T (100 gauss) should be adequate. At room temperature the minimum field would be more like 0.1 T. At 4 K optical excitation of carriers would be necessary.

3.6 $$I = I_0[\exp(eV/k_B T) - 1],$$

and $(eV/(k_B T) = 5.8$ for 0.15 V at 300 K.

The reverse current is therefore essentially I_0, so that the forward current is

$$I = e^{5.8} \times 5\,\mu\text{A}$$

$$= 1.66\,\text{mA}.$$

3.7

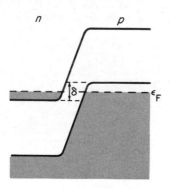
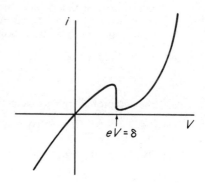

For reverse bias and for forward bias less than δ/e electrons can tunnel between the n side conduction band and the p side valence band. For forward bias greater than δ/e tunnelling cannot occur and the normal diode characteristic is obtained, so that there is negative slope resistance for $eV \sim \delta$. The hole-tunnelling characteristic is similar.

4.1 From Fig. 4.6

$$C = (2.08T + 2.57T^3)\,\text{mJ mole}^{-1}\,\text{K}^{-1}$$

$$= \frac{\pi^2}{2}Nk_B\frac{T}{T_F} + \frac{12\pi^4}{5}Nk_B\left(\frac{T}{\Theta_D}\right)^3,$$

whence

$$T_F = 1.97 \times 10^4\,\text{K},$$

$$\Theta_D = 91\,\text{K}.$$

4.2 $k_F^3 = (3\pi^2 N/V) = (3\pi^2/a^3)$ for a simple cubic structure, so that $k_F = 3.09/a$.

The Brillouin zone is a cube of side $(2\pi/a)$, so that the Fermi sphere is just contained within it; the distance of closest approach is only $(0.05/a)$.

Because the free electron Fermi sphere is so close to the zone boundary, the periodic potential would almost certainly be strong enough to cause contact with the zone boundary, as in Fig. 4.14. However, there would be more unoccupied states near the zone corner than in Fig. 4.14 and there would not be enough electrons to occupy any states in the second zone. In three dimensions such contact with the zone boundary gives a multiply connected Fermi surface; compare copper, Fig. 10.13.

4.3 The width of the K emission band is given by ε_F. From Eq. (4.2) and $(N/V) = (2/a^3)$ for a body-centred cubic structure

$$\varepsilon_F = \frac{\hbar^2}{2ma^2}(6\pi^2)^{2/3}$$

$$= 7.6 \times 10^{-19}\,\text{joule}$$

$$= 4.7\,\text{eV}.$$

The width will increase slightly with temperature as conduction electrons are excited above the Fermi level.

4.4 Since the density of states is proportional to $\varepsilon^{1/2}$

$$E_0 = \frac{N\int_0^{\varepsilon_F} \varepsilon^{3/2}\,d\varepsilon}{\int_0^{\varepsilon_F} \varepsilon^{1/2}\,d\varepsilon} = \frac{3}{5}N\varepsilon_F.$$

$$p = -\frac{\partial E_0}{\partial V} = -\frac{E_0}{V}\frac{\partial \ln E_0}{\partial \ln V} = \frac{2}{3}\frac{E_0}{V} = \frac{2}{5}\frac{N\varepsilon_F}{V}.$$

$$B = -V\frac{\partial p}{\partial V} = -p\frac{\partial \ln p}{\partial \ln V} = \frac{5}{3}p = \frac{2}{3}\frac{N\varepsilon_F}{V}.$$

From the preceding problem, for lithium

$$B = 2.4 \times 10^{10}\,\text{N m}^{-2}.$$

The experimental value is $1.2 \times 10^{10}\,\text{N m}^{-2}$.

4.5 For sodium

$$R_H = -\frac{1}{ne} = -2.45 \times 10^{-10}\,\text{m}^3\,\text{C}^{-1}.$$

For InSb, from Eq. (3.15)

$$n = 2\left(\frac{2\pi k_B T}{h^2}\right)^{3/2} (m_e m_h)^{3/4} e^{-E_G/2k_B T}$$

$$= 0.86 \times 10^{22} \, m^{-3},$$

so that $R_H = -7.2 \times 10^{-4} \, m^3 \, C^{-1}$.

Electrons are the only effective carriers if their mobility is much larger than that of the holes; see answer to next problem.

If a Hall voltage V_H is generated across a sample of width w and thickness t

$$\frac{V_H}{w} = R_H B \frac{i}{wt},$$

so that

$$V_H = R_H Bi/t,$$

independent of w. Thus for sodium

$$V_H = 2.45 \times 10^{-9} \, V = 2.45 \, nV$$

and for InSb

$$V_H = 7.2 \, mV.$$

This illustrates how the small number of carriers in semiconductors makes them suitable for measuring magnetic fields by Hall effect.

4.6 Since the valence band lies below the conduction band it tends to be narrower; inner shell electrons have very narrow bands because they are little influenced by neighbouring atoms.

From Eq. (3.3) a narrower band tends to have a high effective mass. Also

$$\mu = \frac{e\tau}{m^*}$$

and, apart from explicit velocity dependence of the collision cross section,

$$\tau \propto (1/\bar{v}) \propto (m^*)^{1/2}.$$

Therefore, other things being equal

$$\mu \propto (m^*)^{-1/2}$$

and hole mobility tends to be less than electron mobility.

5.1 The average moment per ion, M/N, is obtained by weighting the states of moment $\pm \mu_B$ with appropriate Boltzmann factors:

$$\frac{M}{N} = \frac{\mu_B e^{+\mu_B B/k_B T} - \mu_B e^{-\mu_B B/k_B T}}{e^{+\mu_B B/k_B T} + e^{-\mu_B B/k_B T}}$$

$$= \mu_B \tanh (\mu_B B/k_B T).$$

The energy of the spin system is

$$E = -\mathbf{M} \cdot \mathbf{B} = -N\mu_B B \tanh (\mu_B B/k_B T),$$

so that the specific heat per mole at constant field is

$$C_B = \left(\frac{\partial E}{\partial T}\right)_B = R\left(\frac{\mu_B B}{k_B T}\right)^2 \cosh^2\left(\frac{\mu_B B}{k_B T}\right).$$

Thus for $k_B T \gg \mu_B B$

$$C_B \approx R(\mu_B B/k_B T)^2,$$

and for $k_B T \ll \mu_B B$

$$C_B \approx 4R(\mu_B B/k_B T)^2 \exp\left[-2\mu_B B/k_B T\right].$$

At intermediate temperatures the specific heat has a maximum value of order R; for $CuSO_4$ the maximum occurs in a field of order 1 T at a temperature of 1 K. In comparison, the electronic specific heat of copper metal is of order $10^{-4}R$ at these temperatures. For $B = 0.5T$, temperatures above 1 or 2 K are 'high' for $CuSO_4$.

5.2 From Eq. (4.4) for the specific heat of a free electron gas, and the relation $g(\varepsilon_F) = (3N/2\varepsilon_F)$, we obtain

$$C_v = \frac{\pi^2}{3} g(\varepsilon_F) k_B^2 T.$$

In this form the result applies to any independent particle model, not just a free electron gas. This is because at low temperatures only the occupation of states near ε_F changes, so the heat capacity can depend only on $g(\varepsilon_F)$.

Also from Eq. (5.5)

$$\chi = \mu^2 g(\varepsilon_F),$$

so that

$$\frac{C_v}{\chi} = \frac{\pi^2 k_B^2 T}{2\mu^2}.$$

This result depends on the assumption that the local field is the applied field; any effective field due to exchange interaction is ignored. Also the total measured susceptibility includes Landau diamagnetism of conduction electrons, not just Pauli paramagnetism.

5.3 From Eq. (5.16)

$$\chi = -\frac{ne^2}{4m}\langle r^2 \rangle,$$

where

$$n = \text{number of relevant electrons in unit volume}$$

$$= 6 \times 6.02 \times 10^{23} \times \frac{0.88}{78} \times 10^6 \text{ m}^{-3},$$

since there are 6 relevant electrons per molecule.

For randomly oriented planar molecules

$$\langle r^2 \rangle = \tfrac{2}{3}(1.4 \text{ Å})^2,$$

whence $\chi = -3.4 \text{ A m}^{-1} \text{T}^{-1}$.

The total observed susceptibility is $-6.2\,\mathrm{A\,m^{-1}\,T^{-1}}$; the relatively large contribution of the electrons in extended orbitals is due to the large $\langle r^2 \rangle$ for these electrons.

5.4 The major assumption of the Weiss model is that the energy of a spin depends *only* on the average magnetization of the sample. Specifically, the spins have energies

$$E_\pm = \mp \mu_B B_{\mathrm{eff}}$$

where $B_{\mathrm{eff}} = H + \mu_0 \lambda M$, and the upper sign denotes spin parallel to the magnetization.

If there are N spins, n of which are aligned parallel, so that $M = \mu_B(2n - N)$, the entropy is

$$S = k \ln \Omega = k \ln \frac{N!}{n!(N - n)!}$$

$$= k[N \ln N - n \ln n - (N - n) \ln (N - n)].$$

To evaluate the internal energy U we use the basic expression of the first law of thermodynamics in statistical mechanics

$$dU = \sum_i n_i\, d\varepsilon_i + \sum_i \varepsilon_i\, dn_i$$

where the first term is the work and the second term the heat. This can be written

$$dU = \sum_i n_i \left(\frac{\partial \varepsilon_i}{\partial H}\right)_{M,N} dH + \sum_i \varepsilon_i \left(\frac{\partial n_i}{\partial M}\right)_{H,N} dM$$

$$= \sum n_\pm (\mp \mu_B)\, dH + \sum \frac{\varepsilon_\pm}{(\pm 2\mu_B)}\, dM$$

$$= -\mathbf{M} \cdot d\mathbf{H} - (\mathbf{H} + \mu_0 \lambda \mathbf{M}) \cdot d\mathbf{M}$$

$$= -d(\mathbf{M} \cdot \mathbf{H} + \tfrac{1}{2}\mu_0 \lambda M^2).$$

Therefore

$$U = -\mathbf{M} \cdot \mathbf{H} - \tfrac{1}{2}\mu_0 \lambda M^2.$$

The equilibrium value of n is determined by minimizing F, i.e. by

$$\frac{\partial F}{\partial n} = \frac{\partial U}{\partial n} - T \frac{\partial S}{\partial n} = 0.$$

For $H = 0$

$$U = -\tfrac{1}{2}\mu_0 \lambda \mu_B^2 (2n - N)^2,$$

so that the condition for minimum F becomes

$$2\mu_0 \lambda \mu_B^2 (2n - N) = k_B T \ln \left(\frac{n}{N - n}\right),$$

whence

$$M = N\mu_B \tanh (\mu_B B_{\mathrm{eff}}/k_B T);$$

this is valid also for $H \neq 0$.

The specific heat in zero field is:

$$C_H = \left(\frac{\partial U}{\partial T}\right)_H = T\left(\frac{\partial S}{\partial T}\right)_H = -\frac{\mu_0 \lambda}{2} \frac{d}{dT}(M^2)$$

and the transition is second order on this model. A more general result valid for $H \neq 0$ is

$$C_H = -B_{\text{eff}} \frac{dM}{dT}.$$

5.5 Assuming N spins on A sites and N on B sites

$$M_A = N\mu \tanh (\mu B_{\text{eff}}^A / k_B T),$$

and

$$\frac{\partial M_A}{\partial B_{\text{eff}}^A} = \frac{N\mu^2}{k_B T} \text{sech}^2 (\mu B_{\text{eff}}^A / k_B T).$$

For $H = 0$

$$M_A = M_0, \qquad M_B = -M_0;$$
$$B_{\text{eff}}^A = \mu_0 \lambda M_0, \qquad B_{\text{eff}}^B = -\mu_0 \lambda M_0.$$

In a small field δH parallel to M_0

$$\delta B_{\text{eff}}^A = \delta H - \mu_0 \lambda \, \delta M_B$$

$$= \delta H - \mu_0 \lambda \frac{\partial M_B}{\partial B_{\text{eff}}^B} \delta B_{\text{eff}}^B$$

$$= \delta H - \alpha \, \delta B_{\text{eff}}^B \quad \text{(say)}.$$

Similarly

$$\delta B_{\text{eff}}^B = \delta H - \alpha \, \delta B_{\text{eff}}^A,$$

since

$$|B_{\text{eff}}^A| = |B_{\text{eff}}^B| \qquad \text{for } H = 0.$$

For parallel susceptibility we are interested in a solution with $\delta M_A = \delta M_B > 0$. Therefore

$$\delta B_{\text{eff}}^A = \delta B_{\text{eff}}^B = \frac{\delta H}{1 + \alpha}$$

and

$$\delta M_A = \delta M_B = \frac{1}{\mu_0 \lambda} \frac{\alpha}{1 + \alpha} \delta H,$$

so that

$$\chi_{\parallel} = \frac{\delta M_A + \delta M_B}{\delta H} = \frac{1}{\mu_0 \lambda} \left(\frac{2\alpha}{1 + \alpha} \right) = \frac{C}{T_N} \left(\frac{2\alpha}{1 + \alpha} \right);$$

note that

$$\alpha = \frac{T_N}{T} \text{sech}^2 \left(\frac{\mu \mu_0 \lambda M_0}{k_B T} \right),$$

so that $\alpha \to 0$ as $T \to 0$ and $\alpha \to 1$ as $T \to T_N$.

The perpendicular susceptibility is easier to calculate. Since δH is perpendicular to M_0, $|B_{\text{eff}}|$ is not changed. All that happens is that B_{eff}^A and B_{eff}^B rotate in opposite directions through an angle $(\delta H/2B_{\text{eff}})$, in order to satisfy Eqs. (5.25). The component of magnetization parallel to δH is therefore

$$\delta M = 2 \times M_0 \frac{\delta H}{2B_{\text{eff}}} = \frac{\delta H}{\mu_0 \lambda},$$

so that

$$\chi_\perp = \frac{C}{T_N},$$

independent of temperature below T_N.

5.6 Eq. (5.38) gives

$$\gamma \, \delta B_e + \frac{\hbar}{2m^*} \frac{\pi^2}{d^2} \delta(2n+1)^2 = 0,$$

and from Fig. 5.10, $(2n+1) = 5$ to $(2n+1) = 15$ (corresponding to $\delta(2n+1)^2 = 200$) gives $\delta B_e \approx 0.16$ T. With the electron spin value for $\gamma = e/m$

$$\frac{m^*}{m} = \frac{\hbar}{2e} \left(\frac{\pi}{d}\right)^2 \frac{\delta(2n+1)^2}{\delta B_e} \approx 11.$$

5.7 The straight line shows that

$$C = aT^{3/2} + bT^3.$$

The first term indicates a specific heat due to magnons with $E \propto k^2$. The intercept of the straight line will therefore give m^* for magnons and the slope will give the Debye temperature Θ_D.

6.1

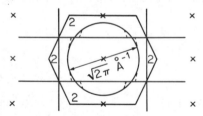

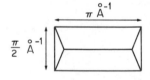

2nd zone reduced.

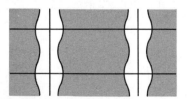

1st zone repeated.

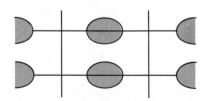

2nd zone repeated.

6.2 In terms of Cartesian axes for a conventional cubic unit cell of side a the primitive translation vectors are

$$\mathbf{a} = \tfrac{1}{2}a(\mathbf{i} + \mathbf{j})$$
$$\mathbf{b} = \tfrac{1}{2}a(\mathbf{j} + \mathbf{k})$$
$$\mathbf{c} = \tfrac{1}{2}a(\mathbf{k} + \mathbf{i}),$$

whence

$$\mathbf{G} = \frac{2\pi}{a}\{(l - m + n)\mathbf{i} + (l + m - n)\mathbf{j} + (-l + m + n)\mathbf{k}\}.$$

In units of $(2\pi/a)$ the lengths of the four reciprocal lattice vectors given are $\sqrt{3}$, $2\sqrt{2}, 2, 2\sqrt{5}$.

6.3 From Problem 6.2, the shortest reciprocal lattice vector for fcc is $\sqrt{3}(2\pi/a)$ and k_M is half this. Since there are four electrons in the cubic unit cell

$$k_F^3 = 3\pi^2 N/V = 12\pi^2/a^3.$$

Thus

$$\frac{k_F}{k_M} = \frac{(12\pi^2)^{1/3}}{\pi\sqrt{3}} = 0.901.$$

For bcc the primitive translation vectors are

$$\mathbf{a} = \tfrac{1}{2}a(\mathbf{i} + \mathbf{j} - \mathbf{k})$$
$$\mathbf{b} = \tfrac{1}{2}a(\mathbf{i} - \mathbf{j} + \mathbf{k})$$
$$\mathbf{c} = \tfrac{1}{2}a(-\mathbf{i} + \mathbf{j} + \mathbf{k})$$

so that by the method of Problem 6.2

$$\mathbf{G} = -\frac{2\pi}{a}\{(n + l)\mathbf{i} + (l + m)\mathbf{j} + (m + n)\mathbf{k}\},$$

of which the shortest value is $\sqrt{2}(2\pi/a)$ so that

$$k_M = \sqrt{2}(\pi/a).$$

With two electrons per unit cell

$$k_F^3 = 6\pi^2/a^3,$$

so that

$$\frac{k_F}{k_M} = \frac{(6\pi^2)^{1/3}}{\pi\sqrt{2}} = 0.879.$$

The lower value of this ratio for bcc accounts (at least in part) for the fact that sodium has an almost spherical Fermi surface, but that of copper touches the hexagonal faces of the Brillouin zone (see Fig. 10.12).

7.1 Wavevector conservation gives

$$\mathbf{k}_1 = \mathbf{k}_2 + \mathbf{q}$$

or

$$q^2 = k_1^2 + k_2^2 - 2k_1k_2 \cos\theta$$
$$\approx (k_1 - k_2)^2 + k_1k_2\theta^2$$

for small θ. Energy conservation gives

$$\hbar qc = \frac{\hbar^2}{2m}(k_1^2 - k_2^2),$$

where c is the velocity of sound and m the neutron mass. Elimination of q gives

$$(k_1 - k_2)^2 \left[\left\{ \frac{\hbar(k_1 + k_2)}{2mc} \right\}^2 - 1 \right] = k_1 k_2 \theta^2.$$

For 0.02 eV neutrons $k_1 = 3.1 \times 10^{10}$ m (section 7.12) so that

$$\frac{\hbar k_1}{mc} \approx 6.8,$$

which implies $k_1 \approx k_2$ and

$$q \approx k_1 \theta$$

$$\Delta E \approx \hbar k_1 \theta c$$

$$= 1.71 \times 10^{-22}\,\text{J} = 1.07 \times 10^{-3}\,\text{eV}.$$

Initial time of flight is 5.2 ms. Since $(\Delta E/\varepsilon) = -(1.07/20)$, $(\Delta\tau/\tau) = +(1.07/40)$, and time of flight is increased by 0.14 ms.

7.2 If $\hbar k_1 < mc$ a small angle solution to the equations of Problem 7.1 with $k_1 \approx k_2$ cannot be found, and there is no scattering near the origin of reciprocal space. But if we choose a larger scattering angle so that

$$\mathbf{k}_1 - \mathbf{k}_2 \approx \mathbf{G}$$

a phonon of small wavevector and hence small energy can be created. This is why inelastic neutron scattering experiments are usually done near reciprocal lattice points other than the origin.

7.3 If $(00n)$ Bragg reflections are examined with neutrons they will be affected by a spiral magnetic structure, because the magnetic scattering amplitude of the crystal planes will be modulated with the period of the spiral. This gives satellite reflections much as in Eq. (7.13), except that for a static modulation of the structure $\omega = 0$, so that all the scattering is elastic. The separation of the satellites from the normal Bragg reflections gives the period of the spiral, and hence the turn angle.

8.1 (a) The maximum conductivity may be obtained approximately from the formula applicable below the maximum, where conduction is limited by boundary scattering of phonons

$$K \approx \tfrac{1}{3} dc C_v$$

$$= \tfrac{1}{3} \times 3 \times 10^{-3} \times 10^4 \times 10^{-1}(30)^3$$

$$= 2.7 \times 10^4\,\text{W m}^{-1}\,\text{K}^{-1}.$$

(b) From the maximum to liquid N_2 temperatures

$$K \propto T^3 \exp[\Theta_D/bT],$$

where $b \approx 2$. Thus

$$\frac{K_{80}}{K_{30}} = \left(\frac{80}{30}\right)^3 e^{-10.4}$$

$$K_{80} \approx 15\,\text{W m}^{-1}\,\text{K}^{-1}.$$

But note that this estimate is *very* rough; K_{80} could easily be 100 times smaller if b were less than 2, so the T^3 factor is almost irrelevant.

9.1 The equation of motion is

$$\frac{d\mathbf{v}}{dt} + \frac{\mathbf{v}}{\tau} = \frac{e\mathbf{E}}{m^*};$$

operating with $(-\rho_0 \, \text{div})$ and assuming $(\rho - \rho_0) \ll \rho_0$ gives

$$\frac{\partial^2 \rho}{\partial t^2} + \frac{1}{\tau}\frac{\partial \rho}{\partial t} = -\frac{\rho_0 e}{m^*}\,\text{div}\,\mathbf{E} = -\frac{\rho_0 e}{m^* \varepsilon_0}(\rho - \rho_0)$$

$$\left[\frac{\partial^2}{\partial t^2} + \frac{1}{\tau}\frac{\partial}{\partial t} + \omega_p^2\right](\rho - \rho_0) = 0.$$

This is the equation of a damped harmonic oscillator, critically damped for $\omega_p \tau = \frac{1}{2}$, overdamped for smaller ω_p. Substituting for ω_p, the critical electron concentration is

$$n = \frac{m^* \varepsilon_0}{4e^2 \tau^2}$$

$$\approx 10^{20}\,\text{m}^{-3} = 10^{14}\,\text{cm}^{-3}.$$

When $\omega_p \tau \ll 1$ electron inertia (the $\partial^2/\partial t^2$ term) is negligible and charge inequalities decay with an exponential time constant

$$\frac{1}{\omega_p^2 \tau} = \frac{\varepsilon_0}{\sigma}.$$

This is often the relevant situation in semiconductors, except that screening by valence band electrons modifies the time constant to $\varepsilon_r \varepsilon_0/\sigma$.

9.2 The Mott transition to the metallic state occurs when the screening distance is greater than the mean distance between electrons,

$$\frac{v_F}{\omega_p} \gtrsim n^{-1/3}$$

$$n^{1/3} \gtrsim \frac{\omega_p}{v_F} = \left(\frac{ne^2}{m^* \varepsilon_0}\right)^{1/2}\frac{m^*}{\hbar}(3\pi^2 n)^{-1/3},$$

so that

$$n^{1/6} \gtrsim \frac{e}{3\hbar}\left(\frac{m^*}{\varepsilon_0}\right)^{1/2}.$$

Therefore, for a metal

$$\frac{n}{k_F} = \frac{n}{(3\pi^2 n)^{1/3}} \approx \frac{n^{2/3}}{3} \gtrsim \frac{e^4 m^{*2}}{240 \hbar^4 \varepsilon_0^2}.$$

But the Bohr radius is given by

$$a_0 = \frac{4\pi \varepsilon_0 \hbar^2}{m^* e^2}$$

so that the condition for a metal becomes

$$\frac{nA}{k_F} \gtrsim \frac{A}{a_0^2} \sim 1.$$

10.1 From Eqs. (10.4) and (10.3)

$$\delta\left(\frac{1}{B}\right) = \frac{2\pi e}{\hbar^2 \omega} \frac{d\varepsilon}{dA_k} = \frac{e}{\omega m_c},$$

and from the data given $\delta(1/B) = 0.345 \, \text{T}^{-1}$, whence

$$m_c = 1.08 \times 10^{-30} \, \text{kg}.$$

10.2 The energy level spacing is $\hbar\omega_c$, with $\omega_c = eB/m_c$; this must be large compared with kT for a good effect. In other words

$$T \ll \frac{eB\hbar}{m_c k_B} \approx 13 \, \text{K};$$

a temperature of 1.3 K, giving a ratio of 10, should therefore be satisfactory.

It is also necessary that collision broadening should be comparably small, $\omega_c\tau \gg 1$, or

$$\frac{1}{\tau} = \frac{n}{10^{14}} \ll \omega_c$$

$$n \ll 1.75 \times 10^{26} \, \text{m}^{-3}.$$

A good effect should therefore be observed for impurity concentrations less than $10^{25} \, \text{m}^{-3} = 10^{19} \, \text{cm}^{-3}$.

10.3 It is intended as a compliment to A. B. Pippard and D. Shoenberg, from whom I have learnt so much physics.

Index

Page numbers in **bold** type give the location where a term is introduced or explained, and the term is printed in **bold** type on the cited page. The other entries are mainly confined to topics whose location is not obvious from the Contents list.